Numerical Methods for Engineering

The ACES Series on Computational and Numerical Modelling in Electrical Engineering

Andrew F. Peterson, PhD – Series Editor

The volumes in this series encompass the development and application of numerical techniques to electrical and electronic systems, including the modelling of electromagnetic phenomena over all frequency ranges and closely related techniques for acoustic and optical analysis. The scope includes the use of computation for engineering design and optimization, as well as the application of commercial modelling tools to practical problems. The series will include titles for senior undergraduate and postgraduate education, research monographs for reference, and practitioner guides and handbooks.

Published Books in the ACES Series:

W. Yu, X. Yang, and W. Li, "VALU, AVX and GPU Acceleration Techniques for Parallel FDTD Methods," 2014.

A.Z. Elsherbeni, P. Nayeri, and C.J. Reddy, "Antenna Analysis and Design Using FEKO Electromagnetic Simulation Software," 2014.

A.Z. Elsherbeni and V. Demir, "The Finite-Difference Time-Domain Method in Electromagnetics with MATLAB® Simulations, 2nd Edition," 2015.

M. Bakr, A.Z. Elsherbeni, and V. Demir, "Adjoint Sensitivity Analysis of High Frequency Structures with MATLAB®," 2017.

O. Ergul, "New Trends in Computational Electromagnetics," 2019.

D. Werner, "Nanoantennas and Plasmonics: Modelling, design and fabrication," 2020.

ACES

Numerical Methods for Engineering

An introduction using MATLAB® and computational electromagnetics examples

2nd Edition

Karl F. Warnick

The Institution of Engineering and Technology

Published by SciTech Publishing, an imprint of The Institution of Engineering and Technology, London, United Kingdom

The Institution of Engineering and Technology is registered as a Charity in England & Wales (no. 211014) and Scotland (no. SC038698).

First published 2011

Second Edition published 2020

The Institution of Engineering and Technology
Michael Faraday House
Six Hills Way, Stevenage
Herts, SG1 2AY, United Kingdom

www.theiet.org

British Library Cataloguing in Publication Data
A catalogue record for this product is available from the British Library

ISBN 978-1-83953-073-9 (hardback)
ISBN 978-1-83953-074-6 (PDF)

Typeset in India by MPS Limited

Contents

About the Author

Karl F. Warnick is a professor in the Department of Electrical and Computer Engineering at Brigham Young University, USA, and a fellow of the IEEE. He has published widely on electromagnetics theory, numerical methods, antenna applications, and high sensitivity phased arrays for satellite communications and radio astronomy.

Preface

This book evolved out of my experience that well-prepared undergraduate engineering students, as well as all beginning graduate students, can learn to understand and implement powerful numerical techniques if the topic is presented with clear explanations, good examples, and problem sets that move from simple algorithms to complex codes with real-world capabilities. I undertook the challenge of developing a good set of class notes and refined them based on significant feedback from students with a variety of specializations, research interests, and career objectives. After nearly a decade of iterations on the course notes followed by external reviews from instructors who class-tested advance versions of the book, I feel confident that the finished book will be a welcome self-study tool and course text.

The book has two major objectives:

1. To impart the skill of taking a mathematical prescription for a numerical method and translating it into a working, validated software code.
2. To help users of numerical modeling and design software understand the relative merits, ranges of validity, domains of applicability, expected accuracy, and underlying properties of several popular classes of numerical algorithms for engineering problems.

Since design work often requires the creation of specialized software tools, engineers must have the ability to implement numerical solution algorithms based on a mathematical model for a system or physical problem. Even when using commercial design packages, it is important to understand how to use the software to obtain sufficiently accurate results within an acceptable amount of computation time.

The examples used in the book are chosen primarily from wave propagation and electromagnetics for a reason. Of all the engineering fields, electromagnetic analysis, and design tends to use the most sophisticated algorithms. This focus ensures that the book touches the state-of-the-art in numerical methods for engineering. In keeping with the emphasis on computational electromagnetics, a secondary goal of the book is to help students better understand the electromagnetic theory by implementing numerical solution methods for the fundamental equations of electromagnetics. One of the best ways to become competent in working with the laws and principles of electromagnetics, as well as to gain insight into the behavior of field and wave solutions, is to translate the governing equations of electric and magnetic fields into a computational algorithm and then to use the algorithm as a tool to visualize fields and waves in conjunction with practical examples.

Although the emphasis is on computational electromagnetics, numerical tools are developed in this book for basic building block tasks such as solving differential equations, evaluating integrals, and solving linear systems. This means that much of the material can be applied to all branches of engineering. General principles of numerical analysis such as efficiency, accuracy, and ill-posed problems will be treated within the context of computational electromagnetics, but these also have broad application in computational science and engineering.

As a textbook, the material works best at the senior year after a first course in basic electromagnetics or as a first-year graduate course. This book bridges the gap between typical one-term introductory electromagnetics and the advanced computational electromagnetics textbooks that are not very accessible to seniors or beginning graduate students. The balance of electromagnetics-specific content with general topics is intended to serve engineering students specializing in electromagnetics, wireless communications, optics, and related fields as well as students concentrating in other areas who desire an exposure to numerical methods.

The electromagnetics content of the book is fairly self-contained. To avoid too much distraction from the central topic of numerical methods, topics from electromagnetic theory are presented in a condensed form without lengthy development. Chapter 2 includes a review of the basic concepts of electromagnetics. More advanced material on wave propagation, radiation integrals, Green's functions, propagation in materials, waveguides, and scattering theory are later introduced as needed throughout the book.

Chapters 3–8 cover finite difference methods, numerical integration, the method of moments for integral equations, linear system solution algorithms, variational methods, and the finite element method. Chapters 9–10 on optimization and inverse problems may be omitted or briefly surveyed in a senior undergraduate numerical methods course. These topics are included to enrich the material for more advanced graduate students.

Homework problems in the book are programmed, in the sense that groups of consecutive problems often build incrementally toward a completed code. It is common for the core of an algorithm to be easy to implement in relation to the preprocessing required to create a grid or mesh and postprocessing used to compute-derived physical quantities. Consequently, one or two homework problems typically deal with the core algorithm. Several successive problems add additional pre- and postprocessing capability. This incremental approach eases the development of complex codes and implicitly teaches the vital debugging skill of verifying a small portion of the code before adding another major functional block.

Another important technique used in the homework problems is cross-code comparison. Many of the physical problems are repeated from one chapter to the next, and homework problems are included to compare numerical results from multiple algorithms for the same physical system. This helps students to understand the relative merits of different algorithms in terms of accuracy, efficiency, and ease of implementation. It is possible to explain in words why the method of moments for surface integral equations can, in some cases, be more efficient than the finite difference time-domain method. It is a more powerful learning experience to run codes for both methods and see results from one of them that are more accurate and take less computation time.

Looking beyond the immense practical value of numerical methods for engineering work, numerical codes, and algorithms are wonderful tools for learning engineering electromagnetics. Assessment data at my university indicate that our beginning electromagnetics course is one of the most difficult in the electrical engineering program. This can be attributed to the large amount of content that must be packed into a single semester course and to the abstract nature of electromagnetic theory in relation to, say, circuit analysis. I have observed many students who leave the beginning electromagnetics course with a marginal understanding of the principles of electromagnetics make their way through a course on computational electromagnetics based on the material in this book. After implementing numerical algorithms on their own and seeing the results appear in graphical form, the light turns on, and they finally feel that they understand something of what the equations and theory of electromagnetics are all about.

Some of the numerical methods covered in the book require considerable mathematical detail. To help the student distinguish supporting material from the primary formulas needed for homework solutions, in this second edition, key results required to implement the methods in code are set off using colored boxes. To further aid the student, implementation helps; summaries of algorithmic steps, code fragments, and examples are also included in colored boxes.

I appreciate the many outstanding students who have used the earlier versions of this textbook in courses and have helped in many ways to refine the material, including particularly Clayton Davis and Derek Hudson. I thank my colleagues who class-tested preliminary drafts and the reviewers for insightful comments, corrections, and valuable suggestions. Jakob Kunzler did a remarkably thorough reading of an early draft of the second edition, even locating errors that had escaped many readers over years of using the first edition. I welcome and appreciate the readers' and instructor's feedback to continue improving the text.

Karl F. Warnick
Brigham Young University
Provo, Utah, USA

Chapter 1

Introduction

This book is about numerical methods that allow the differential equations, integral equations, and other mathematical relationships that govern wave propagation, electromagnetic (EM) fields, and other physical systems to be solved using computational algorithms. The progress of numerical methods has followed the rise of low-cost, widely available computational power. Numerical methods are now fundamentally important in engineering work, since almost all branches of engineering rely on software tools, design packages, and numerical models. Circuit design, antenna design, microwave systems, optical systems, remote sensing, control theory, signal processing, and microelectronics are just a few of the fields that are heavily impacted by software tools and computer-aided engineering. The ability to understand, use, and create numerical methods is already crucial for engineers and will continue to grow in importance.

In this book, we will survey general numerical methods as well as specialized algorithms, with a focus on examples from computational electromagnetics. General numerical tools and techniques to be covered include numerical integration, differentiation, the solution of large linear systems, and methods for discretizing and solving partial differential equations and integral equations. Finite difference methods, the method of moments, and the finite element method will be treated in detail. While the emphasis is on wave propagation and electromagnetics, the principles and techniques of numerical analysis that are illustrated throughout the book can be applied to many branches of engineering.

The goals are to teach the reader how to design algorithms that solve a given set of equations or analyze a particular system and to help users of computer-aided engineering software tools understand how to choose the best analysis technique, interpret numerical results, and optimize the efficiency and accurate of the solution method. Another expected learning outcome is a deeper understanding of the characteristic behaviors of solutions to the differential and integral equations of electromagnetic field theory. Studying numerical methods has a double benefit for engineering students, since they not only become competent at creating and using numerical techniques but also gain insight into the fundamental laws and equations that govern the systems used by engineers in creating new technologies.

The development of numerical methods has become an academic area in its own right. When the emphasis is on modeling physical systems, the field is referred to as scientific computation, and the theoretical study of the properties and behavior of computational algorithms using rigorous mathematics is numerical analysis. We will begin our thinking about numerical methods by considering the relationships between these fields.

1.1 Scientific Computation, Numerical Analysis, and Engineering

Science has historically been divided into two major areas of inquiry: theory and experiment. In the last few decades, a third area of effort has become prevalent. This new area is the application of numerical algorithms implemented in computer code to model physical systems and is commonly referred to as scientific computation. Scientific computation combines aspects of theoretical analysis, since it involves the solution of differential equations or other descriptive frameworks, and experimental science, since a system can be modeled under a variety of stimuli as would be done when conducting an empirical study. Computational modeling is a growth industry in nearly all fields of science, from simulation of quantum field theories for subatomic particles to numerical weather prediction.

Since engineering is the application of physical principles to the design of new technologies, the methods and techniques used to analyze physical systems from a scientific perspective become the tools used by engineers. Numerical methods have become vital in engineering analysis and design, and in recent years engineering applications have driven a flowering of the power, capability, and breadth of algorithms and commercially available software packages for analysis and design of uncounted types of technologies.

As scientific computation and computer-aided engineering have increased in importance, a new branch of mathematics dealing with the development and improvement of solution algorithms has grown and matured. This field is referred to as numerical analysis. Numerical analysis is concerned with creating algorithms that are coupled with rigorous theoretical proofs of theorems that bound the computation time required to obtain a solution for a given class of problems and that guarantee the accuracy of computed solutions.

Historically, there has been a fair amount of interplay between engineering applications of computational algorithms and numerical analysis, but for the most part the gap between the two areas has been unfortunately wide. Developers working in application areas are driven by market forces to develop new numerical methods more rapidly than rigorous theory can keep up, so the tools used in engineering work typically can be quite sophisticated yet unsupported by careful theoretical proofs of validity and accuracy of solutions. Rigorous mathematical analysis often leads to insights that drive the state-of-the-art forward, but theoretical solution convergence proofs often apply only to simplistic problems and some mathematically sophisticated numerical methods can be inadequate for complex, real-world applications. Another barrier between numerical analysis and the engineering world is abstract mathematical notation, which can disguise simple problems and methods that should be familiar to the engineer and hinders the dissemination of new methods to application-oriented users.

To bridge this gap, this book attempts to balance theoretical ideas from numerical analysis with practical examples from scientific computation and engineering.

Interestingly, the more the engineer is focused on applications, the more the study of numerical methods should focus on theory. An engineer doing design work typically uses a commercial software package, so the basic work of implementing mathematical equations as a computer algorithm has already been done. What the practicing engineer needs is a feel for the expected accuracy of the results and insight into the best way to use the software to get optimal results. Gaining this kind of understanding requires only a modest amount of experience in writing numerical codes, and the emphasis should be on accuracy analysis, efficiency, and comparisons of the performance of different algorithms for the same problem. For most engineers, the right mix is an equal combination of numerical analysis (theory) and practical algorithms (applications).

1.2 Computational Electromagnetics

The development and application of numerical methods for problems involving electromagnetic fields and waves is referred to as the field of computational electromagnetics (CEMs). To place CEM techniques in a historical context, we will briefly survey the development of solution methods for wave propagation and electromagnetic systems.

As with most areas of physics, engineering, and applied science, the basic trend in solution methods for electromagnetics applications over the last century is from "pencil and paper" analysis techniques to powerful, sophisticated software packages. In the early days of radar and wireless communications, engineers relied on the analytical techniques—series expansions, canonical problems, and perturbation methods—developed in the late 1800s and the early 1900s. In spite of the great cleverness of researchers in applying these methods to an expanding range of problems with increasing complexity, eventually the wide range of EM applications required the development of more flexible and powerful techniques.

Asymptotic methods based on the high- and low-frequency limits provided a way to analyze problems involving structures and geometries that are too complex for analytical solutions. Prominent among these developments was Ufimtsev's creation of the uniform theory of diffraction [1], which enabled advances in low-observable stealth technology for aircraft. Asymptotic methods are more flexible than analytical solutions and are vital for many engineering applications, but the accuracy of the approximation is difficult to improve for structures that are geometrically complex or otherwise fall outside the range of validity of the asymptotic approach.

More recently, the explosion of wireless technologies, medical imaging, optics, terahertz devices, local area networks, mobile handheld devices, and a host of other applications have required still more powerful methods for analysis and design, and analytical and asymptotic methods have been augmented or replaced by CEM tools. In the mid-1960s, the three pillars of numerical analysis in EMs were introduced: the method of moments (MoM), the finite difference time-domain (FDTD) method, and the finite element method (FEM). These breakthrough algorithms precipitated the present era of computational electromagnetics and are part of the broad trend toward joining simulation with theory and experiment throughout science and engineering.

Analytical, asymptotic, and numerical methods that have been developed for electromagnetic modeling include the following:

Early 1900s: Analytical Methods

- Canonical problems and separable geometries (e.g., spheres, cylinders, rectangles, and cubes)
- Modal solutions, orthogonal functions, and series expansions
- Restricted to a few classes of geometries

Mid-1900s: Asymptotic Methods

- Low-frequency regime (quasistatic approximations)
- High-frequency regime
  - Geometrical optics and ray tracing
  - Physical optics
  - Geometrical theory of diffraction (GTD), physical theory of diffraction (PTD), uniform theory of diffraction (UTD)
  - Wentzel–Kramers–Brillouin (WKB) approximation
  - Incremental length diffraction coefficients (ILDCs)

Late 1900s–Present: Numerical Methods

- Mode matching ("piecing together" analytical solutions)
- Finite difference time-domain method (FDTD), Yee, 1966 [2]
- Method of moments (MoM), Richmond, 1966 [3]
- Finite element method (FEM), 1968 [4,5]
- Hybrid methods that combine "full-wave" numerical methods like FDTD and MoM with asymptotic methods

Numerical methods are highly flexible and can be used to analyze and design an unlimited variety of microwave systems, antennas, components, and structures, but methods of all types, including analytical techniques, are very important and widely used today, and there is much overlap among these three broad areas. Even with powerful computational methods, analytical and asymptotic solutions are very important. They provide qualitative physical insight for complex problems, lead to simple formulas that are useful in design work, and, as will be seen in later chapters, are used as test cases to validate numerical methods.

1.2.1 Applications of CEM Tools

Software analysis and design tools based on CEM algorithms are useful for reducing product development costs by replacing costly device testing with computational simulation, design optimization, and integration of electromagnetic phenomena with other system aspects in a multiphysics environment. A key goal is to replace the design–build–test cycle with a design–simulate cycle, to reduce fabrication costs, make more efficient use of finite resources, and shorten the time to market for product development.

CEM methods are used in application areas such as the following:

Applications of Computational Electromagnetics

- Microwave circuit and system design
- Antenna analysis and design
- Wireless communications
- Propagation of electromagnetic waves in various environments (e.g., indoor, outdoor, urban locations, underground, over the ocean surface, in atmospheric layers, embedded in biological tissues or in the body)
- Optics and photonics
- Design and analysis of metamaterials
- Radar systems
- Stealth and low observable technology
- Remote sensing of oceans, ice caps, land, and other planets
- Electromagnetic compatibility and interference (EMC/EMI)

Since most CEM tools are essentially solution methods for systems of differential equations or integral equations, the basic techniques can be applied to many other fields beyond those directly associated with electromagnetic theory and its applications. CEM methods are related to those used in such broad-ranging fields as structural analysis, mechanics, fluid dynamics, acoustics, microelectronics, weather and climate models, galactic dynamics, simulations of colliding black holes, and financial modeling.

1.2.2 Types of CEM Methods

There are four main categories of numerical methods used in electromagnetics: differential equation methods, integral equation methods, mode matching, and asymptotic approximations. Differential equation methods solve Maxwell's equations directly by transforming the continuous differential equations into discrete difference equations. Integral equation methods use the equivalence principle, Green's functions, and boundary conditions to transform Maxwell's equations into an integral equation for an unknown current. Asymptotic approximations are based on the high- or low-frequency limiting behaviors of fields and waves, and mode matching is based on piecing together analytical solutions for simple, canonical problems to analyze more complex structures.

Prominent algorithms in these classes of methods are outlined as follows:

CEM Algorithms

Differential equation methods

- Finite difference (FD) method
  - Finite difference time-domain (FDTD) method
- Finite element method (FEM)
- Finite integration method
- Transmission line matrix (TLM) method
- Method of lines

Integral equation methods

- Method of moments (MoM)

Mode matching

Numerical methods based on asymptotic approximations

- Ray tracing (geometrical optics)
- GTD, PTD, and UTD codes and hybridizations with other algorithms such as MoM

The methods that we will focus on in this book are emphasized in color.

Numerical methods are also specialized to solving single-frequency (time-harmonic) problems and broadband (time-domain) problems. The major methods in these categories are given in Table 1.1. The MoM–time-domain integral equation (MoM-TDIE) approach is a relatively new development, because only recently have instability problems been overcome sufficiently to allow for useful implementations. In this book, we will consider both time-domain methods (FDTD) and frequency-domain methods (MoM, FEM).

1.2.3 Mesh and Grid Generation

CEM methods require some kind of discrete computational representation of the problem to be solved. Typically this is a regular grid of points or mesh of triangular patches, cubes, or other simple shapes, together with parameter values that specify the physical properties in the material in each mesh element. As will be seen in

Table 1.1 Algorithms for time- and frequency-domain problems

Algorithm/Domain	Frequency domain	Time domain
Differential equation methods	FD, FEM	FDTD
Integral equation methods	MoM	MoM-TDIE

later chapters, the mesh representation must be fine enough to represent oscillatory currents or fields.

For many applications, CEM software must accept meshes generated for non-EM purposes such as mechanical drawings. Mechanical mesh representations typically require preprocessing before they are adequate for use with a CEM code. Since the generation of a smooth, regular mesh for a complex object can be as challenging as an electromagnetic analysis, mesh generation is an important subfield of scientific programing in its own right. Various approaches to mesh generation are developed in this book in connection with the MoM and FEM algorithms in Chapters 6 and 8.

1.3 Accuracy and Efficiency

Two key questions for any numerical method are as follows:

> How good is the solution (accuracy)?
>
> How long does it take to compute the solution (efficiency)?

For the developer, difficulty of implementation and modification is also an issue, but from the user's point of view, accuracy and efficiency are paramount. Both empirical and theoretical approaches can be used to answer these questions. An extensive treatment of accuracy and efficiency for CEM methods can be found in [6].

On the empirical side, a new computational algorithm can be validated by checking results for a few benchmark problems for which analytical solutions or measured data are available. Another empirical validation technique is cross-comparison of numerical results from two different numerical methods. These types of empirical validation methods have been used with great success by the computational electromagnetics community. Essentially all existing numerical methods have been developed, tested, and validated using empirical techniques. The main limitation of empirical validation is that a method that works well for a test case may fail for more complicated, real-world problems.

On the other end of the spectrum is the use of rigorous theoretical techniques to create new and more efficient algorithms and to prove that a numerical method leads to accurate results. This is the domain of numerical analysis. The disadvantages of the theoretical approach are that it often lags behind progress in methods development and can lead to abstract results that have limited practical value for users of numerical models.

Approaches to studying accuracy for numerical methods can be classified in terms of increasing rigor:

Approaches to Studying Accuracy

1. Use a "tried-and-true" code that has been checked by others for many similar types of problems.

2. Verify the qualitative behavior of the solution by comparison with an idealized problem with a known solution or by determining whether the solution is physically reasonable in terms of its order of magnitude or another large-scale property.
3. Compare the computed solution with results from another code (cross-code validation), measured data, or exact or approximate solutions for test problems.
4. Use heuristic, physics-based considerations to gain insight into the range of problem parameters (e.g., physical size, frequency, or geometrical curvature) for which the numerical method is expected to be accurate.
5. Apply rigorous mathematical theory to obtain an error estimate or error bound.

In this book, we will consider all these approaches to the study of accuracy and efficiency.

1.4 Programing Languages

Numerical methods are typically implemented in programing languages such as Fortran or a later version such as Fortran 90, C or C++, or an interactive numerical programing environment such as MATLAB® (Mathworks, Inc.). While Fortran is no longer a widely used programing language, it remains important in scientific computation because of its simplicity and execution speed, but other languages are typically used for commercial codes.

In this book, examples and code fragments will be given in the MATLAB programing language. Relative to other programing languages such as C or Fortran, MATLAB is easier to learn and is more foolproof in the sense that pointers and other low-level computational constructs are hidden from the user. This makes MATLAB a convenient tool for learning the skill of numerical algorithm implementation. MATLAB routines can use more memory and run slower than the same routine implemented in C or Fortran. Some core MATLAB functions are highly optimized and can be quite fast however, so the actual performance of an algorithm depends strongly on the details of how it is implemented. For our purposes, the learning curve, code implementation time, and ease of debugging with MATLAB are much faster than other languages, which makes up much of the difference in efficiency. The MATLAB environment also provides helpful plotting commands, disk input and output routines, and other supporting functions. A tutorial on using MATLAB is given in Section 3.1.

1.5 Writing and Debugging Numerical Codes

A large part of this book is devoted to describing the basic numerical algorithms of computational electromagnetics so that they can be implemented by the student. This can be challenging and time-consuming, particularly for first-time numerical programers. Good programing practices save time in the long run.

Suggestions for writing codes in ways that help one to avoid mistakes include the following:

Good Programing Practices

- Prepare a write-up of an algorithm with key equations manipulated into a form that is as close as possible to the actual code that will be implemented. Map out on paper how the algorithm works and derive key lines of code. This step is often bypassed to save time, but writing a code without a clear understanding of the algorithm can make debugging more difficult.
- Divide a code into logical sections, such as (1) physical parameter definitions, (2) problem parameters, (3) computational domain parameters, (4) mesh generation, (5) matrix filling, (6) linear system solution, and (7) postprocessing.
- Do not include numerical constants such as the number of unknowns or physical dimensions multiple times in the code. Define these as variables in a parameters section at the beginning of the code.
- Indent loops and conditional statements for visual clarity and avoid long lines of code.
- Comment the code liberally. Describe in a few words what each variable represents and the function of each line or group of a few lines. Put in clear dividers between major code sections so the high-level structure of the algorithm is readily apparent. Write out equations with additional detail in the comments. Document tricky minus signs and constants.

Once a code is written, debugging can take many hours if proper practice is not followed. For many of the codes developed in this book, there is a core algorithm that implements the basic solution technique, along with pre- and postprocessing routines that prepare the mesh and problem setup and transform the solution into useful physical quantities. Often, the core algorithm may be fairly simple to translate into code, but the preprocessing and postprocessing are much more difficult to implement and debug. It is critical that the preprocessing steps be verified before attempting to debug the core algorithm, and similarly, the core algorithm should be verified before debugging the postprocessing. Debug a long, complex code in small sections, rather than all at once.

To reduce the time required to obtain working codes, here are several recommendations:

Tips for Debugging Codes

- The most time-consuming (and time-wasting) debugging practice is to change numbers or commands blindly and repeatedly run a code until the results are correct. This is occasionally useful in some situations such as fixing a minus sign but should not be overused.

- The best debugging technique is to break the code into sections and to test each section separately. It is extremely difficult to simultaneously debug more than one complex code segment.
- Once a problem has been isolated to a specific part of the code, inspect the values of the variables in that part of the code, and check to see if each is correct. Often the bug can be a minor detail, such as an incorrect sign or value for a frequency, wave number, or other constant in a formula.
- Check the code against an analytical solution or a problem with a known qualitative behavior.
- If a code is not working properly, one can often get a hint as to the cause of the problem by looking at the results for a test case. Devise simple experiments for which the answer is known to localize the problem to a specific section of code.
- When debugging, it is helpful to watch a computation evolve by plotting results during a simulation. Outputting results every loop cycle slows the simulation considerably, so plotting results every tenth or hundredth cycle is more efficient.
- Take a break and let the code sit for a while, and then come back and read over it slowly. Errors that were missed earlier can become more readily apparent.
- As a last resort, rewrite the code from scratch. This can actually be quicker than finding an obscure bug in a long code.

References

[1] P. Y. Ufimtsev, *Fundamentals of the Physical Theory of Diffraction*. Hoboken, NJ: Wiley-Interscience, 2007.

[2] K. S. Yee, "Numerical solution of initial boundary-value problems involving Maxwell's equations in isotropic media," IEEE Trans. Antennas Propag., vol. 14, pp. 302–307, 1966.

[3] J. Richmond, "A wire-grid model for scattering by conducting bodies," IEEE Trans. Antennas Propag., vol. 14, no. 6, pp. 782–786, 1966.

[4] P. Arlett, A. Bahrani, and O. Zienkiewicz, "Application of finite elements to the solution of Helmholtz's equation," Proc. IEE, vol. 115, no. 12, pp. 1762–1766, 1968.

[5] Z. Csendes and P. Silvester, "Numerical solution of dielectric loaded waveguides: I-finite-element analysis," IEEE Trans. Microwave Theory Tech., vol. 18, no. 12, pp. 1124–1131, 1970.

[6] K. F. Warnick, *Numerical Analysis for Electromagnetic Integral Equations*. Norwood, MA: Artech House, 2008.

Chapter 2
Fundamentals of Electromagnetic Field Theory

As the algorithms to be developed in this book will be applied to example problems from computational electromagnetics, we will briefly review here the governing equations, field and source quantities, and notation and conventions used in electromagnetic field theory. The purpose of this section is to provide a convenient, self-contained reference on notation, terminology, and basic formulas. If additional tutorials on electromagnetic theory are needed, several references are given at the end of the chapter.

The philosophy of this book is to develop numerical methods early rather than after a long treatment of field theory. In keeping with this philosophy, we use a just-in-time approach to avoid unnecessary delays before moving to basic numerical algorithms, and only a few basic principles of field theory are covered here. Later chapters include a significant amount of additional material on Green's functions, radiation integrals, equivalent theorems, and other topics.

2.1 Electromagnetic Field and Source Quantities

Maxwell's equations for the electromagnetic field are expressed in terms of electric and magnetic vector fields $\overline{E}$, $\overline{D}$, $\overline{H}$, and $\overline{B}$, and the source quantities $\overline{J}$ and ρ. These vector fields and one scalar quantity are defined in Table 2.1 along with the units of each. Other systems of units are available, but the International System of Units convention is used throughout this book.

Table 2.1 The field and source quantities of electromagnetics

Name	Symbol	Dimension
Electric field intensity	$\overline{E}$	V/m
Magnetic field intensity	$\overline{H}$	A/m
Electric flux density	$\overline{D}$	C/m^2
Magnetic flux density	$\overline{B}$	Wb/m^2
Electric current density	$\overline{J}$	A/m^2
Electric charge density	ρ	C/m^3

The field quantities can be grouped into two flux densities $\overline{D}$ and $\overline{B}$ and two field intensities $\overline{E}$ and $\overline{H}$, so that the electric and magnetic fields are each described in terms of two vector fields. This approach facilitates the introduction of material models through the constitutive relations between the two field intensities and the two flux densities.

2.2 Maxwell's Equations

The evolution of the electromagnetic field in time is governed by Maxwell's equations. Maxwell's equations in point or derivative form are given next.

Maxwell's Equations in Point Form

$$\nabla \times \overline{E} = -\frac{\partial \overline{B}}{\partial t} \tag{2.1a}$$

$$\nabla \times \overline{H} = \frac{\partial \overline{D}}{\partial t} + \overline{J} \tag{2.1b}$$

$$\nabla \cdot \overline{D} = \rho \tag{2.1c}$$

$$\nabla \cdot \overline{B} = 0 \tag{2.1d}$$

In these equations $\nabla \times \overline{E}$ is the curl of the vector field $\overline{E}$, and $\nabla \cdot \overline{D}$ is the divergence of the vector field $\overline{D}$. These operations are discussed further in Section 2.4.

The first equation is Faraday's law, the second is Ampère's law, and the second pair are Gauss's law for the electric and magnetic fields, respectively. Maxwell's equations can also be expressed in terms of line, surface, and volume integrals, but the point or derivative forms of Maxwell's equations are most useful for the numerical methods to be considered in this book.

Physically, Faraday's law implies that time-varying magnetic flux produces an electric field. By Ampère's law, a magnetic field is produced by time-varying electric flux or electric current. Gauss's law shows that charges lead to nonzero divergence for the electric flux density. Visually, if the electric flux density vector field is represented by lines of flux, the flux lines either begin or end on electric charges. Since there are no isolated magnetic charges, the magnetic flux density must be divergence-free.

2.2.1 Constitutive Relations

In free space (vacuum), the flux densities and field intensities are related by the constitutive relations:

$$\overline{D} = \varepsilon_0 \overline{E} \tag{2.2a}$$

$$\overline{B} = \mu_0 \overline{H} \tag{2.2b}$$

where $\varepsilon_0 \simeq 8.854 \times 10^{-12}$ F/m is the permittivity of free space, and $\mu_0 = 4\pi \times 10^{-7}$ A/m is the permeability of free space. Using two quantities for the electric field and two quantities for the magnetic field provides a convenient way to introduce models for the effect of materials such as dielectrics. In an inhomogeneous dielectric, the permeability and permittivity are functions of position, so that

$$\overline{D} = \varepsilon(x, y, z)\overline{E} \tag{2.3a}$$

$$\overline{B} = \mu(x, y, z)\overline{H} \tag{2.3b}$$

Relative permittivity ε_r and relative permeability μ_r are defined according to

$$\varepsilon = \varepsilon_r \varepsilon_0 \tag{2.4a}$$

$$\mu = \mu_r \mu_0 \tag{2.4b}$$

In a conductive material, the induced current density is

$$\overline{J} = \sigma(x, y, z)\overline{E} \tag{2.5}$$

where σ is the conductivity of the material.

2.2.2 Impressed and Induced Currents

In a physical sense, all electric currents are induced by electric fields according to (2.5). For engineering work, however, it is impractical to model all currents using (2.5). The electric fields inside the signal generator that drive a current in a transmission line, for example, are nearly completely shielded from the fields radiated by an antenna, so it makes little sense to model the antenna, transmission line, and signal generator as a complete system. It is much more efficient to simply consider the current source that excites the antenna to be fixed. A fixed current that excites a system is referred to as an impressed current. Mathematically, an impressed current is a given source term in Ampère's law (2.1b), or a forcing function in differential equation terminology. The total current is the sum of the induced and impressed currents

$$\overline{J} = \overline{J}_{\text{ind}} + \overline{J}_{\text{imp}}$$

where $\overline{J}_{\text{ind}} = \sigma \overline{E}$.

2.2.3 Magnetic Currents

When using the equivalence principle (Section 2.13.7), it is common to introduce fictitious impressed magnetic current sources $\overline{M}$ as well as electric currents. The magnetic current is included in Faraday's law as a source term, so that Faraday's law is modified to

$$\nabla \times \overline{E} = -\frac{\partial \overline{B}}{\partial t} - \overline{M} \tag{2.6}$$

Magnetic currents apparently do not exist in nature, so the magnetic current source $\overline{M}$ is a mathematical convenience used when applying the equivalence principle as a tool in solving electromagnetic radiation and scattering problems.

2.3 Coordinate Systems

To give explicit numerical values for the sources and fields and to compute the derivatives in Maxwell's equations, a coordinate system is required. The most basic coordinate system is the rectangular coordinate system. More generally, curvilinear coordinate systems can be defined, which simplifies the analysis of electromagnetics problems involving curved geometries such as spheres and cylinders.

2.3.1 Rectangular Coordinates

The rectangular coordinate system is shown in Figure 2.1. Using this coordinate system, a point in a three-dimensional (3-D) space can be represented by its distances x, y, and z from the origin of the coordinate system along each of three orthogonal axes.

We define the three orthogonal unit vectors $\hat{x}$, $\hat{y}$, and $\hat{z}$, which point in the directions of each of the three coordinate axes, respectively, and which have unit length, so that $\|\hat{x}\| = \|\hat{y}\| = \|\hat{z}\| = 1$. Using these unit vectors, an arbitrary vector $\overline{A}$ can be expressed in terms of its length along each of the axes as

$$\overline{A} = A_x\hat{x} + A_y\hat{y} + A_z\hat{z} \tag{2.7}$$

where A_x, A_y, and A_z are the components of the vector with respect to the rectangular coordinate system.

The dot product of two vectors can be represented as

$$\overline{A} \cdot \overline{B} = A_xB_x + A_yB_y + A_zB_z \tag{2.8}$$

and the cross-product is

$$\overline{A} \times \overline{B} = \hat{x}(A_yB_z - A_zB_y) + \hat{y}(A_zB_x - A_xB_z) + \hat{z}(A_xB_y - A_yB_x) \tag{2.9}$$

The magnitude or length of the vector $\overline{A}$ is

$$\|\overline{A}\| = (\overline{A} \cdot \overline{A})^{1/2} = (A_x^2 + A_y^2 + A_z^2)^{1/2} \tag{2.10}$$

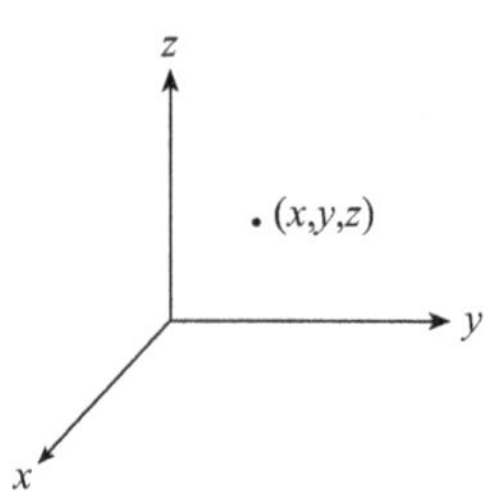

Figure 2.1 The rectangular coordinate system

If the components of the vector are complex, then the magnitude is

$$\|\overline{A}\| = (\overline{A} \cdot \overline{A}^*)^{1/2} = (|A_x|^2 + |A_y|^2 + |A_z|^2)^{1/2} \tag{2.11}$$

2.3.2 Cylindrical Coordinates

The circular cylindrical coordinate system is shown in Figure 2.2. The coordinates ρ, ϕ, z are related to the rectangular coordinates x, y, z by the formulas

$$x = \rho \cos\phi \tag{2.12a}$$

$$y = \rho \sin\phi \tag{2.12b}$$

$$z = z \tag{2.12c}$$

and

$$\rho = \sqrt{x^2 + y^2} \tag{2.13a}$$

$$\phi = \tan^{-1}\frac{y}{x} \tag{2.13b}$$

$$z = z \tag{2.13c}$$

The unit vectors for this coordinate system are $\hat{\rho}$, $\hat{\phi}$, and $\hat{z}$. Unlike the unit vectors associated with the rectangular coordinate system, which are fixed vectors and independent of position, the vectors $\hat{\rho}$ and $\hat{\phi}$ change directions depending on the angle ϕ. An arbitrary vector is expressed in terms of its cylindrical components as

$$\overline{A} = A_\rho\hat{\rho} + A_\phi\hat{\phi} + A_z\hat{z} \tag{2.14}$$

A vector is a coordinate-independent object, so a given vector can be represented in any given coordinate system by transforming its components in one coordinate system to a new set of components with respect to the basis vectors of another coordinate system. Tables of transformations among the representations of vector components in rectangular, cylindrical, and spherical coordinates can be found in many references, including [1].

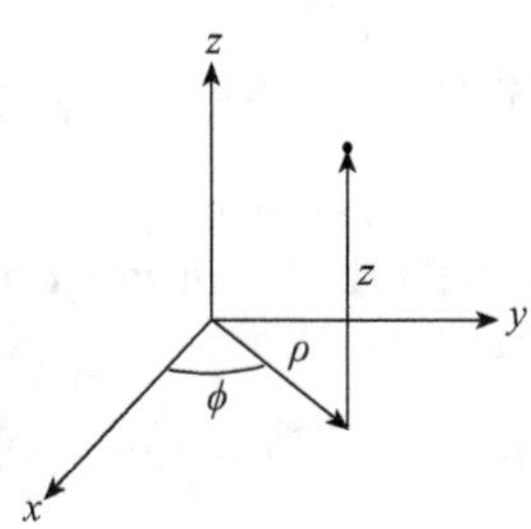

Figure 2.2 The cylindrical coordinate system

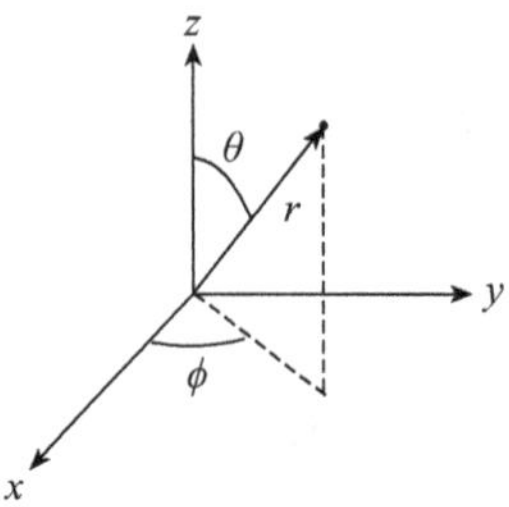

Figure 2.3 The spherical coordinate system

2.3.3 Spherical Coordinates

The spherical cylindrical coordinate system is shown in Figure 2.3. The coordinates r, θ, ϕ are related to the rectangular coordinates x, y, z via

$$x = r \sin\theta \cos\phi \tag{2.15a}$$

$$y = r \sin\theta \sin\phi \tag{2.15b}$$

$$z = r \cos\theta \tag{2.15c}$$

and

$$r = \sqrt{x^2 + y^2 + z^2} \tag{2.16a}$$

$$\theta = \tan^{-1} \frac{\sqrt{x^2 + y^2}}{z} \tag{2.16b}$$

$$\phi = \tan^{-1} \frac{y}{x} \tag{2.16c}$$

The unit vectors for this coordinate system are $\hat{r}$, $\hat{\theta}$, and $\hat{\phi}$. An arbitrary vector can be expressed as

$$\overline{A} = A_r \hat{r} + A_\theta \hat{\theta} + A_\phi \hat{\phi} \tag{2.17}$$

2.4 Gradient, Curl, and Divergence

The fields and source quantities in Maxwell's equations depend on position, and are represented as vector fields. A vector field evaluates to a vector at each point in space, so the coefficients are functions of position:

$$\overline{A}(x, y, z) = A_x(x, y, z)\hat{x} + A_y(x, y, z)\hat{y} + A_z(x, y, z)\hat{z} \tag{2.18}$$

Vector derivative operators such as the curl and divergence operations that appear in Maxwell's equations can be used to quantity how a vector field changes with position.

In the rectangular coordinate system, the vector derivative operator ∇ can be represented as

$$\nabla = \frac{\partial}{\partial x}\hat{x} + \frac{\partial}{\partial y}\hat{y} + \frac{\partial}{\partial z}\hat{z} \tag{2.19}$$

This object is an operator, and not a vector, since the "coefficients" are partial derivatives, but it can be applied to a scalar or combined with a vector field using the dot or cross-products as if it were a vector, to produce the gradient, curl, and divergence operations.

The gradient of the function $f(x, y, z)$ is the vector field

$$\nabla f = \frac{\partial f}{\partial x}\hat{x} + \frac{\partial f}{\partial y}\hat{y} + \frac{\partial f}{\partial z}\hat{z} \tag{2.20}$$

The curl of the vector field $\overline{A}$ is

$$\nabla \times \overline{A} = \dot{x}\left(\frac{\partial A_z}{\partial y} - \frac{\partial A_y}{\partial z}\right) + \dot{y}\left(\frac{\partial A_x}{\partial z} - \frac{\partial A_z}{\partial x}\right) + \dot{z}\left(\frac{\partial A_y}{\partial x} - \frac{\partial A_x}{\partial y}\right) \tag{2.21}$$

The divergence of $\overline{A}$ is the scalar function

$$\nabla \cdot \overline{A} = \frac{\partial A_x}{\partial x} + \frac{\partial A_y}{\partial y} + \frac{\partial A_z}{\partial z} \tag{2.22}$$

For a vector field expressed in terms of its cylindrical components,

$$\nabla \times \overline{A} = \hat{\rho}\left(\frac{1}{\rho}\frac{\partial A_z}{\partial \phi} - \frac{\partial A_\phi}{\partial z}\right) + \hat{\phi}\left(\frac{\partial A_\rho}{\partial z} - \frac{\partial A_z}{\partial \rho}\right) + \hat{z}\frac{1}{\rho}\left(\frac{\partial}{\partial \rho}\rho A_\phi - \frac{\partial A_\rho}{\partial \phi}\right) \tag{2.23}$$

$$\nabla \cdot \overline{A} = \frac{1}{\rho}\frac{\partial \rho A_\rho}{\partial \rho} + \frac{1}{\rho}\frac{\partial A_\phi}{\partial \phi} + \frac{\partial A_z}{\partial z} \tag{2.24}$$

In the spherical coordinate system,

$$\begin{aligned}\nabla \times \overline{A} = {} & \frac{\hat{r}}{r\sin\theta}\left[\frac{\partial}{\partial \theta}(A_\phi \sin\theta) - \frac{\partial A_\theta}{\partial \phi}\right] + \frac{\hat{\theta}}{r}\left[\frac{1}{\sin\theta}\frac{\partial A_r}{\partial \phi} - \frac{\partial}{\partial r}(rA_\phi)\right] \\ & + \frac{\hat{\phi}}{r}\left[\frac{\partial}{\partial r}(rA_\theta) - \frac{\partial A_r}{\partial \theta}\right]\end{aligned} \tag{2.25}$$

$$\nabla \cdot \overline{A} = \frac{1}{r^2}\frac{\partial}{\partial r}r^2 A_r + \frac{1}{r\sin\theta}\frac{\partial}{\partial \theta}\sin\theta A_\theta + \frac{1}{r\sin\theta}\frac{\partial}{\partial \phi}A_\phi \tag{2.26}$$

The gradient, curl, and divergence satisfy the identities

$$\nabla \times \nabla f = 0 \tag{2.27}$$

$$\nabla \cdot (\nabla \times \overline{A}) = 0 \tag{2.28}$$

as long as the function f and the components of the vector field $\overline{A}$ are smooth enough that partial derivative operators can be interchanged without changing the result.

2.5 Laplacian

The Laplacian operator is

$$\nabla^2 = \frac{\partial^2}{\partial x^2} + \frac{\partial^2}{\partial z^2} + \frac{\partial^2}{\partial z^2} \tag{2.29}$$

When applied to a vector field $\overline{A}$, the Laplacian operator is related to the gradient, curl, and divergence operators by

$$\nabla^2 \overline{A} = -\nabla \times \nabla \times \overline{A} + \nabla \nabla \cdot \overline{A} \tag{2.30}$$

For a scalar quantity, this simplifies to

$$\nabla^2 f = \nabla \cdot \nabla f \tag{2.31}$$

2.6 Wave Propagation

Many of the simulation examples we will consider deal with wave propagation. Accordingly, we will review the basic solution procedure for the wave equation and its propagating solutions.

The first step is to combine the system of first-order Maxwell's equations (2.1a)–(2.1d) into a single second-order partial differential equation. Taking the curl of both sides of Faraday's law (2.1a) and using the constitutive relation (2.4b) yields

$$\nabla \times \nabla \times \overline{E} = -\frac{\partial}{\partial t} \mu \nabla \times \overline{H} \tag{2.32}$$

We have assumed that the permeability μ is invariant of position (the medium or material in which the wave propagates is spatially homogeneous), which allowed the curl operator to be moved past the factor of μ on the right-hand side of this expression. Substituting Ampère's law into this expression with $\overline{J} = 0$ leads to

$$\nabla \times \nabla \times \overline{E} = -\frac{\partial^2}{\partial t^2} \mu \varepsilon \overline{E} \tag{2.33}$$

Here, we have assumed that the material parameters are constant, which allowed a time derivative to be moved past the factor of ε.

We now wish to apply the identity (2.30). The Laplacian applied to $\overline{E}$ in rectangular coordinates is

$$\nabla^2 \overline{E} = \frac{\partial^2 \overline{E}}{\partial x^2} + \frac{\partial^2 \overline{E}}{\partial y^2} + \frac{\partial^2 \overline{E}}{\partial z^2} \tag{2.34}$$

which is much simpler than the double curl operation in (2.33). Using Gauss's law for the electric field, and assuming that there are no sources ($\rho = 0$) and the medium is homogeneous, we have

$$\nabla \cdot \overline{E} = \frac{1}{\varepsilon} \nabla \cdot \overline{D} = 0 \tag{2.35}$$

so that the identity (2.30) applied to $\overline{E}$ becomes $\nabla \times \nabla \times \overline{E} = -\nabla^2 \overline{E}$. Using this to simplify (2.33) produces the wave equation

$$\nabla^2 \overline{E} = \varepsilon\mu \frac{\partial^2}{\partial t^2} \overline{E} \tag{2.36}$$

In the rectangular coordinate system, this wave equation expands to

$$\frac{\partial^2 \overline{E}}{\partial x^2} + \frac{\partial^2 \overline{E}}{\partial y^2} + \frac{\partial^2 \overline{E}}{\partial z^2} = \frac{1}{c^2} \frac{\partial^2 \overline{E}}{\partial t^2} \tag{2.37}$$

where the constant c is the speed of wave propagation,

$$c = \frac{1}{\sqrt{\mu\varepsilon}} \tag{2.38}$$

In a vacuum, $c \simeq 2.998 \times 10^8$ m/s.

Considering just one component of the electric field intensity, such as $u = E_x$, the wave equation reduces to

$$\frac{\partial^2 u}{\partial x^2} + \frac{\partial^2 u}{\partial y^2} + \frac{\partial^2 u}{\partial z^2} = \frac{1}{c^2} \frac{\partial^2 u}{\partial t^2} \tag{2.39}$$

where u represents any one of the components of the electric field intensity (E_x, E_y, or E_z). For a wave propagating in the x direction, the wave equation has solutions of the form

$$u(x,t) = u^+(x - ct) + u^-(x - ct) \tag{2.40}$$

where $u^+(x)$ and $u^-(x)$ represent arbitrary pulse shapes. The first term, $u^+(x - ct)$, represents a wave traveling in the $+x$ direction, and the second term, $u^-(x + ct)$, represents a wave in the $-x$ direction. More generally, the solutions to the wave equation can propagate in an arbitrary direction in the 3-D space. Wave solutions with arbitrary directions will be considered in Section 2.9.

2.7 Electromagnetic Boundary Conditions

At any two-dimensional (2-D) surface S, the following boundary conditions must hold:

Electromagnetic Boundary Conditions

$$\hat{n} \times (\overline{E}_1 - \overline{E}_2) = 0 \tag{2.41a}$$

$$\hat{n} \times (\overline{H}_1 - \overline{H}_2) = \overline{J}_s \tag{2.41b}$$

$$\hat{n} \cdot (\overline{D}_1 - \overline{D}_2) = \rho_s \tag{2.41c}$$

$$\hat{n} \cdot (\overline{B}_1 - \overline{B}_2) = 0 \tag{2.41d}$$

The subscripts 1 and 2 denote fields on either side of the surface S. $\hat{n}$ is a unit vector that is normal to S and points into region 1. $\overline{J}_s$ is the electric surface current density and has units A/m, and ρ_s is the surface charge density with units C/m^2.

Good conductors are often approximated as perfect electric conductors (PEC). Inside a PEC object, all time-varying fields must be zero (only a static magnetic field can exist). If the boundary conditions are applied at a PEC surface, the fields inside the object vanish, and the boundary conditions reduce to

$$\hat{n} \times \overline{E} = 0 \tag{2.42a}$$

$$\hat{n} \times \overline{H} = \overline{J}_s \tag{2.42b}$$

$$\hat{n} \cdot \overline{D} = \rho_s \tag{2.42c}$$

$$\hat{n} \cdot \overline{B} = 0 \tag{2.42d}$$

where $\hat{n}$ is normal to the PEC surface.

2.8 Time- and Frequency-Domain Representations

The computational electromagnetics examples we will consider in this book include both time-domain and frequency-domain problems. For frequency-domain problems, sources and fields are sinusoidal or time harmonic, and we assume that the system has reached a steady-state condition. Time-harmonic field and source quantities will be represented using phasors. Faraday's law in time-varying form, for example, is

$$\nabla \times \overline{E} = -\frac{\partial \overline{B}}{\partial t} \tag{2.43}$$

whereas the phasor form is

$$\nabla \times \underline{\overline{E}} = -j\omega \underline{\overline{B}} \tag{2.44}$$

The time-domain and phasor forms of the fields are related by

Phasor to Time-Domain Relationship

$$\overline{E}(x,y,z,t) = \mathrm{Re}[\underline{\overline{E}}(x,y,z)e^{j\omega t}] \tag{2.45}$$

In some treatments, particularly in the physics literature, the time variation is of the form $e^{-i\omega t}$, in which case phasors are complex conjugated relative to the convention of (2.45).

In (2.45), an underline is used to distinguish the phasor quantity from the time-varying field intensity, but in later chapters the underline will be omitted. Some texts use a different typeface to represent time-domain fields and phasors (e.g., $\overline{\mathscr{E}}(x,y,z,t)$

in the time domain and $\overline{E}(x, y, z)$ for phasors). In the remainder of this book, to reduce notational complexity, we will use the same typeface for time-domain and phasor fields, and the distinction should be clear from context or by explicit indication of the time dependence in the argument of the scalar or vector field.

In the frequency domain, Maxwell's equations are

$$\nabla \times \overline{E} = -j\omega\overline{B} \tag{2.46a}$$

$$\nabla \times \overline{H} = j\omega\overline{D} + \overline{J} \tag{2.46b}$$

$$\nabla \cdot \overline{D} = \rho \tag{2.46c}$$

$$\nabla \cdot \overline{B} = 0 \tag{2.46d}$$

where all field and source quantities are complex-valued vectors or scalars in the phasor signal representation. The time derivatives in (2.1a)–(2.1d) are no longer present and have been replaced by factors of $j\omega$.

2.9 Plane Waves

Plane waves are the most fundamental type of solution to Maxwell's equations in a source-free region. Since the fields associated with a plane wave are time-harmonic, it is convenient to use the frequency-domain representation. In terms of the phasor electric field intensity, the wave equation becomes

$$\left(\nabla^2 + k^2\right)\overline{E} = 0 \tag{2.47}$$

where

$$k = \omega\sqrt{\mu\varepsilon} \tag{2.48}$$

and $\omega = 2\pi f$. This PDE is referred to as the Helmholtz equation. Using the method of separation of variables [1], a solution to this equation can be shown to have the form

$$\overline{E}(x, y, z) = \overline{E}_0 e^{-jk_x x} e^{-jk_y y} e^{-jk_z z} \tag{2.49}$$

The constants k_x, k_y, and k_z determine the direction of propagation of the wave and are constrained by the dispersion relation

$$k_x^2 + k_y^2 + k_z^2 = k^2 \tag{2.50}$$

where $k^2 = \omega^2\mu\varepsilon$. $\overline{E}_0$ is a constant vector with complex coefficients that determines the polarization of the plane wave.

2.9.1 Wave Vector

It is common when working with plane waves to use vector notation to simplify the expression (2.49). The wave vector is defined by

$$\overline{k} = k_x\hat{x} + k_y\hat{y} + k_z\hat{z} \tag{2.51}$$

This vector points in the direction of wave propagation. It is sometimes useful to introduce the unit vector

$$\hat{k} = k^{-1}\overline{k} \tag{2.52}$$

which also points in the direction of propagation but has unit length. The length of the vector $\overline{k}$ is

$$k = \sqrt{k_x^2 + k_y^2 + k_z^2} \tag{2.53}$$

By the dispersion relation (2.50), k is equal to $\omega\sqrt{\mu\varepsilon}$.

2.9.2 Position Vector

We also define the position vector

Position Vector

$$\overline{r} = x\hat{x} + y\hat{y} + z\hat{z} \tag{2.54}$$

The vector $\overline{r}$ is nothing more than a compact notation for the point with coordinates (x, y, z). This vector extends from the origin of the coordinate system to (x, y, z). It is often convenient to express the coordinates x, y, and z in terms of the spherical angles θ and ϕ, so that the position vector becomes

$$\overline{r} = r\sin\theta\cos\phi\hat{x} + r\sin\theta\sin\phi\hat{y} + r\cos\theta\hat{z} \tag{2.55}$$

We also define the unit vector $\hat{r} = \overline{r}/r$, which is identical to the vector $\hat{r}$ in the orthogonal triple $(\hat{r}, \hat{\theta}, \hat{\phi})$ of unit vectors associated with the spherical coordinate system.

Many of the problems we will consider in this book are two-dimensional. To distinguish 2-D and 3-D formulations, we use the notation

2-D Position Vector

$$\overline{\rho} = x\hat{x} + y\hat{y} \tag{2.56}$$

for a point in the plane.

2.9.3 Plane Wave in Vector Notation

Using the wave and position vectors, the phasor electric field in (2.49) can be simplified to

Plane Wave in Phasor Representation

$$\overline{E}(\overline{r}) = \overline{E}_0 e^{-j\overline{k}\cdot\overline{r}} \tag{2.57}$$

The Helmholtz equation (2.47) places no constraint on the constant vector $\overline{E}_0$. This means that the Helmholtz equation admits both longitudinal ($\overline{E}$ parallel to $\overline{k}$) and transverse wave solutions ($\overline{E}$ orthogonal to $\overline{k}$). In a homogeneous, isotropic medium, however, Gauss's law for the electric field implies that the direction of wave propagation $\overline{k}$ and the electric field direction $\overline{E}_0$ are orthogonal, which means that propagating waves must be of the transverse type.

Example: x-Directed Plane Wave

For a plane wave propagating in the $+x$ direction and linearly polarized in the z direction, (2.57) becomes

$$\overline{E}(x) = E_0 \hat{z} e^{-jkx} \tag{2.58}$$

where E_0 is a complex constant that determines the amplitude and phase of the plane wave. Using (2.45) and assuming that E_0 is real, the time-dependent electric field intensity is

$$\overline{E}(x,t) = E_0 \hat{z} \cos(\omega t - kx) \tag{2.59}$$

The wavelength of the plane wave is

$$\lambda = \frac{2\pi}{k} = \frac{c}{f} \tag{2.60}$$

Example: Plane Wave in the x–y Plane

In this book, we will use the 2-D electromagnetic scattering problem to illustrate many numerical methods (see Section 2.13.6). For this type of problem, the incident field is often a plane wave propagating in the x–y plane at some angle ϕ^{i}. The wave vector direction for a wave arriving from the angle ϕ^{i} is

$$\hat{k} = -\hat{x}\cos(\phi^{\mathrm{i}}) - \hat{y}\sin(\phi^{\mathrm{i}}) \tag{2.61}$$

For the case of the electric field polarized in the z direction, the phasor plane wave is

$$\overline{E}(x) = \hat{z}E_0 e^{jk[x\cos(\phi^{\mathrm{i}})+y\sin(\phi^{\mathrm{i}})]} \tag{2.62}$$

If the amplitude E_0 is real, then the time-varying form of the electric field is

$$\overline{E}(x,t) = \hat{z}E_0 \cos\left\{\omega t + k[x\cos(\phi^{\mathrm{i}}) + y\sin(\phi^{\mathrm{i}})]\right\} \tag{2.63}$$

2.9.4 Characteristic Impedance

Using Faraday's law (2.46a), the magnetic field intensity associated with the plane wave in (2.58) can be found to be

$$\overline{H}(x) = -\frac{E_0}{\eta}\hat{y}e^{-jkx} \tag{2.64}$$

where the ratio of the electric and magnetic field intensities associated with a plane wave, or the characteristic wave impedance, is

$$\eta = \sqrt{\frac{\mu}{\varepsilon}} \tag{2.65}$$

This quantity is also referred to as intrinsic impedance. In a vacuum, $\eta \simeq 376.7\ \Omega$.

2.10 Propagating, Standing, and Evanescent Waves

There are three basic types of waves: propagating, standing, and evanescent. The plane wave in (2.49) represents a single wave propagating in one direction. More generally, the solution for a component of the electromagnetic field is a combination of plane waves propagating in both the positive and negative directions for each coordinate. Thus, the general solution for a component of the electromagnetic field expressed in rectangular coordinates is

$$E(x,y,z) = \left[A_m e^{-jk_x x} + B_m e^{jk_x x}\right]\left[C_m e^{-jk_y y} + D_m e^{jk_y y}\right]\left[E_m e^{-jk_z z} + F_m e^{jk_z z}\right] \tag{2.66}$$

where E represents E_x, E_y, or E_z. Using Euler's theorem, it is possible to express this solution in terms of sine and cosine functions:

$$\begin{aligned} E(x,y,z) = {} & \left[A'_m \sin(k_x x) + B'_m \cos(k_x x)\right]\left[C'_m \sin(k_y y) + D'_m \cos(k_y y)\right] \\ & \times \left[E'_m \sin(k_z z) + F'_m \cos(k_z z)\right] \end{aligned} \tag{2.67}$$

If the coefficients A'_m, B'_m, and k_x are real, then the x-dependent factor in this expression represents a standing wave in the x direction. A standing wave can be viewed as the superposition of two plane waves of equal amplitude propagating in opposite directions. A standing wave is formed when a wave bounces between two parallel conducting plates, so waveguide modes have the form of a standing wave in the coordinates transverse to the axis of the waveguide.

In certain situations, it is possible for one or more of the constants k_x, k_y, and k_z to be imaginary. If $k_x = j\alpha$, for example, then the first term in (2.66) becomes

$$f(x) = A_m e^{\alpha x} \tag{2.68}$$

If α is positive, as x increases the real exponential $e^{\alpha x}$ increases, and the wave grows in magnitude. If α is negative, then the wave decreases in amplitude for large x. A solution to Maxwell's equations with an exponential amplitude factor of this form is known as an evanescent wave. Evanescent waves are produced when a plane wave strikes a dielectric interface at an angle of incidence beyond the critical angle and undergoes total internal reflection or when a mode is excited in a waveguide at a frequency that is lower than the modal cutoff frequency.

2.11 Bessel Functions

Electromagnetic fields can be represented as combinations of complex exponentials or plane waves propagating in arbitrary directions. For structures that naturally match the rectangular coordinate system, such as a rectangular waveguide, plane waves form a convenient basis for field solutions. For geometries that are cylindrical, however, it is easier to represent field solutions in terms of cylindrical waves. A cylindrical wave is a combination of complex exponentials and Bessel functions.

Bessel functions are solutions to Bessel's differential equation, which is

$$\rho^2 \frac{d^2 f(\rho)}{d\rho^2} + \rho \frac{df(\rho)}{d\rho} + (\rho^2 - m^2) f(\rho) = 0 \tag{2.69}$$

The general solution to this differential equation is

$$f(\rho) = AJ_m(\rho) + BY_m(\rho) \tag{2.70}$$

where A and B are arbitrary coefficients, $J_m(x)$ is the Bessel function of order m, and $Y_m(x)$ is the Neumann function of order m. The Bessel function $J_m(x)$, Neumann function $Y_m(x)$, and the Hankel functions (as well as modified Bessel functions and several other variants not discussed here) are all referred to collectively as Bessel functions, since all of them are solutions to various forms of Bessel's differential equation.

The Bessel and Neumann functions are similar to the sine and cosine functions. Like the trigonometric functions, as well as many other tabulated special functions, the Bessel and Neumann functions can be defined using a power series with a given set of coefficients. The Bessel and Neumann functions are also oscillatory, but the amplitude of the oscillation decays as $x^{-1/2}$ as x becomes large. Graphs of the Bessel and Neumann functions of the first few orders are shown in Figures 2.4 and 2.5. As $x \to 0$, the Neumann functions tend to $-\infty$, due to a logarithmic singularity at $x = 0$. Several handbooks of mathematical functions (e.g., [2]) and online Internet references provide more details on special functions, including the Bessel and Neumann functions.

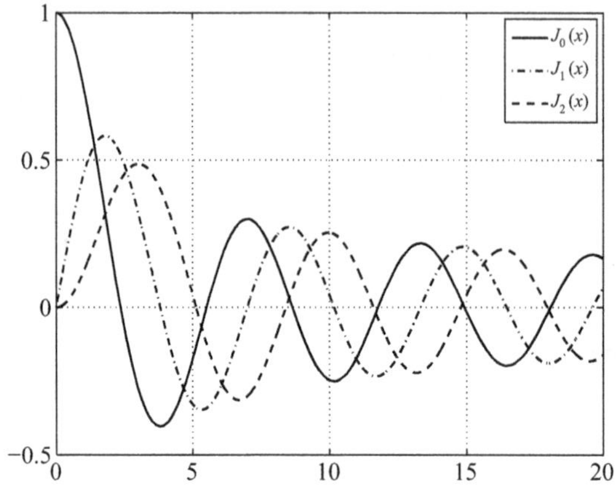

Figure 2.4 Bessel functions of orders 0, 1, and 2

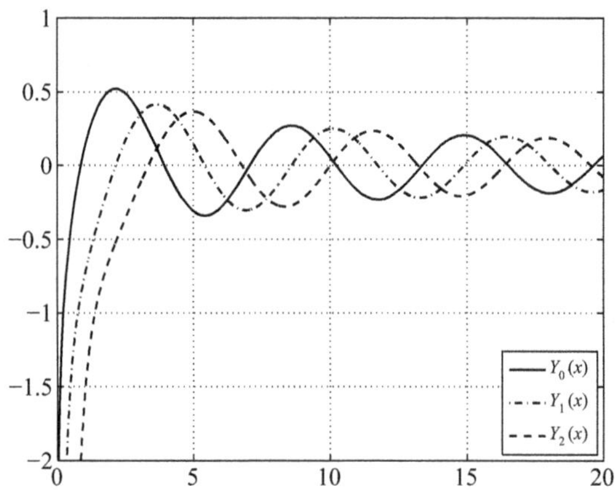

Figure 2.5 Neumann functions of orders 0, 1, and 2

2.11.1 Hankel Functions

The Hankel functions are related to the Bessel function $J_m(x)$ and the Neumann function $Y_m(x)$ by

$$H_m^{(1)}(x) = J_m(x) + jY_m(x) \tag{2.71a}$$

$$H_m^{(2)}(x) = J_m(x) - jY_m(x) \tag{2.71b}$$

These relationships are similar in form to Euler's theorem. Since Euler's theorem relates the real functions sine and cosine to the complex exponential, this implies that Hankel functions are analogous to complex exponentials. Physically, the Bessel and

Neumann functions $J_m(x)$ and $Y_m(x)$ represent standing waves, whereas the Hankel functions represent propagating waves.

When the argument is large, the Hankel functions can be approximated by

$$H_m^{(1)}(x) \simeq \sqrt{\frac{-2j}{\pi x}} e^{jx - m\pi/2} \tag{2.72a}$$

$$H_m^{(2)}(x) \simeq \sqrt{\frac{2j}{\pi x}} e^{-jx + m\pi/2} \tag{2.72b}$$

As with the Bessel and Neumann functions, the amplitude of the Hankel functions decays as $x^{-1/2}$ as x becomes large, and the phase of the oscillation depends on the order m. If the asymptotic forms of the Hankel functions are used in (2.45), it can be seen that the first-kind Hankel function $H_m^{(1)}(k_\rho \rho)$ represents a wave propagating in the $-\rho$ direction (i.e., an ingoing wave), and the second-kind Hankel function $H_m^{(2)}(k_\rho \rho)$ is an outgoing wave and propagates in the $+\rho$ direction.

2.11.2 Cylindrical Waves

Bessel functions can be used as building blocks for a type of field solution known as a cylindrical wave. For a cylindrical wave solution, the ρ dependence of the wave is given by a Bessel function. Using the method of separation of variables [1], it can be shown that any component of the electric field intensity can be expressed in the form

$$\begin{aligned} E(\rho, \phi, z) = {} & \sum_{m=0}^{\infty} \left[A_m J_m(k_\rho \rho) + B_m Y_m(k_\rho \rho)\right] \left[C_m e^{jm\phi} + D_m e^{-jm\phi}\right] \\ & \times \left[E_m e^{jk_z z} + F_m e^{-jk_z z}\right] \end{aligned} \tag{2.73}$$

where E represents E_ρ, E_ϕ, or E_z. Each term in the summation is a cylindrical wave of order m.

For a given electromagnetic boundary value problem, the coefficients A_m, B_m, C_m, D_m, E_m, and F_m are determined by the boundary conditions and source. The coefficients k_ρ and k_z are constrained by the dispersion relation

$$k_\rho^2 + k_z^2 = k^2 \tag{2.74}$$

where $k = \omega\sqrt{\mu\varepsilon}$ is the wave number of the fields. This dispersion relation is similar to (2.50), except that $k_x^2 + k_y^2$ is replaced by k_ρ^2.

Depending on the values of the coefficients C_m, D_m, E_m, and F_m, the cylindrical wave can either be standing or propagating in the ρ, ϕ, and z directions. If A_m and B_m are real, then the mth term represents a standing wave in the ρ direction. If A_m or B_m is imaginary, then the wave may propagate in the $+\rho$ or $-\rho$ direction. For propagating cylindrical waves, the Bessel and Neumann functions are typically combined to form Hankel functions.

2.12 Power and Energy

From Maxwell's equations, a conservation of energy relationship for electromagnetic fields can be derived:

$$\oint_S (\overline{E} \times \overline{H}) \cdot d\overline{S} = -\frac{\partial}{\partial t} \int_V \left[\frac{1}{2}\mu \|\overline{H}\|^2 + \frac{1}{2}\varepsilon \|\overline{E}\|^2 \right] dV - \int_V \sigma \|\overline{E}\|^2 dV - \int_V \overline{E} \cdot \overline{J}\, dV \tag{2.75}$$

This is known as Poynting's theorem. The terms in Poynting's theorem are

$$\oint_S (\overline{E} \times \overline{H}) \cdot d\overline{S} = \text{Power leaving the volume through the surface } S$$

$$\frac{1}{2}\mu \|\overline{H}\|^2 = \text{Stored magnetic energy density inside } V$$

$$\frac{1}{2}\varepsilon \|\overline{E}\|^2 = \text{Stored electric energy density inside } V$$

$$\frac{\partial}{\partial t} \int_V \left[\frac{1}{2}\mu \|\overline{H}\|^2 + \frac{1}{2}\varepsilon \|\overline{E}\|^2 \right] dV = \text{Rate of increase of stored energy inside } V$$

$$\int_V \sigma \|\overline{E}\|^2\, dV = \text{Power lost to heat inside } V$$

$$\int_V \overline{E} \cdot \overline{J}\, dV = \text{Power supplied by the impressed source } \overline{J}$$

where V is a 3-D region in space and S is the surface or boundary of V.

Since the first term of (2.75) represents the flow of energy through the surface S, the integrand

$$\overline{S}(\overline{r}, t) = \overline{E}(\overline{r}, t) \times \overline{H}(\overline{r}, t) \tag{2.76}$$

can be interpreted as the power flux density and has units W/m^2. This quantity is the instantaneous Poynting power flux density vector. The time-average Poynting vector

$$\overline{S}_{\text{av}} = \frac{1}{T} \int_0^T \overline{S}(\overline{r}, t)\, dt, \quad T = \frac{2\pi}{\omega} \tag{2.77}$$

represents the average real power density carried by the electric and magnetic fields.

For time-harmonic fields in phasor form, complex power flux density can also be defined, so for time-harmonic fields, there are three related vector power quantities:

Vector Power Quantities

1. Instantaneous Poynting vector (time domain):

   $\overline{S}(\overline{r}, t) = \overline{E}(\overline{r}, t) \times \overline{H}(\overline{r}, t)$

2. Time-average Poynting vector (time-harmonic fields):

   $\overline{S}_{\text{av}}(\overline{r}) = \frac{1}{T}\int_0^T \overline{S}(\overline{r}, t)\, dt,\ T = \frac{2\pi}{\omega}$

3. Complex Poynting vector (phasor domain):

   $\overline{S}(\overline{r}) = \overline{E}(\overline{r}) \times \overline{H}^*(\overline{r})$

The time-average power flux density associated with a time-harmonic electromagnetic field can be obtained from the complex Poynting vector using the relationship

Time-average Power Flux Density

$$\overline{S}_{\text{av}}(\overline{r}) = \frac{1}{2}\text{Re}\{\overline{E}(\overline{r}) \times \overline{H}^*(\overline{r})\} \tag{2.78}$$

2.13 Initial Value Problems and Boundary Value Problems

CEM methods can be used to solve hundreds of different types of problems that arise in engineering practice. To expose their essential unity and provide a logical framework for seemingly disparate types of electromagnetics problems, it is helpful to look at these problems from a more abstract, mathematical point of view. For frequency-domain analyses, the fundamental concept is the boundary value problem (BVP). A BVP consists of two ingredients:

Boundary Value Problem

1. *PDE:* A partial differential equation (PDE) defined on some region V of space governs the field solution for the problem. The PDE may be a single differential equation or a system of equations with multiple unknowns. The PDE may also include forcing functions that represent sources such as electric charges or currents.
2. *BC:* A boundary condition (BC) determines the behavior of the field solution on the boundary S of the region of interest V. Typical boundary conditions are vanishing tangential electric field at the surface of a conductor, or a radiation boundary condition for an infinite region. If V is unbounded, instead of specifying the value of a field quantity on a finite surface S, a radiation boundary condition specifies the asymptotic behavior of the field at infinity.

A PDE alone has many solutions. Together with forcing functions (source or excitations) and a properly posed boundary condition, a PDE has a unique solution. This important mathematical property corresponds to the fundamental physical property that two field measurements under the same conditions must yield the same result. Because of the importance of this property in mathematical physics, much effort over the history of the theory of BVPs has been devoted to proving the existence and uniqueness of a solution for various types of PDEs.

If the PDE includes time as an independent variable, one of the boundary conditions is an initial condition at a given time for the fields in the region of interest. Consequently, time-dependent PDE problems are referred to as initial value problems.

2.13.1 Modes

For some boundary value problems, the forcing function or source in the PDE is left unspecified. In the language of differential equations, this is called a homogeneous differential equation problem. Typically, the homogeneous PDE alone has a continuum of solutions, whereas if the BC confines the problem to a finite region, the solutions are countable. Solutions to a BVP without forcing functions are called modes. If a forcing function is specified, then the unique solution consists of a particular linear combination of the modes.

Consider the example of a perfectly conducting rectangular waveguide. The fields in the waveguide satisfy Maxwell's equations, and at each point inside the waveguide the solution can therefore be expressed as a combination of plane waves, of which there are an uncountably infinite number. Adding the boundary condition at the waveguide walls restricts the solution to a countable infinity of modal solutions, in which the fields behave like sine and cosine functions along a cross section of the waveguide, and have a complex exponential form in the longitudinal direction. Adding a forcing function (placing a driving current source in the waveguide) completes the BVP, uniquely specifying the amplitudes of each waveguide mode and leading to a unique solution for the fields inside the waveguide.

Numerical methods are most commonly used to find the unique solution to a BVP, although numerical methods are also available that find the modes of a structure without a specified forcing function or source. Typically, finding the unique solution to a BVP corresponds to solving a linear system with the right-hand side vector determined by the source, whereas characterizing the modes of a structure generally requires finding the eigenvalues and eigenvectors of a matrix.

2.13.2 1-D, 2-D, and 3-D Boundary Value Problems

Electromagnetic boundary value problems and Maxwell's equations inherently involve three spatial dimensions plus time. To simplify the mathematical and numerical treatment, it is common to consider problems for the sources and structures have one or more directions of translational invariance. The field solutions inherit the symmetry of the boundary value problem, and one or more of the coordinate variables can be eliminated from the governing partial differential equations. In this book, we will consider one-dimensional (1-D), two-dimensional (2-D), and three-dimensional (3-D) BVPs.

For 1-D and 2-D problems, numerical algorithms can be developed more easily than for 3-D problems. This allows the basic principles of finite difference methods, the method of moments, and other numerical methods to be learned without dealing with multidimensional arrays and BCs that become more complex for higher dimensional problems. 1-D and 2-D numerical methods are limited in capability but can often be used in practice to approximate 3-D problems and obtain insight and even quantitative results much more quickly than would be required with a more computationally expensive 3-D algorithm. For real-world analysis and design problems with tight design requirements, approximate solutions can be inadequate and 3-D numerical methods are generally required.

2.13.2.1 1-D Problems

If the sources and structures in an EM boundary value problem are infinite in two dimensions and only vary in one direction, the problem is said to be one-dimensional. This is illustrated in Figure 2.6. An example of a 1-D problem is the wave equation considered in Section 4.2. The field varies only in the x direction in this case and is invariant in the y and z directions. Maxwell's equations reduce to a 1-D wave equation that can be solved with a simple finite difference algorithm. The 1-D finite difference time domain (FDTD) algorithm is developed in Section 4.2.

Another way to reduce the dimensionality of an EM problem is to use the thin-wire integral equations discussed in Section 6.2.5. The boundary value problem remains three-dimensional, but the unknown quantity in the integral equation is a current that is approximated as a 1-D filament along a thin antenna or other conductor.

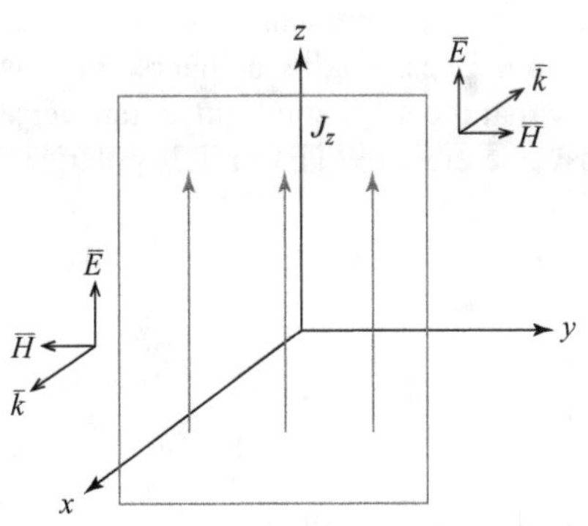

Figure 2.6 One-dimensional EM problem. An infinite planar time-harmonic current sheet in the y–z plane radiates plane waves propagating in the +x and −x directions. The plane waves are polarized in the same direction as the current flow. When solving such a 1-D problem numerically, the code only needs to model the x dependence of the sources and fields

2.13.2.2 2-D Problems and the Transverse Electric (TE) and Transverse Magnetic (TM) Polarizations

For some boundary value problems, the material structures are long cylinders that can be approximated as infinite in length. Examples include circular waveguides, rectangular waveguides, coaxial cable, wires, and optical fibers. If a cylindrical structure and the sources that radiate the fields that excite the structure can be approximated as infinite in length, a three-dimensional boundary value problem can be reduced to a two-dimensional boundary value problem.

Typically, the "long" dimension of the material geometry is along the z-axis of the rectangular coordinate system. If the cross-sectional shape of the material geometry and sources are invariant in the z direction, then the field solution is also independent of the z coordinate. In the PDE that governs the fields, all z derivatives therefore evaluate to zero, and the z coordinate drops out of the problem almost entirely.

For electromagnetics problems, even though the material geometry, sources, and fields do not depend on the z coordinate, it is possible for the field vectors to have a z component. The fields radiated by a general z-invariant source $\overline{J}(x,y) = J_x(x,y)\hat{x} + J_y(x,y)\hat{y} + J_z(x,y)\hat{z}$ can be decomposed into a combination of the fields radiated by the first two terms, with the electric field vector in the x–y plane, and by the third term, with electric field in the z direction. If the materials and structures in the problem are infinite in the z direction, then the field radiated by the part of the source in the x–y plane has no z component. Similarly, the field radiated by the z component of the current has no x or y component.

For this reason, we distinguish between two types of 2-D problems: one for which the electric field intensity vector $\overline{E}$ is in the z direction, so that $\overline{E} = E_z\hat{z}$; and another for which the magnetic field is in the z direction, so that $\overline{H} = H_z\hat{z}$. In the first case, the symmetry of the problem dictates that the magnetic field is confined to the x–y plane. Since the magnetic field has no z component, this case is referred to as the transverse magnetic (TM) polarization. Occasionally, a superscript z is added to the notation TM^z to indicate explicitly that the magnetic field is transverse to the z direction. The second case is the transverse electric (TE) or TE^z polarization. The two cases are illustrated in Figure 2.7.

For the two possible polarizations for 2-D problems, the electric and magnetic field intensity vectors are of the form

$$\left.\begin{aligned} \overline{E} &= E_z\hat{z} \\ \overline{H} &= H_x\hat{x} + H_y\hat{y} \end{aligned}\right\} \quad \text{TM Polarization} \tag{2.79}$$

$$\left.\begin{aligned} \overline{E} &= E_x\hat{x} + E_y\hat{y} \\ \overline{H} &= H_z\hat{z} \end{aligned}\right\} \quad \text{TE Polarization} \tag{2.80}$$

Since an infinitely long, cylindrical object made of an isotropic material cannot convert an incident wave of one polarization to a scattered field of the other polarization, the polarization for a given problem is dictated by that of the incident field. In general, it is possible for the fields in the problem to be a combination of the TM and TE polarizations, but the two cases are nearly always solved separately and then combined if needed.

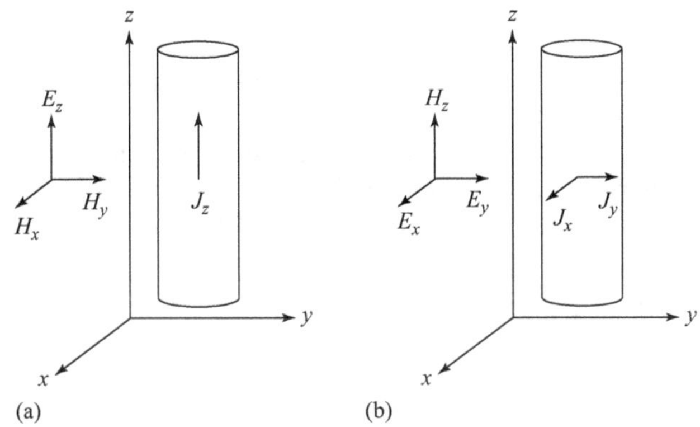

Figure 2.7 Two-dimensional EM problems. (a) For the TM^z polarization, current sources flow in the z direction, radiating z-directed electric field, and magnetic field in the x–y plane. (b) For the TE^z polarization, current sources flow in the x–y plane, radiating z-directed magnetic field, and electric field in the x–y plane. In both cases, all source distributions, materials, interfaces, and structures are infinite in the z direction

Numerical methods for two-dimensional problems are usually easier to implement and less computationally intensive than methods for three-dimensional problems. For this reason, we will place a significant emphasis on 2-D boundary value problems in this book. Further discussion of 2-D problems can be found in Section 4.4.1.

2.13.2.3 3-D Problems

If the sources and materials associated with an EM boundary value problem have no invariant direction, the problem is said to be three-dimensional. In this case, the fields vary in x, y, and z, and all three spatial coordinates must be considered in a numerical solution algorithms. Numerical algorithms for 3-D boundary value problems are treated in Sections 4.8 and 6.8.

2.13.3 Radiation and Scattering Problems

Two important classes of boundary value problems are radiation and scattering problems. Radiation and scattering problems are distinguished by the type and location of the source or excitation. For either type of problem, the region of interest is typically unbounded, and the BC is a radiation condition at infinity.

Radiation problems involve materials such as conductors and dielectrics near a radiating source. Antenna analysis is the most prominent type of radiation problem. Commonly, the source is a transmission line or feed gap corresponding to a driving excitation at the terminals of the antenna.

Scattering problems model the fields scattered by a material object illuminated by a known incident field from a distant source. Since the source is far from the object, the incident field can typically be approximated as a plane wave. The source itself is assumed to be decoupled from the scatterer and is not included within the domain of the problem solution. The canonical scattering problem involves finding the radar signal reflected by a metallic target, although many other physical problems from applications quite different from radar target analysis can be treated as scattering problems. In the BVP viewpoint, the radiating source for a radiation problem or the incident field for a scattering problem becomes part of the forcing function or boundary condition.

Another related BVP classification is that of interior and exterior problems. For interior problems, the source is inside an impenetrable object, such as a PEC cavity, and the solution domain is finite. Exterior problems have an unbounded solution domain and usually involve a radiation boundary condition on the fields as they propagate to infinity.

2.13.4 Inverse Problems

Radiation and scattering problems are forward problems, in the sense that a material structure and excitation or source are specified, and the goal is to find the fields that satisfy Maxwell's equations for the given configuration. An inverse scattering problem involves the determination of the shape and material properties of an unknown structure from measurements of fields scattered by the structure for one or more incident sources. Radar is a very simple inverse problem for which only the location of an object and possibly limited size information are determined from scattered fields. Ground penetrating radar and medical imaging of biological tissues using microwaves or ultrasound are examples of more sophisticated inverse techniques. As a general rule, inverse problems are computationally more challenging than forward problems, ultimately because there are many similar objects that scatter an incident field in nearly the same way.

2.13.5 Green's Functions and Radiation Integrals

Boundary value problems can be solved using many techniques. One approach is to solve the boundary value problem for the field radiated by a point source defined by a delta function located at some point in space. This is known as a Green's function.

In principle, any given boundary problem with arbitrary materials and structures has a Green's function. A closed form solution for the Green's function does not exist in general for complicated boundary value problems, but the Green's function can be found analytically for free space and a few other types of boundary value problems with simple, canonical geometries for the structures involved.

If Green's function is known for a given boundary value problem, the field radiated by an arbitrary, distributed source can be found with a radiation integral. A radiation integral is a solution for the fields radiated by a current source. The integral includes the Green's function weighted by the current source in the integrand. If the boundary value problem is shift-invariant (e.g., free space or a homogeneous

Table 2.2 Scalar Green's functions for time-harmonic problems of various dimensions

Dimension	Green's function	Source	Solution type
1-D:	$g(x,x') = -\frac{1}{2jk}e^{-jk\lvert x-x'\rvert}$	Plane current	Plane wave
2-D:	$g(\overline{\rho},\overline{\rho}') = -\frac{j}{4}H_0^{(2)}(\lvert\overline{\rho}-\overline{\rho}'\rvert)$	Line current	Cylindrical wave
3-D:	$g(\overline{r},\overline{r}') = \frac{e^{-jk\lvert\overline{r}-\overline{r}'\rvert}}{4\pi\lvert\overline{r}-\overline{r}'\rvert}$	Point source	Spherical wave

medium), then the radiation integral is a convolution of the Green's function and current source.

A list of scalar, free-space Green's functions is given in Table 2.2. A scalar Green's function is a solution to the Helmholtz partial differential equation with a delta function type source. The scalar Green's function can be used to find the electric or magnetic field radiated by a 1-D, 2-D, or 3-D source in free space. By convolving the scalar Green's function with a current source distribution $\overline{J}(\overline{r}')$, we can find the magnetic vector potential $\overline{A}(\overline{r}')$. From the magnetic vector potential, we can compute the electric or magnetic field radiated by the source. This set of calculations can be combined into one expression, which is referred to as the free-space radiation integral. Green's functions and the free-space radiation integral are covered in Sections 4.6.2.1 and 4.6.2.2 for 2-D problems. The 3-D Green's function and radiation integral are given in Section 6.8.

2.13.6 Formulating and Solving Boundary Value Problems

A given boundary value problem has different formulations, or ways in which the solution can be expressed. A BVP generally is defined using a basic set of partial differential equations. If the PDEs are a coupled set of first-order differential equations, like Maxwell's equations, the PDEs can be used to derive a second-order partial differential equation. The PDEs can be transformed into an integral equation. The BVP also has a variational formulation. If the solution to the BVP is known for one or more sources or excitations, coefficients in the PDE can be determined numerically. This is an inverse problem. Each of these ways of expressing a boundary value problem leads to a different numerical method. All these approaches will be considered in this book.

To illustrate these formulations, we will consider an example BVP, 2-D electromagnetic scattering of a TM^z polarized incident field by a conducting or dielectric object. This BVP will be treated from various perspectives throughout this book. We summarize the following various ways in which this BVP can be formulated and solved:

- Coupled set of first-order PDEs (Maxwell's equations). This is the most direct solution method. The first-order PDEs can be solved by using the Yee cell construction to derive the 2-D FDTD method (Section 4.4).

- Integral equation formulation. Using a solution to the free-space BVP with a point source, or a Green's function, the 2-D scattering problem can be transformed to an integral equation. This integral equation is solved using the method of moments (MoM) in Section 6.4.
- Second-order wave equation or Helmholtz equation formulation. From the coupled first-order frequency domain form of Maxwell's equations, we can derive a second-order Helmholtz equation (in the time domain, this would be a wave equation).[1]
- Variational principle formulation. The Helmholtz equation can be solved by applying the Rayleigh–Ritz procedure to a variational principle for the Helmholtz equation to derive the finite element method (FEM) as in Chapter 8.
- Inverse scattering problem. From measurements of the scattered field, the dielectric constant in the Helmholtz equation or the shape of the scatterer can be determined (Chapter 10).

We have seen that the same BVP can be solved using different mathematical formulations and numerical methods. It is also true that many different BVPs can be reduced or related to a mathematical problem involving the same differential operator. This means that the core algorithm for a solver for one differential operator can be used to solve different types of physical problems.

2-D electromagnetic scattering problems and the analysis of waveguide modes, for example, can be formulated to involve the Laplacian operator. In the former case, we discretize the Laplacian and solve a matrix equation to find fields scattered by an object. For waveguide mode analysis, we discretize the Laplacian with a BC on the waveguide surface and find the eigenvalues of the resulting matrix. From the eigenvalues, we can compute the waveguide mode cutoff frequencies. The connection between modal problems and scattering problems allows complex algorithms to be developed in stages in Chapters 4 and 8, where the core solver for the Laplacian is tested first for the simpler waveguide problem and then combined with additional processing needed for the more challenging 2-D electromagnetic scattering problem.

2.13.7 The Equivalence Principle

A powerful tool in electromagnetic theory for transforming and simplifying problems of many kinds is the equivalence principle. The equivalence principle is applied when deriving the near to far field transformation used in postprocessing for the FDTD method (Section 4.6.2) and in developing integral equations for electromagnetic boundary value problems (Section 6.2).

The goal when using the equivalence principle to a boundary value problem is usually to replace a complicated system or structure with an equivalent current source in free space. The equivalence principle can be used to transform a boundary

[1] For 3-D problems, the wave equation and Helmholtz equation allow for longitudinal waves that do not satisfy Maxwell's equations, so the second-order formulation cannot be used directly as a basis for a numerical solution method.

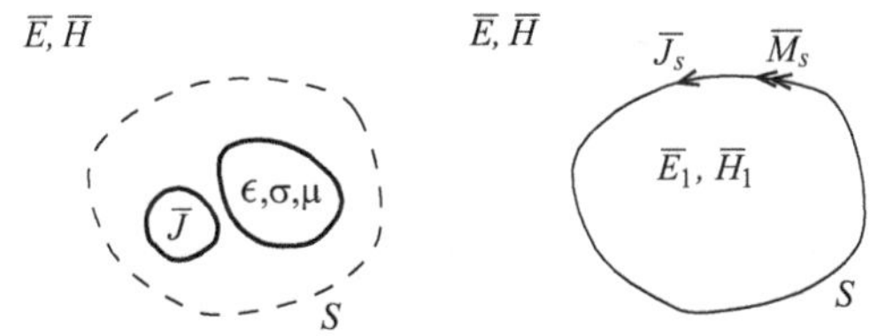

Figure 2.8 Left: original boundary value problem with sources and materials inside the mathematical surface S. Right: equivalent problem with impressed surface currents on the surface S that radiate the original fields $\overline{E}$ and $\overline{H}$ outside S and fields $\overline{E}_1$ and $\overline{H}_1$ inside S

value problem from one involving material and structures to a new problem with an equivalent source radiating in free space so that the free-space Green's function and radiation integral can be used to find the radiated fields. With the equivalence principle, it is common to introduce fictitious impressed magnetic current sources $\overline{M}$ as well as electric currents (see Section 2.2.3).

The equivalence principle is illustrated in Figure 2.8. A mathematical surface S is introduced into the original problem. The region of interest is outside the surface S. When using the equivalence principle, all sources and materials are usually inside S and outside S is free space. Equivalent sources $\overline{J}_s$ and $\overline{M}_s$ are impressed on the surface. The fields outside the surface S are unchanged. The fields inside the surface are changed by the radiation from the surface currents to $\overline{E}_1$ and $\overline{H}_1$. $\overline{E}_1$ and $\overline{H}_1$ are arbitrary and can be chosen for convenience to suit various applications of the equivalence principle.

If there are sources or materials inside S, they can be removed using the equivalence principle. The new, equivalent boundary value problem then only includes current sources radiating in free space. The radiated fields can be found using the free-space radiation integral. The equivalence sources on S are given by

$$\overline{M}_s = -\hat{n} \times (\overline{E} - \overline{E}_1) \tag{2.81}$$

$$\overline{J}_s = \hat{n} \times (\overline{H} - \overline{H}_1) \tag{2.82}$$

where $\overline{J}_s$ and $\overline{M}_s$ are electric and magnetic surface currents with units A/m and V/m, respectively, and $\hat{n}$ is the outward pointing normal vector on the surface S. Surface currents can be modeled as volume currents with a delta function that confines the current flow to an infinitely thin sheet. The sources $\overline{J}_s$ and $\overline{M}_s$ radiate the fields $\overline{E}$ and $\overline{H}$ outside S and $\overline{E}_1$ and $\overline{H}_1$ inside S.

A common choice for the fields inside S is $\overline{E}_1 = \overline{H}_1 = 0$. This is known as Love's equivalent. With Love's equivalence principle, the fields in the region of interest remain unchanged and the fields outside the region of interest are zero. The equivalent sources become

$$\overline{M}_s = -\hat{n} \times \overline{E} \tag{2.83}$$

$$\overline{J}_s = \hat{n} \times \overline{H} \tag{2.84}$$

Love's equivalence principle is used in modeling aperture antennas, deriving the near to far transformation used in Section 4.6.2, deriving integral equations for electromagnetic radiation and scattering, and other applications.

2.14 Other Topics

This chapter is intended to review a few basic concepts from electromagnetic field theory and to establish notation and conventions that will be used throughout this book. Other important topics from electromagnetics, including electrostatics, anisotropic materials, Green's functions, and integral equations, will be introduced on an as-needed basis in later chapters.

If a more in-depth treatment is desired, there are numerous comprehensive treatments of electromagnetics, including introductory textbooks [1,3–8], more advanced references [9–11], and freely available online resources.

Problems

2.1 Find the gradient of the function $f(x,y,z) = x^2 + y^2 + z^2$. How does the vector field ∇f relate to the graph of the function?

2.2 Compute the curl and divergence of the following vector fields:
(a) $x^2\hat{y} + y^2\hat{x}$
(b) $\rho\hat{\phi}$
(c) $\sin\theta\hat{\phi}$

2.3 Check the following identities:
(a) $\nabla \times (\nabla\phi) = 0$
(b) $\nabla \cdot (\nabla \times \overline{A}) = 0$
(c) $\nabla^2\overline{A} = -\nabla \times (\nabla \times \overline{A}) + \nabla(\nabla \cdot \overline{A})$
To simplify the calculations, this can be done for a simple vector field with only an x component.

2.4 Write down (a) the electric field intensity vector and (b) the magnetic field intensity vector in phasor form for a plane wave with frequency 2.4 GHz, propagating in the $\hat{x} + \hat{y}$ direction, linearly polarized in the z direction, and carrying a time-average power density of 2 W/m^2. (c) Give the time-varying electric and magnetic fields. Provide numerical values for all constants and coefficients in the expressions. (d) Is there more than one electric field that meets the given criteria?

2.5 Starting with Maxwell's equations and the constitutive relations, derive a wave equation for the magnetic field intensity $\overline{H}$.

2.6 Find the (a) complex Poynting vector and (b) time average Poynting vector for a plane wave. (c) What is the electric field strength corresponding to 10 W/m^2 time average power density?

2.7 Find an expression for the electric field radiated by the planar current source $\overline{J}(\overline{r}) = \hat{y}J_0 \cos(\omega t)\delta(x)$. Hint: This planar current distribution radiates plane waves in the $+x$ and $-x$ directions. Write expressions for the two plane waves, and use the magnetic field boundary condition at $x = 0$ to find the amplitudes of the fields.

2.8 A plane wave with electric field intensity 1 V/m strikes an infinite perfect electric conductor (PEC) plane at a normal incidence angle. (a) Write down expressions for the incident and reflected plane waves. (b) Use the electromagnetic boundary conditions to find the amplitude of the reflected electric field relative to the amplitude of the incident field (reflection coefficient). (c) Use the magnetic field boundary condition to find the current induced on the PEC plane.

References

[1] C. A. Balanis, *Advanced Engineering Electromagnetics*. New York, NY: John Wiley & Sons, 1989.

[2] M. Abramowitz and I. Stegun, *Handbook of Mathematical Functions with Formulas, Graphs and Mathematical Tables*. New York, NY: John Wiley & Sons, 1993.

[3] M. N. O. Sadiku, *Elements of Electromagnetics*. Orlando, FL: Saunders, 1989.

[4] R. S. Elliott, *Electromagnetics—History, Theory, and Applications*. New York, NY: IEEE Press, 1991.

[5] Z. Popović and B. D. Popović, *Introductory Electromagnetics*. Englewood Cliffs, NJ: Prentice Hall, 2000.

[6] D. A. De Wolf, *Electromagnetics for Engineering*. Cambridge: Cambridge University Press, 2001.

[7] N. Ida, *Engineering Electromagnetics*. Berlin: Springer, 2003.

[8] K. E. Lonngren, S. V. Savov, and R. J. Jost, *Fundamentals of Electromagnetics With MATLAB*, 2nd edn. London: SciTech, 2007.

[9] R. F. Harrington, *Time-Harmonic Electromagnetic Fields*. New York, NY: McGraw-Hill, 1961.

[10] W. C. Chew, *Waves and Fields in Inhomogeneous Media*. New York, NY: IEEE Press, 1995.

[11] P. Russer, *Electromagnetics, Microwave Circuit and Antenna Design for Communications Engineering*, 2nd edn. Norwood, MA: Artech House, 2006.

Chapter 3
Basic Numerical Tasks

As a warmup before developing more sophisticated numerical methods, we will first develop in this chapter some of the basic routines that are commonly used in scientific computation and numerical modeling. These include numerical differentiation, integration, interpolation, and curve fitting. Since numerical differentiation and integration are building blocks for algorithms like the finite difference method and the method of moments, we will return to these two key topics in later chapters with more rigorous treatments.

Since the algorithm examples given throughout this book are based on the MATLAB® programing language, we will also introduce the MATLAB programing environment and some of the basic commands that will be used throughout the book. The simple algorithms introduced in this chapter are intended as an easy way to learn MATLAB programing. Many textbooks, monographs, and online references on numerical analysis are available that provide greater depth on these topics (e.g., [1,2]).

3.1 Introduction to MATLAB Programing

In this tutorial, we will assume that the reader is able to run the MATLAB user interface. Since MATLAB is built around an interactive programing environment, commands can be entered and executed directly at the prompt indicated by the symbol ». To assign a value to a variable, enter the command

```
» a = 3
```

and press the <Enter> key to execute the assignment. MATLAB executes the command and displays the result `a = 3`. To suppress the interactive output of the result, end the line with a semicolon:

```
» a = 3;
```

MATLAB executes the command and returns to the command prompt without displaying the result of the operation. Further commands can use the variable `a` in formulas and other operations. The command

```
» a^2
```

returns the result

```
ans = 9
```

The variable `ans` is used by MATLAB to store the result of any operation that is not explicitly assigned to a variable. The MATLAB interpreter is case sensitive, so `a` and `A` represent different variables. A list of defined variables can be displayed using `who`.

3.1.1 Vectors and Arrays

MATLAB is particularly suited for manipulating vectors and multidimensional arrays. An empty matrix can be created using

```
A = zeros(10,20);
```

This command creates an array that has 10 rows and 20 columns. With one argument, the command `B = zeros(10);` creates a square matrix with 10 rows and 10 columns. Predefined matrix types include the following:

Predefined Matrix Types

```
B - zeros(M,N);          % M by N matrix of zeros
C = ones(M,N);           % M by N matrix of ones
D = eye(M);              % M by M identity matrix
E = rand(M,N);           % Random matrix (uniform PDF)
F = randn(M,N);          % Random matrix (normal PDF)
G = spalloc(M,N,100)     % Sparse matrix data type
H = diag(1:10);          % Diagonal matrix
I = zeros(M,N,P);        % Three dimensional array
```

The `%` symbol indicates a comment that is not executed. Elements of a matrix can be accessed using `A(m,n)`, which returns the matrix element A_{mn}. A 2-D array can also be treated as a vector with one argument, so that `A(m)` returns the *m*th element of the array counting along rows (row scanned). Vectors and arrays need not be preallocated before values are added. If the variable `J` is undefined, then executing the command `J(3) = 1` produces the row vector `[0,0,1]`.

3.1.1.1 Colon Operator

When creating vectors, the colon operator is particularly useful. The command `1:10` forms a vector of length 10 with elements $[1, 2, \ldots, 10]$. The command

```
x = a:step:b
```

forms a vector with smallest element `a` and adjacent elements increasing by `step` until the upper value `b` is reached but not exceeded. An array of values can be passed as an argument to a function.

The following code illustrates the use of the colon operator to display a graph of the sine function:

```
» x = 0:.01:10*pi;
» plot(x,sin(x))
```

3.1.1.2 Vector and Array Operations

Vectors and arrays can be multiplied, divided, or transformed using many different MATLAB functions. The previously created vector `x` can be used as the argument of a more complication function such as $x \sin(x)$:

```
» y = x.*sin(x);
» plot(x,y)
```

The `.*` operator causes MATLAB to multiply the values stored in the arrays `x` and `sin(x)` on an element-by-element basis. Arrays combined using the operator `.*` and other similar operators such as `./` must have identical dimensions. Using vector and array operations rather than a loop to compute many values at once is known as "vectorizing" a code and typically leads to much faster execution.

Other useful array operations include the following:

Array Operations

```
A'          % Hermitian or conjugate transpose
A.'         % Transpose without complex conjugate
A^2         % Matrix square A*A
A.^2        % Element-by-element square A(m,n)*A(m,n)
A*B         % Matrix product AB
A.*B        % Element-by-element product
sqrt(A)     % Element-by-element square root
sqrtm(A)    % Matrix square root
eig(A)      % Eigenvalues of A
inv(A)      % Matrix inverse
x = A\b     % Solve the linear system A*x = b
```

3.1.2 Working with Plots

The command `help plot` displays documentation for the `plot` function. MATLAB help can also be accessed through the graphical user interface. Commands are available to change attributes of the plot such as the range of x or y values (`xlim` and `ylim`) or the axis labels (`xlabel` and `ylabel`). The range of x values in the previously created plot can be adjusted by entering the command

```
» xlim([0 10*pi])
```

The command `plot(y)` creates a graph that is similar to `plot(x,y)` but with the independent values on the horizontal axis equal to the integer indices of the elements of the vector `y`.

When comparing numerical results to analytical or expected data, it is helpful to overlay multiple curves on one plot. This can be done using

```
» plot(x,sin(x),'b',x,cos(x),'r')
```

Another way to overlay curves is

```
» plot(x,sin(x),'b')
» hold on
» plot(x,cos(x),'r')
» hold off
```

The strings `'b'` and `'r'` set the line colors to red and blue, respectively. Other colors are listed in the `help plot` documentation entry. To add markers to a line, use

```
» plot(x,sin(x),'bs-')
```

The string `'s'` adds squares to the line at each data point. If the line type is `'bs'`, square markers are plotted without a line. Plots with a logarithmic scale on the x-axis, y-axis, or both can be created using `semilogx`, `semilogy`, or `loglog`, respectively.

MATLAB handles can be used to control many aspects of a plot's appearance. The command

```
» get(gca)
```

displays a list of plot attributes that can be changed using the `set` command. The function `gca` returns the handle of the current selected plot. The font size, for example, can be adjusted for a more pleasing appearance using

```
» set(gca,'fontsize',18)
```

Each line in a plot also has attributes that can be adjusted. The command

```
get(gca,'children')
```

returns the handles of lines on the current selected plot. The command

```
» set(get(gca,'children'),'linewidth',2)
```

thickens the lines on the current plot.

Plots can be saved, printed, or converted to a variety of graphical file formats. To save a plot as a portable document format (PDF), use

```
print -dpdf filename.pdf
```

The current selected plot is saved as a PDF file in the working directory. See `help print` for other formats. To create a new figure window or select existing figure windows, use the `figure` command. With an argument, `figure(1)` selects figure 1 and brings it to the top of other figure windows.

3.1.2.1 3-D Plots

MATLAB can also produce density plots, contour plots, 3-D plots, and other specialized plots of various types. To view a comprehensive list of plotting commands, use the commands `help graph2d`, `help graph3d`, and `help specgraph`.

For plotting 2-D field solutions, useful routines include `image` and `imagesc`. The latter is particularly convenient, since the data array is automatically scaled to span the range of available colors before it is plotted. If `A` is a real-valued matrix, `imagesc(A)` produces a 2-D density plot with pixel colors chosen based on the values of the elements of `A`. Typically, the values in the matrix represent field solution data at points in the x–y plane. If `x` and `y` are vectors of the coordinates of values in `A`, then `imagesc(x,y,A)` produces a plot with the x and y coordinate values labeled on the plot. To show the color map used to generate the plot, use the command `colorbar`.

The MATLAB `image` and `imagesc` commands display arrays in the standard way with rows horizontal. For arrays representing samples in the x–y plane, however, it is better to have the first array index represent the x coordinate of the physical domain and the second index represent y, with the (1,1) element corresponding to the lower left corner of the physical domain. To view an array of x–y plane data in the proper orientation, rotate the array using the command

```
imagesc(x,y,A.')
```

The command

```
set(gca,'ydir','normal')
```

adjusts the vertical axis so that values increase toward the top of the plot, according to the convention for x–y plots, rather than from top to bottom as used when writing out the elements of a matrix with row 1 at the top.

The command `plot3(x,y,z)` produces a 3-D scatter plot of data in the vectors or arrays `x`, `y`, and `z`. Other useful 3-D plotting routines include `mesh`, `surf`, `waterfall`, and `slice`. The `view` command can be used to rotate the viewpoint of a 3-D plot. The MATLAB figure window interface also includes tools that can be used with a mouse to rotate the view and otherwise transform the plot.

3.1.3 Scripts

While all MATLAB commands can be accessed from the command line, complex routines are typically saved as text files or scripts and executed by entering the name of the script at the command prompt. MATLAB scripts are also referred to as *m*-files. To create a script, enter the command

```
» edit scriptexample.m
```

at the command prompt. This command runs a text editor in which MATLAB commands can be entered and stored. If `scriptexample.m` contains the text

scriptexample.m

```
x = 0:.01:10*pi;
y = x.*sin(x);
plot(x,y)
xlim([0 10*pi])
```

and the file `scriptexample.m` is saved in the MATLAB current working directory, entering the command

```
» scriptexample
```

at the command prompt executes the code in the script to generate a plot of the function $y = x \sin x$.

3.1.3.1 Comments

Comments can be added to scripts using the `%` symbol. The script in the previous section with comments added is

scriptexample.m with comments

```
% scriptexample.m
%
% Display a plot of the function x*sin(x)
%

% Generate an array of x coordinate values
x = 0:.01:10*pi;

% Evaluate the function x*sin(x)
y = x.*sin(x);

% Display the plot
plot(x,y)
xlim([0 10*pi])
```

The `help` command returns all commented lines at the beginning of a script or function until the first blank line or uncommented MATLAB command, so `help scriptexample` can be used to display the leading comments in the script in the MATLAB command window. When creating scripts and functions, it is important to use comments liberally to aid in debugging and modifying the code and to help other users understand how to use the script or function.

3.1.3.2 Parameter Values and Plot Commands

Scripts are useful when a numerical algorithm must be rerun several times with different parameter values. Since a given algorithm is almost always modified and used in slightly different ways for a number of problems, it is tempting to copy a script and save multiple versions of the file. Making multiple copies of a code has the disadvantage that if a change is made or a bug fixed the change must be propagated to all the versions of the code. A better approach to managing multiple sets of parameter values is to keep only a single copy of the code and embed parameters and short sections of code for each type of problem inside `if 1 ... end` and `if 0 ... end` statements that can be turned on and off as needed, as illustrated in the following example:

Changing Parameter Values in Scripts

```
% First set of parameter values
if 1
   a = 0;
   b = pi;
end

% Second set of parameter values
if 0
   a = 0;
   b = 10;
end

% Generate an array of x coordinate values
x = a:.01:10*b;

% Evaluate the function x*sin(x)
y = x.*sin(x);

% Display the plot
plot(x,y)
xlim([a b])
```

The first set of parameter values is used by the script as written. By changing `if 0` to `if 1`, the second set of parameter values can be activated.

Commands used to create and format plots should also be embedded in scripts, rather than entered interactively. A rule of thumb is that a given plot of a numerical result will be regenerated five to ten times before the result is validated as correct. When debugging or using an algorithm, to avoid having to reenter plotting commands many times over include the commands used to plot results from the algorithm at the end of a code or in a separate script rather than entering them interactively.

3.1.4 Functions

A function is a script that includes a defined interface for passing and returning variables. Unlike a script, a function does not have access to variables stored in the MATLAB workspace memory. Only globally defined variables such as predefined constants (e.g., `pi`) and variables passed to the function as an argument are available for manipulation by commands in the function. The first line of a function after comments should include the MATLAB keyword `function` and a list of the input and output variables. The function `rms.m`, for example, returns the root mean square (RMS) value of numbers stored in the input vector `x`:

rms.m (Function Example)

```
% function y = rms(x)
%
% Compute RMS value of the elements of the vector x
% Input: x = real or complex vector
% Output: y = RMS value of elements of x
%
function y = rms(x)

N = length(x);
y = sqrt(sum(abs(x).^2)/N);
```

If this function is stored in the file `rms.m` in the current working directory or in a directory listed in the current MATLAB path, then the function can be called from the command line, a script, or another function. When entered at the MATLAB prompt, the command `rms(1:5)` returns the value 3.3166. The function definition is repeated in the leading comments so that the command `help rms` returns the function definition as well as initial comments in the function text file.

Creating functions for commonly used operations is a powerful programing technique, but with MATLAB it is sometimes convenient to use scripts rather than functions when debugging. When a script is run from the command line, internal variables are accessible interactively when a program breaks. Local variables used within a function are not available at the command line but can be accessed by embedding the command `keyboard` in the function. When the `keyboard` command is reached, program control returns to the MATLAB interactive prompt in debug mode, and all local variables are available at the command prompt.

Once a program is debugged and converted to a function, it is useful to create a script that stores the function calls for future reference rather than entering them interactively. When using numerical methods, a code is often run many times before the final desired result is obtained. It is easier to edit a script with function calls than to make changes at the command line, and the final parameter values are stored and can be documented within the script.

3.1.5 Arguments and Structure Arrays

As suggested in Section 3.1.4, numerical algorithms can be easier to debug in scripts rather than as functions. Once an algorithm has been validated, it can be converted to a function for convenient reuse. For complex functions with many parameters, it is convenient to use the MATLAB `struct` command to create structure arrays with fields containing parameter values. The structure can be passed as a single variable to a function. The function then has access to all parameter values. Most importantly, when new parameters are added, they can be included as fields in the structure and passed to the function without having to change the function's argument list.

3.1.6 Speeding Up MATLAB Codes

For numerical algorithms, efficiency and execution time are critical. Numerical methods require repeated evaluation of functions, integrals, and other operations. Execution time grows rapidly with the number of grid points, mesh elements, or degrees of freedom. When implementing numerical methods, familiarity with methods for improving run time is important. Some methods for speeding up codes and algorithms include:

Speeding Up MATLAB Codes

- When using `for` loops and other types of loops, assign variables and perform calculations outside of the loop when possible.
- Use vector and array operations (see Section 3.1.1.2).
- Use `parfor` to distribute looped operations across processor cores.
- Plot or image commands used to visualize a numerical solution as it evolves within a loop can slow down the code execution. Use the condition `if mod(n,10)==0` to render the plot every tenth iteration of the loop. The `drawnow` command should also be used to update the figure while the code is running (see Section 4.2.7 for an example).
- Compiled programing languages such as C can be faster than interpreted language implementations like MATLAB. Compiled languages can be integrated with MATLAB using `mex` functions.

3.1.7 Other MATLAB Commands

MATLAB includes many other programing language constructs, operators, built-in functions, and scripts. The `help` command returns a list of categories of help documents. `help general` returns a list of commands used to manage the MATLAB workspace, creating and managing files, accessing operating system functions, and controlling the command window and interactive environment. `help lang` lists commands for loops, controlling program flow and other programing operations. `help ops` returns a list of operators. A variety of additional MATLAB commands and operators will be introduced throughout this book as new algorithms are presented.

3.2 Numerical Differentiation

Numerical methods often require evaluation of derivatives and integrals. Numerical approximations to the exact derivative of a function can be derived from the definition of the derivative. These approximations are the basis for the finite difference method of Chapter 4.

The derivative of the function $f(x)$ can be defined by the limit

$$f'(x) = \lim_{\Delta x \to 0} \frac{f(x + \Delta x) - f(x)}{\Delta x} \tag{3.1}$$

To compute derivatives numerically, we simply choose a small but finite value for Δx rather than passing to the limit $\Delta x \to 0$:

$$f'(x) \simeq \frac{f(x + \Delta x) - f(x)}{\Delta x} \tag{3.2}$$

To increase the accuracy of the computed derivative, this difference formula can be modified so that $f(x)$ is evaluated symmetrically about x according to

Central Difference Rule

$$f'(x) \simeq \frac{f(x + \Delta x/2) - f(x - \Delta x/2)}{\Delta x} \tag{3.3}$$

This is the first-order central difference approximation for the derivative. Another common notation for the small parameter Δx is h.

By evaluating the difference approximations for a polynomial, it is easy to see that the central difference approximation is more accurate than the forward difference approximation in (3.2). Both difference approximations are exact if $f(x)$ is a first-order polynomial. For $f(x) = x^2$, we have

$$\frac{f(x + \Delta x) - f(x)}{\Delta x} = 2x + \Delta x \tag{3.4}$$

$$\frac{f(x + \Delta x/2) - f(x - \Delta x/2)}{\Delta x} = 2x \tag{3.5}$$

which shows that the central difference approximation is exact for second-order polynomials as well. In general, the error for the forward difference approximation is $O(\Delta x)$, whereas the error for the central difference approximation is $O(\Delta x^2)$.

To further illustrate the accuracy of the central difference formula, it is helpful to plot $f(x)$ and compare the graph of the function with the straight line through the point

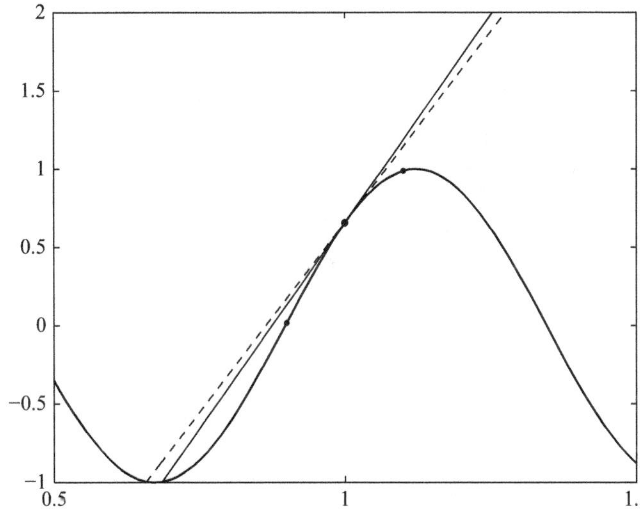

Figure 3.1 *Numerical differentiation of the function $f(x) = \sin(7x)$ at the point $x_0 = 1$, with $\Delta x = 0.2$. The solid straight line is $a(x - x_0) + f(x_0)$ with the slope a equal to the exact derivative $f'(x_0)$, and the dashed line has the slope given by the derivative estimate* (3.3). *The points $[x_0 - \Delta x, f(x_0 - \Delta x)]$, $[x_0 + \Delta x, f(x_0 + \Delta x)]$, and $[x_0, f(x_0)]$ are indicated with markers*

$[x_0, f(x_0)]$ having slope given by (3.3), as shown in Figure 3.1. Estimates of derivatives are noise sensitive, because low amplitude noise with rapid spatial variation can perturb $f'(x)$ significantly. There are many other ways to estimate derivatives, based on rigorous techniques for characterizing and improving the accuracy of finite difference approximations like (3.3).

In the example used in Figure 3.1, the accuracy of the central difference rule can be easily improved by reducing the parameter Δx. Due to rounding error, however, Δx cannot be made arbitrarily small. Δx should be set to a value somewhat larger than the smallest difference between two machine-representable numbers. In MATLAB, this can be done using `dx = 10*eps`. This value works well if the goal is to estimate a derivative accurately. In Chapter 4, the central difference rule will be used to transform a differential equation into a set of difference equations. In that case, the smaller Δx, the larger the number of difference equations, so Δx must be chosen to compromise between accuracy and computation time.

3.2.1 Code Example: Central Difference Rule

To aid the reader in developing good coding practice for the algorithms to be considered throughout this book, a short MATLAB script is presented here that implements the central difference rule illustrated in Figure 3.1.

Central Difference Rule Example Script

```
% Compute a derivative using the central difference rule

% Define a function handle
f = @(x) sin(7*x);

% Define point at which derivative is computed
x0 = 1;

% Difference step size
dx = 0.2;

% Approximate value of derivative
D = (f(x0+dx/2) - f(x0-dx/2))/dx;
% Display the result
disp(['Derivative = ',num2str(D)])
```

The `@()` operator avoids the need for a separate file to define a function and allows the function in the derivative computation to be easily changed. Function handles can be passed as arguments to other functions.

3.3 Numerical Integration

The simplest method for numerical integration is Euler's rule or the Riemann sum. The interval of integration $[a, b]$ is divided into N subintervals of width h, and the area under the graph of the function $f(x)$ is approximated by the sum of the areas of N rectangles of width h and height $f(x_n)$, where x_n is a point in the nth subinterval. As will be proved in Chapter 5, the best choice for x_n is the midpoint of each region, so that $x_n = a + (n - 1/2)h$, $n = 1, \ldots, N$ and $h = (b - a)/N$. The integral is approximated by

$$\int_a^b f(x)dx \simeq \sum_1^N w_n f(x_n) \tag{3.6}$$

where $w_n = h$. This formula is commonly referred to as the midpoint rule. An example is shown in Figure 3.2. Integration is less noise sensitive than differentiation, because zero-mean noise integrates to a small value over large intervals, whereas differentiation amplifies the effect of the noise on the computed result.

For the midpoint rule, the function $f(x)$ is evaluated at evenly spaced points x_n and the weights x_n are all equal. We will see in Chapter 5 that more accurate integration rules can be derived by choosing unevenly spaced integration points and unequal weights.

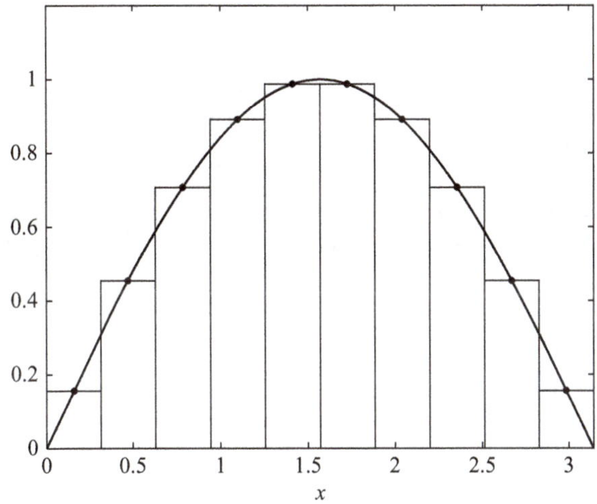

Figure 3.2 Numerical integration of the function f(x) = sin x from 0 to π using the midpoint rule with N = 10 and h = π/10. The total area of the rectangles is 2.0082, and the exact value of the integral is 2

3.3.1 Code Example: Midpoint Integration Rule

In implementing the midpoint integration rule, the MATLAB colon operator can be used to generate a vector of integration points. The actual integration is accomplished using the `sum` function, which adds the elements in a vector and returns the result.

Midpoint Integration Rule Example Script

```
% Define integrand
f = @(x) sin(x);

% Define region of integration
a = 0;
b = pi;

% Number of integration points and step size
N = 10;
h = (b-a)/N;

% Generate integration points
x = (a+h/2):h:(b-h/2);

% Approximate value of integral
I = sum(f(x))*h
```

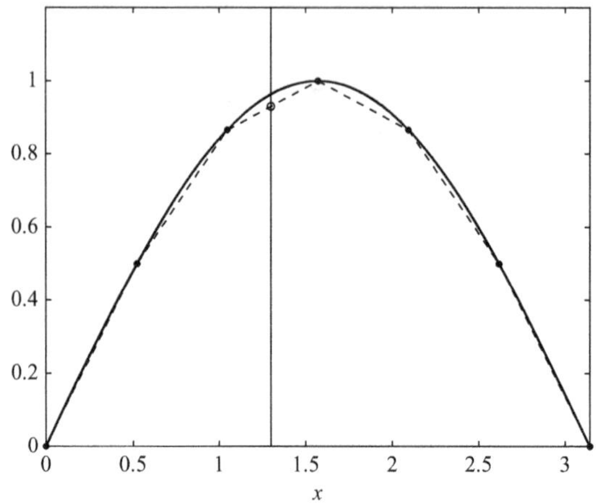

Figure 3.3 Interpolation of $\sin x$ *at* $x = 1.3$*. Solid dots mark the sample points. The interpolated point is marked with a circle. The interpolated value is 0.93, and the exact value is* $\sin 1.3 = 0.96$

3.4 Interpolation

If we have sample values of a function $f(x)$ at a discrete set of points x_n, interpolation is the task of estimating $f(x)$ at a point x lying between the sample points. The simplest method is nearest-neighbor interpolation, which takes $f(x)$ to be $f(x_n)$, where x_n is the sample point closest to x. This can be considered a zeroth-order interpolation method, since the function is approximated as a constant on intervals around the points x_n.

First-order or linear interpolation is most commonly used in practice. $f(x)$ is taken to lie on a straight line between the points $[x_{n-1}, f(x_{n-1})]$ and $[x_n, f(x_n)]$, where x_{n-1} and x_n are the two sample points closest to x, as shown in Figure 3.3. For linear interpolation, $f(x)$ depends only on the values of the function at two adjacent sample points. By allowing the interpolated value $f(x)$ to depend on more than two samples, higher order interpolation methods can be obtained.

Extrapolation is similar to interpolation but with x outside the interval spanned by the sample points. Extrapolation is less accurate and more noise sensitive than interpolation.

One-dimensional nearest neighbor, linear, and higher order interpolation are implemented in the MATLAB function `interp1`. Two-dimensional interpolation can be done using `interp2`, but the locations of the data points must be regular. If the data points are nonuniform, `griddata` or `scatteredInterpolant` can be used instead.

3.5 Curve Fitting

A set of measurements of a physical quantity can often be represented in the form $y_n = f(x_n) + p_n$, $n = 1, 2, \ldots, N$, where $f(x)$ represents the underlying physical behavior of interest, and p_n are samples of a random noise process. To estimate the function $f(x)$, we often want to find the coefficients of a polynomial that is a close fit to the samples y_n. If $f(x)$ is slowly varying with respect to x and the noise samples are statistically independent, then the polynomial can closely approximate $f(x)$ even if the noise samples are large in value in relation to the "signal" $f(x_n)$. Even for noise-free data, it is often valuable to approximate a complex function as a simple polynomial. Curve fitting is closely related to interpolation, since the curve can be used to estimate $f(x)$ at any point x.

To fit a curve to a set of data points, we choose a set of M basis functions $f_n(x)$ that are like $f(x)$ and not like the noise p_n. The function $f(x)$ is often smooth or slowly varying, whereas the noise is uncorrelated from sample to sample and so is rapidly varying with respect to x. In this case, the natural choice for the basis functions $f_n(x)$ are low-order polynomials.

The goal is to find M numbers a_n such that

$$f(x) \simeq \sum_{n=1}^{M} a_n f_n(x) = \hat{f}(x) \tag{3.7}$$

The functions $f_n(x)$ can be called basis functions. One way to choose the a_n is according to the rule

$$a_1, a_2, \ldots, a_M = \underset{a_1, a_2, \ldots, a_M}{\operatorname{argmin}} \sum_{n=1}^{N} \left| \hat{f}(x_n) - y_n \right|^2 \tag{3.8}$$

This leads to the approximation $\hat{f}$ that is closest to the y_n in the sense of least squared distance from the sample values y_n at the sample points x_n.

In matrix notation, we can write

$$\hat{f} = Fa \tag{3.9}$$

where $\hat{f}_n = \hat{f}(x_n)$, $a = [a_1, a_2, \ldots, a_M]^T$, and $F_{mn} = f_n(x_m)$ is an $N \times M$ matrix. The sum of the squared distances is then

$$\sum_{n=1}^{N} \left| \hat{f}(x_n) - f(x_n) \right|^2 = (Fa - y)^T (Fa - y) \tag{3.10}$$

where $y = [y_1, y_2, \ldots, y_N]^T$. If we minimize this expressing by taking the derivative with respect to each of the a_n and setting the result to zero, we obtain

$$0 = 2(F^T F)a - 2F^T y \tag{3.11}$$

Solving for the vector a gives

$$a = (F^T F)^{-1} F^T y \tag{3.12}$$

Since F is typically nonsquare, it cannot be inverted directly. The matrix $(F^T F)^{-1} F^T$ on the right-hand side of this expression is a type of pseudo-inverse that exists for nonsquare matrices for which $F^T F$ is invertible.

3.5.1 Code Example: Polynomial Fitting

For polynomial curve fitting, the basis function are monomials, so that $f_n(x) = x^{n-1}$. The number of functions M is one greater than the order of the fitted polynomial. The most common case is a first-order polynomial fit ($M = 2$), which is commonly referred to as linear regression. In this case, $f_1(x) = 1$, and $f_2(x) = x$. The second-order case ($M = 3$) is a quadratic curve fit. Examples of first- and second-order curve fitting are shown in Figure 3.4.

One issue with polynomial fitting is that as the order increases, the fitted polynomial becomes unstable and can "ring" or oscillate in between data points. To avoid this problem, polynomial fitting is typically used only for low orders (linear, quadratic, or sometimes cubic). If greater smoothness is needed, more sophisticated approaches to curve fitting are available, such as piecewise polynomials fit in subintervals or splines that enforce continuity in the function and its derivatives at subinterval boundaries.

To implement polynomial curve fitting, the MATLAB code fragment

```
F = [];
for n = 1:M,
    F(:,n) = x.^(n-1);
end
```

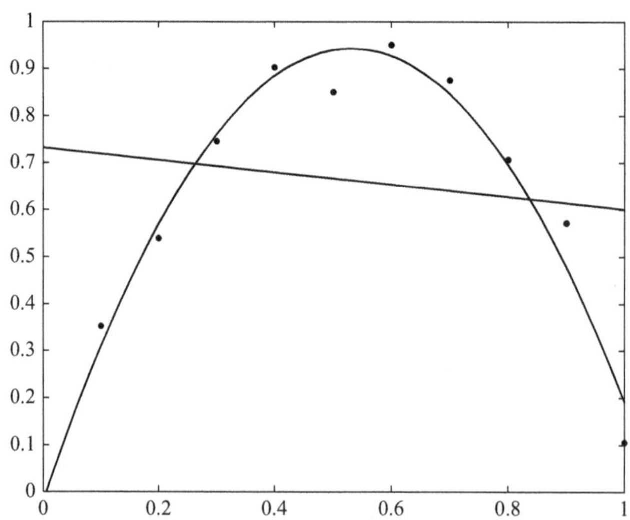

Figure 3.4 First- and second-order polynomial fits to noisy data. The solid lines represent the linear and quadratic functions that minimize the RMS value of the vertical distances between the polynomial curve and the data points

can be used to fill the matrix F. x is a column vector of the locations of data points. The colon symbol allows an entire row or column of an array to be accessed, so that F(:,n) represents the nth column of the matrix F. Each iteration of the loop adds a column to the matrix F. For large numbers of data points, the code will run more quickly if the matrix F is preallocated. This can be done if needed using F = zeros(length(x),M).

If y is a vector of data points, then the polynomial coefficients can be computed using

```
a = (F'*F)\F.'*y.';
```

where y is a column vector of data point values at the locations in the vector x. The backslash operator \ is an efficient way to compute $A^{-1}b$, where A is a square matrix, and b is a vector. The value of the fitted polynomial at a more finely sampled vector of points can be computed using

```
x2 = [0:.001:1];
y2 = polyval(flipud(a),x2);
```

If the data points in x are in a different range, the limits 0 and 1 in x2 can be adjusted accordingly. The flipud command is used to reorder the entries in the vector a because the polyval function expects the coefficient of the highest order monomial to be the first element of a, whereas the definition $F_{mn} = x_m^{n-1}$ implies that the lowest order coefficient is the first element.

The data point values can be plotted using plot(x,y,'o'). The polynomial fit curve can be added to the plot using

```
hold on
plot(x2,y2)
hold off
```

If the code is implemented correctly, then the polynomial curve is close to the data point values. What happens to the polynomial fit as the order $M-1$ of the polynomial is increased (e.g., to 10 or 20)?

3.6 Newton's Method

Occasionally one must compute the zeros of a given function. A convenient and efficient algorithm for this is Newton's method. The algorithm begins with a starting estimate x_0 of the location of a zero of $f(x)$. With the derivative $f'(x)$, it is straightforward to find the equation

$$y = f'(x_0)(x - x_0) + f(x_0) \tag{3.13}$$

of the line that passes through the point $[x_0, f(x_0)]$ and is tangent to the graph of $f(x)$ at $x = x_0$. The x intercept of this line will typically be closer to the actual zero of $f(x)$ than x_0. This provides a second estimate x_1 of the zero. Iterating the algorithm leads to a sequence $x_0, x_1, x_2, \ldots$ that converges rapidly to a zero of $f(x)$. If the function has multiple zeros, the zero found by the algorithm depends on the choice of the initial guess x_0. If the derivative of $f(x)$ is not available analytically, the central difference formula can be used to estimate $f'(x)$. This is the Newton–Raphson method.

Problems

3.1 Become familiar with MATLAB by running the example commands, scripts, functions, and plots in Section 3.1.

3.2 Write a MATLAB script to estimate the derivative of $f(x) = \sin(x)$ at $x = 0.5$ using the central difference approximation. Compare the answer with the analytical derivative for different values of Δx.

3.3 Write a MATLAB script to estimate the value of the integral

$$I = \int_0^1 \sin(x)\, dx \tag{3.14}$$

using the midpoint rule. Compare the answer with the analytical integral for different values of the width Δx of the integration subintervals. How rapidly does the error go to zero as Δx decreases?

3.4 Create 20 or so noisy samples of $f(x) = \sin(x)$ by adding samples of a Gaussian random variable with a standard deviation of about 0.1. The noise samples can be generated using `0.1*randn(size(x))`, where `x` is a vector of sample locations. Choose the sample points to be evenly distributed on the interval $0 \leq x \leq \pi$. Write a code to fit a quadratic polynomial to these samples using the least squares approach. Create a plot of the fitted polynomial superimposed on the noisy data points and the noise-free function $f(x)$. Use different line markers or colors so the curves can be distinguished. Hint: To plot the fitted polynomial and the function $f(x)$, create a more finely spaced vector of sample points using `x1 = 0:.01:pi`.

3.5 Use the MATLAB `interp1` function to interpolate the noisy samples generated in Problem 3.4. Compare the results with polynomial curve fitting by plotting the interpolated values from `interp1` overlaid on a plot of the fitted polynomial from Problem 3.4. How are the results similar? How are they different?

3.6 Use the polynomial fitting code from Problem 3.4 with the polynomial order set to 1. Verify that the resulting line slope and y-intercept agree with the standard formulas for linear regression.

3.7 Implement Newton's method in code and test the routine with the functions $f(x) = x^2$ and $f(x) = \cos x$.

3.8 Implement and test the Newton–Raphson method.

References

[1] R. L. Burden and J. D. Faires, *Numerical Analysis*, 8th edn. Pacific Grove, CA: Brooks Cole, 2005.

[2] J. Kiusalaas, *Numerical Methods in Engineering With MATLAB*. New York, NY: Cambridge University Press, 2010.

Chapter 4

Finite Difference Methods

The basic approach to solving a partial differential equation (PDE) numerically is to transform the continuous differential equation into a finite number of difference equations, which can be solved using a computational algorithm to obtain an approximate solution to the PDE. The general framework for this approach is called the finite difference (FD) method. Since nearly every branch of mathematical physics involves partial differential equations, finite difference methods are used in a vast array of applications, including electromagnetic (EM) wave propagation, optics, fluid dynamics, analysis of structures, acoustics, solid-state physics, and quantum mechanics.

A finite difference algorithm requires a grid or list of points spanning the computational domain, a stencil or difference equation similar to (3.3) that relates sample values of a function at the grid points in a way that approximates a PDE, a boundary condition or difference equation for points on the boundary of the computational domain, a source or excitation, and a method for solving the difference equations for sample values of the unknown function or field at the grid points. Postprocessing is used to compute derived physical quantities from the field solution.

When the PDE includes time as an independent variable, the FD approach is typically referred to as the finite difference time-domain (FDTD) algorithm. In this chapter, we will develop FD and FDTD solvers for a sequence of PDEs of increasing complexity. We will begin with the one-dimensional (1-D) wave equation, and then we will consider Laplace's equation with two spatial dimensions, Maxwell's equations for two-dimensional (2-D) problems, and the full system of three-dimensional (3-D) Maxwell's equations. Algorithm stability and accuracy will be analyzed, and several types of postprocessing will be developed to extract useful physical quantities from computed field solutions. Comprehensive references on the FDTD method include [1–3].

4.1 Basic Components of Finite Difference Solvers

The core of an FD algorithm is a finite difference equation or stencil for approximating the derivative of a continuous function in terms of the value of the function at discrete sample points. In practice, the basic set of finite difference equations is usually easy to derive and translate into code. Supporting functions such as grid generation, boundary conditions, and postprocessing are actually more difficult and generally require many more lines of code than the core finite difference routine, so we will give significant attention to these critical aspects of practical implementations of the FDTD algorithm.

4.1.1 Grid

A grid is a list of points at which the unknown function in the PDE is sampled. A grid is most often regular or evenly spaced, so that in 1-D the grid points can be written in the form

$$x_n = n\Delta x \tag{4.1}$$

where $n = 1, 2, \ldots, N$, and Δx is the distance between grid points or the step size. For 2-D problems, a rectangular grid consists of points for which

$$(x_m, y_n) = (m\Delta x, n\Delta y) \tag{4.2}$$

with an additional offset if needed to properly locate the computational domain in the coordinate plane. A cubic 3-D grid is defined similarly with points given by

$$(x_m, y_n, z_p) = (m\Delta x, n\Delta y, p\Delta z) \tag{4.3}$$

A two-dimensional rectangular grid is shown in Figure 4.1.

For problems involving both space and time coordinates, a temporal grid for the time coordinate is also required. Typically, the temporal grid consists of time points of the form

$$t_n = n\Delta t \tag{4.4}$$

where Δt is the temporal step size.

Other types of grids include polar, hexagonal, and conformal. Conformal grids are used when it is advantageous to have grid points that follow curved edges of a material object. Formulating the difference equations for the FD algorithm is much

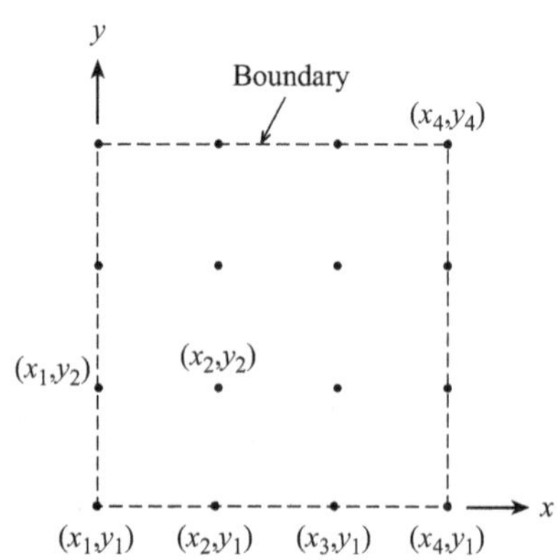

Figure 4.1 Two-dimensional rectangular grid

more complicated for conformal grids than for regular grids but has been implemented in practical commercial codes.

4.1.2 Stencil

A stencil is a finite difference equation that approximates the derivative of a quantity at one grid point in terms of values at neighboring points. The most common stencil for 1-D problems is the first-order central difference formula

$$\frac{du(x)}{dx} \simeq \frac{u(x+\Delta x) - u(x-\Delta x)}{2\Delta x} \tag{4.5}$$

This is similar to (3.3), except that for convenience Δx is defined differently so that it matches the grid points spacing. Using a grid, the central difference formula can be expressed as

$$u'(x_n) \simeq \frac{u(x_{n+1}) - u(x_{n-1})}{2\Delta x} \tag{4.6}$$

We can use a subscript notation to write this in the form of a difference equation:

$$u'_n \simeq \frac{u_{n+1} - u_{n-1}}{2\Delta x} \tag{4.7}$$

where $u_n = u(x_n)$. For a 2-D or 3-D grid, multiple subscripts can be used. For time-domain finite difference methods, the time grid index is sometimes denoted with a superscript to distinguish the temporal and spatial coordinates.

For a given derivative operator, there are many stencils or difference approximations. Higher order stencils can be used to obtain more accurate approximations to derivative operators in terms of sample values at many neighboring points. Often, the goal of higher order stencils is to develop a finite difference algorithm that produces accurate results for large grid spacings, thereby reducing the computational cost of the algorithm relative to low-order finite difference methods. High-order finite difference approximations can provide significant performance improvement over low-order stencils, but high-order methods inherently assume that the solution is smooth and can lead to inaccurate results for problems with nonsmooth or discontinuous solutions.

4.1.3 Boundary Conditions

At the edges of the grid, the stencil applied at points in the interior of the grid typically cannot be used, for the simple reason that the stencil requires solution values at grid points that are outside the simulation domain. A modified difference equation for boundary grid points is required. The modified difference equation is obtained from some type of physical boundary condition on the behavior of the solution at the edges of the simulation domain.

There are several different classes of boundary conditions:

Classes of Boundary Conditions

Dirichlet: $u|_{bd} = 0$, so that the field vanishes at grid points on the boundary. Alternately, the field may be specified by some given nonzero function at the boundary, so that $u|_{bd} = f$, but the zero or homogeneous Dirichlet condition is the most common.

Neumann: $\frac{\partial u}{\partial n}\big|_{bd} = 0$, where n represents the coordinate that is normal to the boundary. Implementing this boundary condition on the FD grid using a forward difference approximation for the derivative leads to the relationship $u(x_1) = u(x_2)$, where x_1 is a point on the grid boundary, and x_2 is the neighboring interior point. This boundary condition constrains the solution to tend to a constant approaching the boundary. The inhomogeneous Neumann condition for which the normal derivative is constrained to be equal to a given function is also used occasionally.

Mixed: A linear combination of u and its normal derivative is set to a constant or a given function.

Absorbing boundary condition (ABC): This type of boundary condition is very important in applications of the finite difference method to wave propagation. Most electromagnetics problems involve unbounded regions, which cannot be modeled computationally without special handling. One option is to use one of the previously given boundary conditions and make the simulation region very large, so that the boundary does not affect the solution in the region of interest. For wave propagation problems, the simulation must be terminated before reflections from the boundary perturb the solution in the region of interest. The drawback of this approach is that the larger the simulation region, the greater the computational cost of the simulation. A better approach is to use a boundary condition that absorbs waves and reflects as little energy as possible. This is the computational analog of an anechoic chamber. There are several types of ABCs, including the following:

- One-way wave equation or Mur ABC. These are easy to implement but imperfect in 2-D and higher dimensions.
- Perfectly matched layer (PML) with loss.
- Surface integral equation on the boundary obtained using the method of moments–time-domain integral equation (MoM–TDIE) algorithm.

Material interfaces: Another type of boundary condition arises at the interface between materials within the computational domain. This type of boundary can often be handled simply by adjusting coefficients in the difference equation or stencil to reflect the material properties at each grid point, but improved accuracy near material interfaces can be obtained for some problems by deriving more sophisticated stencils for grid points lying on the boundary.

4.1.4 Sources

Most finite difference algorithms require a source or excitation that introduces an incident wave or other signal to energize the system being modeled. Sources used in finite difference modeling include the following:

Hard source: A hard source is easy to implement, since one simply sets the value of an unknown field at a grid point or many grid points, equal to a given function of time, such as a sinusoid or Gaussian pulse. Mathematically, a hard source is an inhomogeneous Dirichlet boundary condition. As will be seen for the FDTD method, hard sources have the disadvantage that the source reflects any wave that is incident on it.

Soft source: Soft sources model a forcing function in the PDE and do not reflect incident waves. An impressed electric current $\overline{J}$ is a soft source.

Gap source: To drive a conducting antenna with a small feed gap, we use a hard source in the gap. If the antenna is a dipole oriented in the z direction with a feed gap of length Δz (i.e., equal to one grid step), then we specify the electric field component E_z at the grid cell in the gap with a hard source equal to $v(t)/\Delta z$.

Modal source: A modal sources is used to represent an excitation introduced into a system through transmission lines or waveguides.

Scattered field formulation: The scattered field formulation essentially transforms any material objects in the simulation domain into soft sources.

Nonzero initial condition: Normally with the FDTD method the initial condition for the field variables is zero, but occasionally it is useful to model a system with a nonzero initial condition for the fields.

4.1.5 Solution Method

Applying a stencil at each grid point leads to a system of difference equations that can be solved for the unknown sample values. Various approaches are available for solving the difference equations. Solution methods can be grouped into two categories:

Explicit: Field values at one grid point are updated in terms of neighboring values.

Implicit: The difference equations are arranged into a linear system that can be solved for all of the unknowns at one time.

Typically, time-domain algorithms are explicit, with samples at a future time step updated in terms of past time samples. The finite difference algorithm in Section 4.3 uses an implicit solution method. Another approach is an iterative algorithm for which an initial guess for the solution value at each grid point is updated using the stencil until it converges to within a given tolerance.

4.2 Wave Equation: 1-D FDTD Method

The simplest PDE to which the FDTD algorithm can be applied is the one-dimensional wave equation

1-D Wave Equation

$$\frac{\partial^2 u(x,t)}{\partial x^2} = \frac{1}{c^2}\frac{\partial^2 u(x,t)}{\partial t^2} \tag{4.8}$$

In general, a particular PDE can be used to model many different physical problems. One physical problem that is modeled by the 1-D wave equation is the field radiated by a planar time-domain current source that varies in intensity only in the x direction and is infinite in the y and z directions. Since the source varies only in the x direction, the radiated electric field inherits the same independence of the y and z coordinates and also varies only in the x direction. In view of (2.36), the components $E_y(x,t)$ and $E_z(x,t)$ both satisfy (4.8) in a source-free region.

This PDE has one spatial independent variable and another independent variable that represents time. Since there is only one spatial dimension, we refer to this as the 1-D wave equation, but the solution actually lives in two-dimensional spacetime. Physicists sometimes refer to this as the 1+1 dimensional wave equation, since there is one spatial dimension and one time dimension.

4.2.1 1-D Grid

For the simulation domain $0 \leq x \leq d$, we will choose the grid points to be

$$x_m = (m-1)\Delta x \tag{4.9}$$

where the grid index m ranges from 1 to M. The grid step size is

$$\Delta x = \frac{d}{M-1} \tag{4.10}$$

The time grid points are

$$t_n = (n-1)\Delta t \tag{4.11}$$

The sample value of the solution $u(x,t)$ at one of the grid points is $u(x_m, t_n) = u[(m-1)\Delta x, (n-1)\Delta t]$. As a shorthand notation, we write this as

$$u_m^n = u(x_m, t_n) \tag{4.12}$$

where the subscript indexes the spatial grid point, and the superscript is the time index.

4.2.2 Update Equation for the 1-D Wave Equation

To apply the finite difference method to the wave equation, difference approximations for the derivatives are required. The stencil for the second derivative in x is

$$\frac{\partial^2 u(x,t)}{\partial x^2} \simeq \frac{\partial}{\partial x}\left[\frac{u(x+\Delta x/2,t)-u(x-\Delta x/2,t)}{\Delta x}\right]$$

$$\simeq \frac{u(x+\Delta x,t)-2u(x,t)+u(x-\Delta x,t)}{(\Delta x)^2} \tag{4.13}$$

The stencil for the time derivative operator on the right-hand side of (4.8) is similar. Substituting the difference approximations for the derivatives in the wave equation leads to

$$r^2\left[u(x+\Delta x,t)-2u(x,t)+u(x-\Delta x,t)\right]$$

$$= u(x,t+\Delta t)-2u(x,t)+u(x,t-\Delta t) \tag{4.14}$$

where $r = c\Delta t/\Delta x$.

By introducing the notation (4.12) into (4.14), the difference equation becomes

$$r^2\left(u_{m+1}^n - 2u_m^n + u_{m-1}^n\right) = u_m^{n+1} - 2u_m^n + u_m^{n-1} \tag{4.15}$$

The final step is to solve for the sample value u_m^{n+1} of the solution at the future time t_{n+1} in terms of sample values at time steps n and $n-1$ to obtain an explicit FD update equation:

1-D FDTD Update Equation

$$u_m^{n+1} = r^2\left(u_{m+1}^n - 2u_m^n + u_{m-1}^n\right) + 2u_m^n - u_m^{n-1} \tag{4.16}$$

The algorithm based on this update equation is known as the 1-D finite difference time-domain method (1-D FDTD) method. The algorithm takes sample values at previous time steps and computes a new set of sample values for u at a future time step.

4.2.3 Initial Condition

For time-domain PDEs, boundary conditions in both space and time are required. The boundary condition for the time dimension consists of a specified value for the function $u(x, t_0)$ and its derivative at some initial time t_0, or an initial condition. Since the PDE involves a second-order time derivative, initial conditions for u and the first derivative of u are required. When the PDE is discretized, the second-order difference equations require initial values for the solution at two time steps. This means that the algorithm is initialized by specifying given values for u_m^1 and u_m^2. Most often, the initial condition is simply $u_m^1 = u_m^2 = 0$.

4.2.4 Boundary Conditions for the 1-D FDTD Method

Boundary conditions are required at the left and right grid endpoints x_1 and x_M. The simplest boundary condition is the Dirichlet condition, for which

$$u(0,t) = u_1^n = 0$$
$$u(d,t) = u_M^n = 0$$

The Neumann condition is implemented by setting the endpoint value equal to the point next to it, so that

$$u_1^n = u_2^n$$
$$u_M^n = u_{M-1}^n$$

4.2.4.1 First-Order Mur Absorbing Boundary Condition

An absorbing boundary condition can be obtained by discretizing the one-way wave equation at the endpoints of the simulation domain. This is known as the Mur boundary condition. At the boundary $x = d$, we enforce the PDE

$$\frac{\partial u}{\partial x} = -\frac{1}{c}\frac{\partial u}{\partial t} \tag{4.17}$$

This equation has solutions of the form $u(x,t) = u_0(x - ct)$, which is a wave of arbitrary shape moving to the right as time increases. Using this PDE rather than the 1-D wave equation at the endpoints allows waves to propagate out of the simulation domain without reflection, so that the FDTD algorithm models an infinite domain with a finite number of grid points.

We have to be careful in discretizing this partial differential equation, because the approximations for the spatial and temporal first derivatives in the one-way wave equation need to be evaluated at the same point. For a first derivative, the central difference formula leads to an approximation for the derivative that is halfway between two grid points. As a result, the central difference formulas for the right- and left-hand sides of (4.17) lead to derivatives at different points. To overcome this problem, the central difference formulas for two adjacent pairs of grid points can be averaged according to

$$\frac{1}{2}\left(\underbrace{\frac{u_M^n - u_{M-1}^n}{\Delta x}}_{\text{at }(x_M-\Delta x/2,t_n)} + \underbrace{\frac{u_M^{n+1} - u_{M-1}^{n+1}}{\Delta x}}_{\text{at }(x_M-\Delta x/2,t_{n+1})}\right) = -\frac{1}{2c}\left(\underbrace{\frac{u_M^{n+1} - u_M^n}{\Delta t}}_{\text{at }(x_M,t_n+\Delta t/2)} + \underbrace{\frac{u_{M-1}^{n+1} - u_{M-1}^n}{\Delta t}}_{\text{at }(x_{M-1},t_n+\Delta t/2)}\right) \tag{4.18}$$

Both the right- and left-hand sides of this expression are evaluated at the point $(x_M - \Delta x/2, t_n + \Delta t/2)$. Solving for u_M^{n+1} gives

First-Order Mur Absorbing Boundary Condition

$$u_M^{n+1} = u_{M-1}^n + \frac{r-1}{r+1}\left(u_{M-1}^{n+1} - u_M^n\right) \tag{4.19}$$

The boundary condition at $x = 0$ can be derived similarly. For 1-D problems, the Mur ABC is a perfect absorber, but for 2-D and 3-D problems, waves arriving at oblique incidence angles partially reflect, so more sophisticated ABCs are typically used.

4.2.5 Hard and Soft Sources

The two basic types of sources that can be used in the FDTD method are hard sources and soft sources. A hard source simply sets the value of the field at one or more grid points equal to a specified function of time and therefore is a type of Dirichlet boundary condition. This corresponds to an EM problem in which the electric field at some point is known, and we wish to find the values of the radiated field at other points. One property of a hard source is that waves propagating toward the source are reflected by the source, which can cause modeling error. One solution is to remove the source after launching the incident wave but before reflections arrive at the location of the source.

A soft source corresponds to a forcing function added to the wave equation (4.8). For EM problems, this represents an impressed electric current. To generalize the wave equation to include a forcing function, we will rederive (4.8) from Maxwell's equations. Taking the curl of Faraday's law and substituting the result in Ampère's law leads to

$$\nabla \times \nabla \times \overline{\mathscr{E}} + \frac{1}{c^2}\frac{\partial^2 \overline{\mathscr{E}}}{\partial t^2} = -\mu \frac{\partial \overline{\mathscr{J}}}{\partial t} \tag{4.20}$$

where $c = 1/\sqrt{\mu\varepsilon}$. If we use the vector calculus identity

$$-\nabla \times \nabla \times \overline{\mathscr{E}} + \nabla(\nabla \cdot \overline{\mathscr{E}}) = \nabla^2 \overline{\mathscr{E}} \tag{4.21}$$

and assume that the permittivity is constant and the net electric charge is zero, so that $\nabla \cdot \overline{\mathscr{D}} = \nabla \cdot \overline{\mathscr{E}} = 0$, then we arrive at the wave equation

$$\nabla^2 \overline{\mathscr{E}} - \frac{1}{c^2}\frac{\partial^2 \overline{\mathscr{E}}}{\partial t^2} = \mu \frac{\partial \overline{\mathscr{J}}}{\partial t} \tag{4.22}$$

For a problem in which the current density vector is in one direction only, and the source varies also only in one direction, this reduces to a 1-D wave equation of the same form as (4.8) but with a forcing function determined by the current source:

$$\frac{\partial^2 u(x,t)}{\partial x^2} - \frac{1}{c^2}\frac{\partial^2 u(x,t)}{\partial t^2} = \mu \frac{\partial J(x,t)}{\partial t} \tag{4.23}$$

where u represents the electric field component E_y and J represents the source component J_y.

A time-harmonic plane current of the form $J_y = J_0 \sin(\omega t)\delta(x - x_s)$ launches plane waves traveling away from the source on both sides. The delta function $\delta(x - x_s)$ in the source can be discretized by modeling it as a rect or pulse function of width Δx and height $1/\Delta x$. If (4.23) with a plane current source is discretized using the finite difference method, we arrive at the difference equation

$$u_m^{n+1} = r^2\left(u_{m+1}^n - 2u_m^n + u_{m-1}^n\right) + 2u_m^n - u_m^{n-1} - \frac{c^2(\Delta t)^2\mu}{\Delta x}\frac{\partial J_s(t)}{\partial t}\bigg|_{t=t_n} \tag{4.24}$$

where $J_s(t)$ multiplies the delta function in the plane current source. It is convenient to specify the magnitude of the current source in terms of the amplitude E_0 of the plane waves radiated by the source. This can be done using the solution to Problem 2.7. By combining these results, it can be shown that the 1-D FDTD update equation with a soft source is

$$u_m^{n+1} = r^2\left(u_{m+1}^n - 2u_m^n + u_{m-1}^n\right) + 2u_m^n - u_m^{n-1} - 2r\omega\Delta t E_0 \cos(\omega t_n)\delta_{m,m_s} \tag{4.25}$$

where $x_{m_s} = x_s$ is the grid point at the location of the source.

In deriving (4.25), the initial phase of the time-harmonic current source was such that the time dependence of the source term in (4.25) is of the form $\cos(\omega t)$. If the FDTD simulation starts at $t = 0$, because $\cos(0) = 1$ the value of the source function jumps discontinuously to a nonzero value. To avoid strong field transients caused by the jump in the source function, the phase of the source can be changed so that the time dependence of the source term is of the form $\sin(\omega t)$, in which case the update equation becomes

1-D FDTD Update Equation With Soft Source

$$u_m^{n+1} = r^2\left(u_{m+1}^n - 2u_m^n + u_{m-1}^n\right) + 2u_m^n - u_m^{n-1} - 2r\omega\Delta t E_0 \sin(\omega t_n)\delta_{m,m_s} \tag{4.26}$$

Since the initial phase of the source is unimportant for most problems, this form of the update equation is almost always preferable to (4.25).

4.2.6 *Source Turn-On Functions*

While the sine function in (4.26) reduces the transients that occur when the simulation begins, the first derivative of the source is still discontinuous. To further reduce its singularity at $t = 0$, the source function can be multiplied by a turn-on function that gradually transitions from zero to one over 10–20 simulation time steps.

Another problem specific to a soft source is that it can introduce a nonzero direct current (DC) component in the solution. While accurate field solutions can still be obtained by ignoring or filtering out the DC component, it is more convenient to use a source function that does not excite a DC component. This can be accomplished using the modified source function [4]

$$s(t)\sin(\omega t) \tag{4.27}$$

The function $s(t)$ is a turn-on function defined by

$$s(t) = \begin{cases} 0 & t < 0 \\ 0.5[1 - \cos(\omega t/(2\alpha))] & 0 \le t \le \alpha T \\ 1 & t > \alpha T \end{cases} \tag{4.28}$$

where α is a half-integer, so that its value is chosen from $1/2, 3/2, 5/2, \ldots$ and T is the period of the time-harmonic source.

4.2.7 Code Example: 1-D FDTD Algorithm

A code implementation of the one-dimensional finite difference can be divided into several major blocks: preprocessing, which includes definition of physical constants, definition of the problem to be solved, mesh or grid generation, and initialization; the solution algorithm; and postprocessing. We will illustrate each of these code sections for the 1-D FDTD algorithm.

Physical constants: This portion of the code specifies values for physical constants:

```
% Physical constants
eps0 = 8.854e-12;        % Free space permittivity (F/m)
mu0 = 4*pi*1e-7;         % Free space permeability (H/m)
c0 = 1/sqrt(mu0*eps0); % Speed of light (m/s)
```

Problem definition: This code section defines the parameters of the physical problem to be solved. For the 1-D FDTD method, the problem specification is particularly simple. If we model two dielectric half-spaces with a plane interface, this portion of the code specifies the values of the relative permittivities of the two half-spaces:

```
% Problem definition
epsr1 = 1;              % Relative permittivity region 1
epsr2 = 2;              % Relative permittivity region 2
c1 = c0/sqrt(epsr1); % Speed of light in region 1
c2 = c0/sqrt(epsr2); % Speed of light in region 2
```

Simulation domain definition: This code section specifies the physical size and other parameters of the simulation domain. For the 1-D FDTD method, we must define the length of the simulation domain in x and the total simulation time t:

```
% Simulation domain
D = 1;                    % Simulation domain length (m)
tmax = 4.5e-9;            % Simulation time (s)
```

Grid generation: This code section determines the spatial and temporal grid points that span the simulation domain:

```
% Grid generation
r = 1;                    % FDTD grid parameter
NX = 100;                 % Number of spatial grid points
dx = D/(NX-1);            % Spatial grid step size (m)
dt = r*dx/c0;             % Temporal step size (s)
NT = ceil(tmax/dt);       % Number of time steps
x = 0:dx:D;               % Spatial grid points
```

Source: The type of source and the parameters of the source function must be specified. For time-harmonic problems, the parameter is the frequency of the excitation. For the 1-D FDTD algorithm with a Gaussian pulse excitation, the source is defined as follows:

```
% Constants in source
t0 = 60*dt;                    % Time at pulse center (s)
s = 10*dt;                     % Pulse width (s)
```

Initialization: The variables used to store field samples must be defined and initialized. For the FDTD algorithm, the most common initial value for the field is zero:

```
% Initialization
u1 = zeros(NX,1);           % Field at past time step
u2 = zeros(NX,1);           % Field at current time step
u3 = zeros(NX,1);           % Field at future time step
```

Solution algorithm: This is the core computational routine. For the FDTD algorithm, the solution algorithm consists of a loop over time with an update equation to compute values of the unknown field at a future time step in terms of values at earlier time steps. The source or excitation must also be included in the loop, as well as boundary conditions at the endpoints of the simulation domain:

```
% Constant in update equation
a1 = (c1*dt/dx)^2*ones(NX,1);
idx2 = find(x > D/2);
a1(idx2) = (c2*dt/dx)^2;

% Constant in absorbing boundary condition
a2 = (sqrt(a1(NX))-1)/(sqrt(a1(NX))+1);

% Loop over time
figure(1)
for n = 1:NT,
      % Update equation for interior grid points
      u3(2:NX-1) = a1(2:NX-1).*(u2(3:NX) - ...
      2*u2(2:NX-1) + u2(1:NX-2)) + 2*u2(2:NX-1) - ...
      u1(2:NX-1);

      % Source at left-hand side
      t = n*dt;
      u3(1) = exp(-(t-t0)^2/(2*s^2));

      % Absorbing boundary condition at right-hand side
      u3(NX) = u2(NX-1) + a2*(u3(NX-1)-u2(NX));

      % Plot field
      plot(x,u3)
      ylim([-2 2])
      xlabel('x (m)')
      ylabel('u')
      drawnow

      % Shift field variables in time
      u1 = u2;
      u2 = u3;
end
```

For this 1-D FDTD code, the hard source on the left-hand side of the simulation domain acts as a Dirichlet boundary condition. A pulse propagating to the left will reflect from the source. To avoid this, an `if` statement can be used to switch between the source and an absorbing boundary condition on the left-hand side after the pulse excitation has decayed to a small value but before reflected pulses arrive at the left boundary.

When the field solution is plotted during a loop over time, the figure window may not refresh until after the script completes running. In that case, an update of the figure window can be triggered using the `drawnow` command.

Postprocessing: After the solution algorithm is complete, postprocessing is used to compute required physical quantities from the fields. Often, the field itself is not the final desired result of the simulation. For the 1-D FDTD algorithm, we can compute a reflection coefficient from the amplitude of the reflected pulse:

```
% Postprocessing
idx1 = find(x < D/2);
% Get numerical reflection coefficient
[tmp,idx] = max(abs(u3(idx1)));
R1 = u3(idx)
% Exact reflection coefficient
R2 = (sqrt(1/epsr2) - sqrt(1/epsr1))/ ...
(sqrt(1/epsr2) + sqrt(1/epsr1));
```

4.2.7.1 Numerical Results With the 1-D FDTD Code

Figure 4.2 illustrates the basic operation of the 1-D FDTD algorithm. A hard source is used to launch a Gaussian pulse from the left endpoint of the simulation domain ($x = -0.5$). The pulse propagates to the right and strikes the right endpoint ($x = 0.5$), where a Dirichlet boundary condition is imposed. The pulse reflects from the right endpoint with an amplitude of -1.

If the propagation velocity c steps from one value to a different value at some point in the simulation domain, the 1-D FDTD algorithm models reflection and transmission from a planar material interface. We will assume that the region $x < 0$ is free space and that the region $x > 0$ is filled with a dielectric material. The $z = 0$ plane is the interface between the two regions. If a wave propagates toward the interface in the $+x$ direction, then a reflected wave propagating in the $-x$ direction and a transmitted wave propagating in the $+x$ direction are produced when the pulse strikes the interface. For a unit amplitude incident wave, the amplitudes of the reflected and transmitted waves are the reflection coefficient and transmission coefficient, respectively. Simulated results for reflection and transmission at an interface between materials with relative permittivity $\varepsilon_{r1} = 1$ and $\varepsilon_{r2} = 2$ are shown in Figure 4.3.

From the boundary condition on the incident, reflected, and transmitted electric fields that must be satisfied at the interface, it can be shown that the reflection and transmission coefficients are given by

$$R = \frac{\eta_2 - \eta_1}{\eta_2 + \eta_1} \tag{4.29a}$$

$$T = \frac{2\eta_2}{\eta_2 + \eta_1} \tag{4.29b}$$

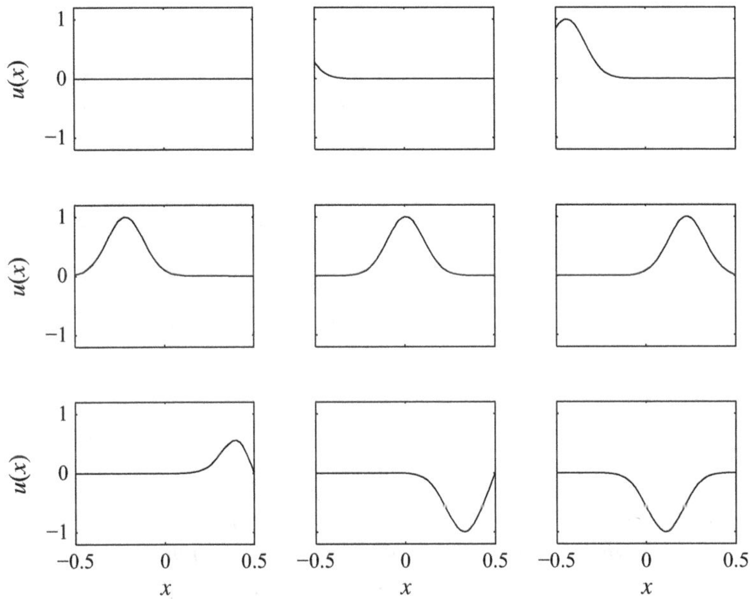

Figure 4.2 Snapshots of an FDTD simulation of a Gaussian pulse as it propagates from left to right, reflects from a Dirichlet boundary condition at the right endpoint of the simulation domain, and returns to the left

where $\eta_1 = \sqrt{\mu_1/\varepsilon_1}$ and $\eta_2 = \sqrt{\mu_2/\varepsilon_2}$ are the characteristic impedances of the materials in the regions $z < 0$ and $z > 0$, respectively. For $\varepsilon_{r1} = 1$ and $\varepsilon_{r2} = 2$, the reflection and transmission coefficients are $R \simeq -0.17$ and $T \simeq 0.83$. These values correspond closely to the amplitudes of the reflected and transmitted pulses in Figure 4.3.

4.2.8 Stability

By running simulations for different values of Δx and Δt, it is easy to see that the FDTD method is not always stable, and the solution blows up as time progresses for some values of the discretization lengths. We can study the solution to the FDTD difference equation analytically to gain insight into this problem.

A single-frequency solution to the 1-D wave equation has the form

$$u(x, t) = \cos(kx \pm \omega t) \tag{4.30}$$

By inserting this solution into the wave equation, it can be seen that $k = \omega/c$. This relationship between time frequency ω and spatial wave number k is the free-space dispersion relation (2.50). If we discretize this solution by sampling it at the grid points and insert it into the 1-D FDTD update equation, we can derive a different dispersion relation for the FDTD algorithm, called the numerical dispersion relation.

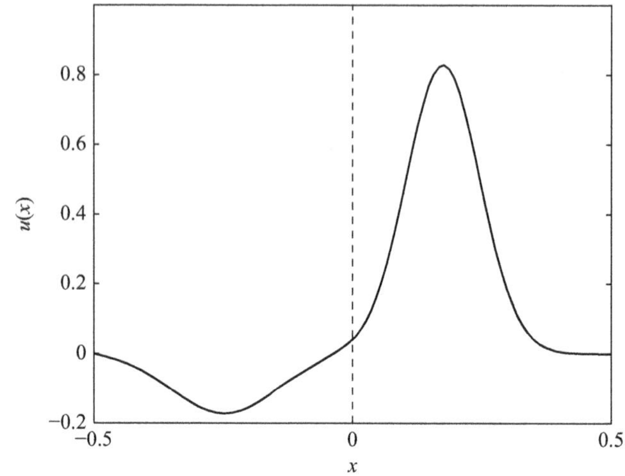

Figure 4.3 *FDTD simulation of reflection of a Gaussian pulse of peak value equal to 1 from an interface between free space and a medium with relative permittivity $\varepsilon_r = 2$. The plot shows the pulse as a function of position after it has reflected from the interface. The reflected pulse has a negative amplitude, and the transmitted pulse is positive. The vertical dashed line marks the location of the dielectric interface*

This numerical dispersion relation provides a tool for understanding the stability properties of the FDTD algorithm.

For convenience in the following analysis, we use the fact that $\sin(kx \pm \omega t)$ is also a solution to put (4.30) into the complex exponential form $e^{jkx+j\omega t}$. Using $x = m\Delta x$ and $t = n\Delta t$ for the space and time coordinates, we can express this solution in terms of the spatial and temporal grid indices as

$$u_m^n = e^{jkm\Delta x + j\omega n\Delta t} \tag{4.31}$$

By inserting this solution into the 1-D FDTD difference equation, we will obtain a dispersion relation for the FDTD algorithm that is different from the free-space dispersion relation. Substituting (4.31) into the difference equation (4.16) leads to

$$\cos(\omega\Delta t) = r^2[\cos(k\Delta x) - 1] + 1 \tag{4.32}$$

This is the numerical dispersion relation for the 1-D FDTD algorithm.

If the right-hand side of (4.32) is greater than 1 in magnitude for some value of k, then ω must have a nonzero imaginary part. If this is the case, then the time exponential in solution (4.31) becomes a real exponential, and the solution blows up as time increases. Restricting the right-hand side of (4.32) to be less than or equal to 1 in magnitude leads to the Courant stability criterion

1-D FDTD Stability Criterion

$$\frac{c\Delta t}{\Delta x} \leq 1 \tag{4.33}$$

for the 1-D FDTD algorithm.

This restriction on Δx and Δt leads to a nice physical picture. If we consider a given grid point (x_m, t_n), the light cone for that point is defined to be all points in the future that can be reached by traveling away from the point at the speed of light. If (4.33) is met, then only one grid point at the $n + 1$ time step is inside the light cone. This is illustrated in Figure 4.4. If the stability condition is not satisfied, then there is more than one grid point at the next time step inside the light cone. If more than one grid point is inside the cone, then in a loose sense, too much energy is transferred from the solution at one time step to the solution at the next time step, and unwanted signal amplification occurs.

4.2.9 Accuracy

The FDTD method is relatively easy to implement and works well for many EM problems. With any numerical method, however, it is important to understand the accuracy of the solution before relying on the results for engineering design and analysis. It is possible for an algorithm that performs well for simple test cases to fail for complex, real-world problems, which means that it is essential to characterize the expected solution accuracy for a wide range of algorithm parameters and problem types.

Section 1.3 listed various approaches to analyzing the accuracy of a numerical method. FDTD can be validated using test cases for which analytical solutions are available, or solutions can be compared with results obtained from other algorithms.

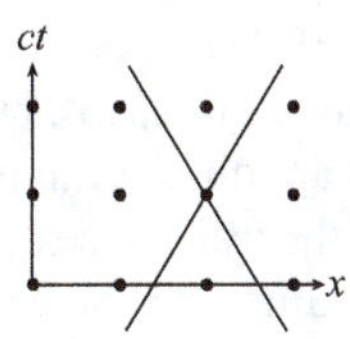

Figure 4.4 Illustration of FDTD stability criterion with the light cone for one grid point indicated by two diagonal lines. The condition $c\Delta t \leq \Delta x$ implies that only one grid point at time $t + \Delta t$ is in the light cone of a grid point at time t. If more than one grid point at time $t + \Delta t$ lies in the light cone, then the algorithm is unstable

Here, we will consider a more rigorous approach based on an analytical derivation of the solution error for a specific type of solution.

For the FDTD method, we have already determined the criterion for stability, but a stable solution does not guarantee that the solution error is accurate. As we will see, the same plane wave solution used in the stability analysis of the previous section can be used to analyze the accuracy of the FDTD algorithm. Solving the numerical dispersion relation (4.32) for the wave number of the discrete plane wave solution (4.31) leads to

1-D FDTD Numerical Dispersion Relation

$$k\Delta x = \cos^{-1}\left(1 + r^{-2}[\cos(\omega\Delta t) - 1]\right) \tag{4.34}$$

For stability, we required that ω was real for all possible values of k. For accuracy, we want to be sure that the value of k determined by (4.34) is as close as possible to the value given by the free-space dispersion relation. If k is different from the free-space value for a given value of ω, then that frequency component travels at a different velocity than the speed of light, and error due to dispersion occurs. We will look at the cases $r = 1$ and $r < 1$ separately.

4.2.9.1 $r = 1$ Case

For $r = 1$, (4.34) becomes

$$k\Delta x = \cos^{-1}[\cos(\omega\Delta t)] = \omega\Delta t \tag{4.35}$$

Using $c\Delta t = \Delta x$, we have

$$k = \frac{\omega}{c} \tag{4.36}$$

which is the free-space dispersion relation. Thus, for $r = 1$, there is no dispersion as the wave propagates. This is a "magic" special case.

For 2-D and 3-D problems, unfortunately, the effective value of r cannot be equal to 1 for all directions of wave propagation, so dispersion error is unavoidable in higher dimensions. Variants of the FDTD algorithm have been developed that reduce dispersion error or even purportedly eliminate it, but these numerical methods either increase the computational cost or are difficult to implement.

Even though (4.34) reduces to the free-space dispersion relation with $r = 1$, there is a subtle difference between the numerical wave (4.31) and the exact wave solution (4.30). The numerical solution consists of discrete values at the FDTD grid points. Samples of a function on a grid with a given step size Δx can represent only a finite range of spatial frequencies. This is implicit in (4.34), since the arccosine function on the right side takes on values in the range $\pm\pi$, which implies that k lies in the interval $|k| \le \pi/\Delta x$. If λ is the spatial period of the wave, then we must have

$$\Delta x \le \frac{\lambda}{2} \tag{4.37}$$

This is the Nyquist criterion for the FDTD grid. If we try to launch a wave with a value of ω large enough that the free-space wavelength is smaller than $2\Delta x$, the wave aliases to a lower spatial frequency.

What is surprising about the FDTD algorithm for the $r = 1$ case is that signals such as a square pulse with infinite frequency content can propagate without distortion. In view of the limited resolving power of samples on a finitely spaced grid, how can this be? The high spatial frequency components of the pulse alias to lower spatial frequencies, which by (4.36) propagate with the correct phase velocity $\omega/k = c$.

4.2.9.2 $r < 1$ Case

For the 1-D FDTD method, we can simply set r to unity and avoid dispersion error entirely, but for 2-D and 3-D problems dispersion cannot be avoided. We can use the $r < 1$ case for the 1-D dispersion relation as a model for the error behavior of FDTD for 2-D and 3-D problems. For $r < 1$, (4.34) deviates from the free-space value $k = \omega/c$. As a wave propagates away from a source, it travels at an incorrect phase velocity and eventually will be out of phase with the exact solution.

One consequence of this is that for some problems, there is an upper bound on the size of the problem that can be simulated. For large objects involving multiple scattering, distant parts of the object do not interact with the correct phase. In some cases, this may not be a significant issue, since only the power contained in the reflected wave may be needed, but for some problems, accumulation of dispersion error leads to unacceptable loss of accuracy.

If we require the phase error for a wave of a given frequency propagating across the simulation region to be less than, say, $\pi/5$, then the physical size of the simulation region is bounded by

$$d < \frac{\pi/5}{|\omega/c - k|} \tag{4.38}$$

where ω is the largest time frequency component of the source signal waveform. Another way to rewrite this expression is in terms of the maximum number of grid points in a simulation for a given time frequency and value of r:

$$N_{\max}(\omega\Delta t, r) = \frac{\pi/5}{\left|\omega\Delta t/r - \cos^{-1}(1 + r^{-2}[\cos(\omega\Delta t) - 1])\right|} \tag{4.39}$$

This upper limit decreases as r becomes smaller than one and as $\omega\Delta t \to \pi$.

We will now consider the solution error analysis from a different point of view. For a given problem, how can one make the solution more accurate? Of course, in 1-D, we could choose the grid spacings such that $r = 1$ and dispersion is eliminated, but in higher dimensions, this simple fix is not possible. To use the 1-D FDTD method as a model for the more important higher dimensional cases, we must understand how the grid spacing affects solution error.

For a plane wave solution, the error after propagating a distance d is

$$\begin{aligned} |E_{\text{exact}} - E_{\text{FDTD}}| &= |e^{jk_0 d} - e^{jkd}|, \quad k_0 = \omega/c \\ &\simeq |k - k_0|d \end{aligned} \tag{4.40}$$

where we have assumed that the error is small. Using the numerical dispersion relation (4.34) to obtain a value for k, the error can be shown to be

$$|E_{\text{exact}} - E_{\text{FDTD}}| \simeq \frac{1}{24}(k_0\Delta x)^2(k_0 d) \tag{4.41}$$

where we have assumed that $\omega\Delta t \ll 1$.

This result has a very intuitive interpretation. The relative error in using the central difference approximation for the second derivative of a plane wave is

$$\begin{aligned}\text{Relative error} &= \frac{\left|\frac{e^{jk_0(x+\Delta x)}-2e^{jk_0x}+e^{jk_0(x-\Delta x)}}{\Delta x^2} - \frac{\partial^2}{\partial x^2}e^{jk_0x}\right|}{\left|\frac{\partial^2}{\partial x^2}e^{jk_0x}\right|} \\ &= \left|1 - \frac{\sin^2(k_0\Delta x/2)}{(k_0\Delta x/2)^2}\right| \\ &\simeq \frac{1}{12}(k_0\Delta x)^2\end{aligned} \tag{4.42}$$

Identifying a similar term in (4.41) leads to

1-D FDTD Numerical Solution Error Estimate

$$|E_{\text{exact}} - E_{\text{FDTD}}| \simeq \frac{1}{2}\quad \underbrace{\frac{1}{12}(k_0\Delta x)^2}_{\substack{\text{Error in using}\\ \text{central difference}}}\quad \underbrace{(k_0 d)}_{\substack{\text{Distance of}\\ \text{error accumulation}}} \tag{4.43}$$

This result implies that to improve error Δx must be reduced. This increases the computational cost of the simulation, because the number of spatial grid points increases. The time step Δt may also need to be decreased in order for the algorithm to be stable.

For the FDTD algorithm, dispersion is not the only source of solution error. In dimensions greater than 1, we must often represent material bodies with curved shapes or continuously varying properties. Representing curved material boundaries on a discrete grid leads to "stairstepping error," which increases solution error. Error can also be introduced by imperfect sources that differ from the actual excitations in a device measurement or experimental setup.

4.3 Laplace's Equation: 2-D Finite Difference Method

We will now consider a PDE with two spatial dimensions and no time dependence. Consider the 2-D boundary value problem (BVP) consisting of Laplace's equation

2-D Laplace's Equation

$$\nabla^2\phi(x,y) = \frac{\partial^2\phi(x,y)}{\partial x^2} + \frac{\partial^2\phi(x,y)}{\partial y^2} = 0 \tag{4.44}$$

with an inhomogeneous (nonzero) Dirichlet condition for the value of the potential ϕ at the boundary of the simulation domain. As with the wave equation, this partial differential equation can model many different physical systems. Prominent examples include the deflection of a stretched membrane or drumhead and electric potential near conducting objects at specified potentials.

To develop the 2-D FD algorithm for Laplace's equation, we will consider the example of a square domain with the boundary condition that $\phi = 10$ V on the top of a square region and $\phi = 0$ on the other three sides. We will discretize the domain using a rectangular grid, so that some of the grid points lie on the edges of the boundary and the rest are inside the region of interest. If the grid is N by N in size, then $4N - 4$ grid points lie on the boundary. With the Dirichlet boundary condition, the potential is known at these boundary grid points. To find the unknown values of the potential at the remaining $N^2 - 4(N - 1)$ grid points, we need a difference equation or stencil for the Laplacian derivative operator in (4.44).

Figure 4.5 shows the two-dimensional grid that we will use to develop a finite difference stencil for the Laplacian. Using the stencil developed earlier for the second derivative operator, the Laplacian can be discretized to obtain

$$\frac{\phi(x+\Delta x,y) - 2\phi(x,y) + \phi(x-\Delta x,y)}{\Delta x^2} + \frac{\phi(x,y+\Delta y) - 2\phi(x,y) + \phi(x,y-\Delta y)}{\Delta y^2} = 0 \tag{4.45}$$

On the grid, if $\Delta x = \Delta y$, the stencil has the form

$$\phi_{i+1,j} + \phi_{i-1,j} + \phi_{i,j+1} + \phi_{i,j-1} - 4\phi_{ij} = 0 \tag{4.46}$$

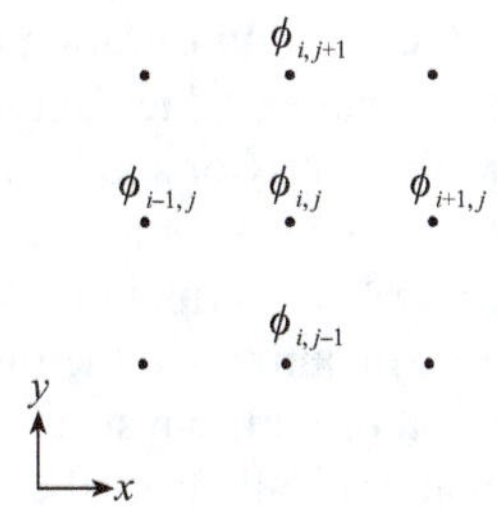

Figure 4.5 Two-dimensional grid used to discretize the Laplacian operator

where i corresponds to the x location of the grid point and j to the y location. This stencil relates the values of the potential ϕ at a given grid point to values at the points to the north, south, east, and west. Many other stencils for the Laplacian operator are available, but the second-order central difference stencil is by far the most commonly used.

The finite difference stencil for the Laplacian provides one linear equation for each interior grid point that is not on the boundary. Together with the boundary conditions at points on the edge of the computational domain, this leads to a set of linear equations that can be solved for the samples of the unknown potential at the interior points. There are several methods for finding the unknown values of the potential:

1. *Iterative method:* Solving the stencil equation for ϕ_{ij} gives

$$\phi_{ij} = \frac{1}{4}(\phi_{i+1,j} + \phi_{i-1,j} + \phi_{i,j+1} + \phi_{i,j-1}) \tag{4.47}$$

   This form of the stencil provides a way to iteratively approach the solution to the Laplace problem starting with an initial guess. The iterative method is initialized by setting boundary grid points to their known values from the boundary condition and all unknown interior values to some starting value such as zero. The interior values are then updated repeatedly by looping through all the interior points and setting the potential at each point equal to the value given by (4.47) until the potential converges to a constant within some error tolerance at all points on the grid. The iterative method can be implemented in a simple way using formulas in a spreadsheet, with settings adjusted to allow circular cell dependencies.

2. *Matrix method:* The linear equations for each interior point can be arranged into a linear system of the form

$$Ax = b \tag{4.48}$$

   where the vector b contains the boundary values of ϕ. This linear system can be solved for the unknown potential values in the vector x.

Interestingly, (4.47) tells us that if ϕ is a solution to Laplace's equation, then the value at one grid point is the average of the four neighboring grid points to north, south, east, and west. This means that the solution $\phi(x,y)$ must vary smoothly and cannot have a sudden jump to a larger or smaller value than nearby values of the potential. This corresponds to a well-known property of solutions to Laplace's equation. In the mathematics literature, solutions to Laplace's equation are called harmonic functions. Harmonic functions satisfy the maximum principle, which implies that the value of a harmonic function must take its maximum value on the boundary of any region. The maximum principle is consistent with physical intuition for the static electric potential, which always varies monotonically from the potential value at one electrode or source to the value at another in the region around the sources. This property also has intriguing connections to the theory of analytic functions in complex analysis, since the real and imaginary parts of an analytic function are harmonic.

4.3.1 Example: 2-D FD Method on a 4-by-4 Grid

For a square grid, the matrix representation of the finite difference equations has a particularly simple form. As an example, we will consider the 4-by-4 grid shown in Figure 4.1. The stencil equations associated with the four interior grid points are

$$\begin{aligned} \phi_{32} + \phi_{12} + \phi_{23} + \phi_{21} - 4\phi_{22} &= 0 \\ \phi_{42} + \phi_{22} + \phi_{33} + \phi_{31} - 4\phi_{32} &= 0 \\ \phi_{33} + \phi_{13} + \phi_{24} + \phi_{22} - 4\phi_{23} &= 0 \\ \phi_{43} + \phi_{23} + \phi_{34} + \phi_{32} - 4\phi_{33} &= 0 \end{aligned} \tag{4.49}$$

Only four of the sample values of ϕ in these equations are unknown, because the other 12 points lie on the boundary. Arranging these equations into a linear system and substituting for the boundary condition leads to

$$\begin{bmatrix} -4 & 1 & 1 & 0 \\ 1 & -4 & 0 & 1 \\ 1 & 0 & -4 & 1 \\ 0 & 1 & 1 & -4 \end{bmatrix} \begin{bmatrix} \phi_{22} \\ \phi_{32} \\ \phi_{23} \\ \phi_{33} \end{bmatrix} = \begin{bmatrix} 0 \\ 0 \\ -10 \\ -10 \end{bmatrix} \tag{4.50}$$

where the values on the right-hand side come from the Dirichlet boundary condition that contains the potential to be 10 V on the top of the square region and 0 on the other three sides. This linear system can be solved for the unknown potential values on the interior of the grid.

4.3.2 Waveguide Modes

We now wish to apply the 2-D FD method to the problem of finding the cutoff frequencies and cross-sectional field patterns of the modes of a waveguide. A waveguide is inherently a 3-D problem, but because of the translational invariance of the waveguide along its axis the waveguide modes can be analyzed using a two-dimensional numerical method. Since the number of grid points required for a 2-D solution is much smaller than the number that would be required for a 3-D grid, this results in a substantial savings in computation time.

The structure to be analyzed is a z-directed, infinite, hollow waveguide with perfect electric conductor (PEC) walls. The shape of the cross-sectional interior boundary is denoted as Γ. It can be shown using Maxwell's equations that the z components of the electric and magnetic fields are solutions to the Helmholtz problems

$$\text{TM modes:} (\nabla^2 + k_t^2)E_z = 0, \qquad E_z|_\Gamma = 0 \quad \text{(Dirichlet BC)} \tag{4.51}$$

$$\text{TE modes:} (\nabla^2 + k_t^2)H_z = 0, \qquad \left.\frac{\partial H_z}{\partial n}\right|_\Gamma = 0 \quad \text{(Neumann BC)} \tag{4.52}$$

where n is a coordinate normal to the boundary. The derivative operator is the 2-D Laplacian

$$\nabla^2 = \frac{\partial^2}{\partial x^2} + \frac{\partial^2}{\partial y^2} \tag{4.53}$$

In the PDEs, k_t is the transverse wave number of the field and is defined by the dispersion relation

$$k_t^2 + k_z^2 = k^2 \tag{4.54}$$

The PDEs are independent of z, because we have assumed that the z dependence of the fields in the waveguide is of the form $e^{\pm jk_z z}$.

The boundary condition for the transverse electric (TE) modes is normally expressed in terms of the tangential component of the electric field intensity vector, which must vanish at the PEC waveguide walls. When transferred to the magnetic field using Maxwell's equations, this boundary condition is equivalent to the Neumann condition in (4.52). To prove that the boundary condition on H_z is of Neumann form, we use the boundary condition on $\overline{E}$ and Ampère's law:

$$\begin{aligned} 0 &= \hat{t} \cdot \overline{E} \\ &= \hat{t} \cdot (\nabla \times \hat{z} H_z) \\ &= \nabla \cdot (H_z \hat{z} \times \hat{t}) + H_z \hat{x} \cdot (\nabla \times \hat{t}) \\ &= \nabla \cdot (\hat{n} H_z) \\ &= \frac{\partial H_z}{\partial n} \end{aligned}$$

where $\hat{t}$ is the tangential to the waveguide wall and orthogonal to the z-axis.

The BVPs in (4.51) and (4.52) are different from the standard boundary value problem, because the constant k_t in the differential operator is unknown. These equations are operator eigenvalue and eigenfunction problems. The correspondence between the operator eigenvalue problem and a matrix eigenvalue problem can be emphasized by rewriting the PDEs in the form

$$\nabla^2 u = \lambda u, \quad u|_\Gamma = 0 \text{ or } \left.\frac{\partial u}{\partial n}\right|_\Gamma = 0 \tag{4.55}$$

where the eigenvalue is $\lambda = -k_t^2$ (the symbol λ is also commonly used for wavelength but here represents an eigenvalue of the Laplacian operator). Solutions u satisfying the PDE and boundary condition for some value of λ are called eigenfunctions of the Laplacian operator for the boundary Γ. This is much like a matrix eigenvalue/eigenvector problem, except that for partial differential operators, a boundary condition on the eigenfunction is required in addition to the eigenvalue equation.

For the case of a Dirichlet condition on a rectangular boundary for the region $0 \le x \le a$, $0 \le y \le b$, the eigenfunctions of the Laplacian operator are of the form

$$u_{mn}(x, y) = \sin\left(\frac{m\pi x}{a}\right) \sin\left(\frac{n\pi y}{b}\right) \tag{4.56}$$

where a and b are the dimensions of the rectangle. The mode indices m and n are positive integers. The corresponding eigenvalues are

$$\lambda_{mn} = -\left(\frac{m\pi}{a}\right)^2 - \left(\frac{n\pi}{b}\right)^2 \tag{4.57}$$

For the Neumann boundary condition, the eigenfunctions involve the cosine function instead of the sine function. For the Neumann case, the mode numbers can be zero.

4.3.3 Spectrum of the Laplacian

The Laplace boundary value problem is of fundamental importance in mathematical physics. Not only does it represent a model for many different physical problems, but it also provides a simplified or approximate way to understand the qualitative behavior of more complex boundary value problems.

Considerable insight can be gained by relating the mathematical properties of the Laplacian to the physical properties of the systems it models. Many of these properties are encapsulated in the set of eigenvalues, or the spectrum, of the operator. Before developing a numerical method to solve the Laplace eigenvalue value problem, we will consider a few general properties of the spectrum of the Laplacian operator.

Discrete spectrum and modal cutoff frequencies:
For a given boundary shape Γ and boundary condition, the Laplacian has a discrete spectrum of eigenvalues λ_n, where $n = 1, 2, \ldots$ is an integer index. Because Laplace's equation corresponds to many different physical problems, the eigenvalues and associated eigenfunctions have a variety of interpretations. For the waveguide problem, the eigenvalues give the cutoff frequencies of the modes of the waveguide. At cutoff, $k_z = 0$, so that by making use of the dispersion relation (4.54) we obtain the relationship

$$-\lambda_n = k_t^2 = k^2 = \omega_c^2 \mu\varepsilon \tag{4.58}$$

where ω_c is the cutoff frequency. The eigenfunction $u_n(x, y)$ corresponds to the x, y dependence of the z component of the electric or magnetic field of the nth waveguide mode.

Negative eigenvalues:
Generally, the eigenvalues of the Laplacian operator are negative. This can be understood intuitively by recognizing that the eigenfunctions are oscillatory, so that where they are positive, the function must decrease and the second derivative must be negative, and where the value is negative, the function must increase, so the second derivative must be positive.

Zero eigenvalue and the DC mode:
If Γ is not simply connected, such as for the case of a coaxial waveguide, we know that the waveguide must support a DC mode. What does this mean about the spectrum of the Laplacian? A DC mode has a cutoff frequency of zero, which implies that the Laplacian has a vanishing eigenvalue.

Shape with the smallest eigenvalue:
For the Dirichlet boundary condition, what simply connected shape of a given area has eigenvalue of smallest magnitude? It has been proved that the answer to this question is the circle. The z component of the electric field for the modes of a circular waveguide is given in terms of the Bessel function $J_m(x)$ as

$$E_z(\rho, \phi) = J_m(k_\rho \rho) \cos(m\phi) \tag{4.59}$$

where $k_\rho = k_t = \sqrt{-\lambda}$ (see Section 2.11 for a review of Bessel functions). If we apply the Dirichlet boundary condition at $\rho = a$, we find that

$$J_m(\sqrt{-\lambda}a) = 0 \tag{4.60}$$

The zeros of the Bessel functions are commonly written as $J_m(\chi_{mn}) = 0$, where χ_{mn} is the nth zero of the mth-order Bessel function. The smallest zero is $\chi_{01} \simeq 2.4048$, so that

$$-\lambda_{\min} = \left(\frac{\chi_{01}}{a}\right)^2 \tag{4.61}$$

$$= \frac{\pi \chi_{01}^2}{A} \tag{4.62}$$

$$\simeq \frac{5.8\pi}{A} \tag{4.63}$$

where A is the area of the circle. Physically, for the waveguide problem, this means that the circular waveguide has a smaller cutoff frequency for its lowest order transverse magnetic (TM) mode than any other simply connected waveguide of the same cross-sectional area. The TM mode with the smallest eigenvalue for a square waveguide has an electric field with x, y dependence of the form $\sin(\pi x/a) \sin(\pi y/a)$, so the eigenvalue is

$$-\lambda_{11} = \left(\frac{\pi}{a}\right)^2 + \left(\frac{\pi}{a}\right)^2 = \frac{(2\pi)\pi}{A} > \frac{5.8\pi}{A} \tag{4.64}$$

From the inequality, it follows that the smallest Laplace eigenvalue for the square has a larger magnitude than the smallest Laplace eigenvalue for a circle with the same area.

Can one hear the shape of a drum?
The vibrational modes of a drumhead of a given shape are governed by (4.55) with the Dirichlet boundary condition. The drum's sound is produced by the motion of the drumhead, which is a superposition of modes of the Laplacian. So, this question can be rephrased as "Is the spectrum of the Laplacian always different for two distinct shapes with equal areas?" This famous question was posed by the mathematician Mark Kac in a classic paper [5]. A few years ago, the question was answered in the negative [6]. Since then, numerous examples of pairs of shapes with identical spectra have been found, many of them constructed from combinations of squares and triangles.

4.3.4 Numerical Implementation

To solve the eigenvalue problem for the Laplacian numerically, we use the matrix method. For TM waveguide modes, the Dirichlet boundary condition requires that E_z be zero at the waveguide wall, so the right-hand side in (4.48) is zero, and it does not make sense to solve the linear system. Instead, we will find the eigenvalues of the matrix and use these to estimate the eigenvalues of the Laplacian operator for the given boundary condition. The eigenvalues in turn determine the waveguide mode cutoff frequencies.

The key elements of a 2-D FD Laplace eigenvalue/eigenfunction solver are as follows:

Key Elements of a 2-D FD Laplace Solver

Grid generation: This portion of the code defines a grid of points that spans the interior of the boundary Γ.

Matrix fill: A finite difference stencil is used to fill a matrix approximation for the Laplacian operator.

Compute eigenvalues: An algorithm such as the MATLAB® `eig` or `eigs` functions can be used to find the eigenvalues of the matrix.

Compute eigenvectors (optional): If the waveguide mode patterns are required, the matrix eigenvectors can also be computed, typically with the same algorithm used to determine the eigenvalues. The eigenvectors represent samples of $E_z(x, y)$ for each mode at the interior grid points.

Determine cutoff frequencies: The operator eigenvalues are related to the waveguide mode cutoff frequencies through (4.58). The matrix eigenvalues are related to the operator eigenvalues by a scale factor, which includes the factor of Δx^2 in the denominator of (4.45).

4.3.5 2-D FD for Transmission Lines with Dielectric Materials

So far, we have considered the FD algorithm only for homogeneous materials (i.e., constant ε and μ). The FD algorithm can be modified to account for inhomogeneous dielectrics using the electric flux density boundary condition (2.41c). To take into account changes in material properties, we need to rederive Laplace's equation from Maxwell's equations, allowing for possible spatial variation of the permittivity.

For static fields, the electric field intensity is related to the scalar electric potential by

$$\overline{E} = -\nabla\phi \tag{4.65}$$

where the electric potential ϕ must be distinguished from the azimuthal coordinate of the cylindrical and spherical coordinate systems by the context. Using the constitutive relationship (2.2a), the electric flux density is

$$\overline{D} = -\varepsilon \nabla \phi \tag{4.66}$$

Using Gauss's law (2.1c) and requiring that no charge accumulates in the simulation region ($\rho = 0$), we arrive at

$$\nabla \cdot \varepsilon \nabla \phi = 0 \tag{4.67}$$

For a 2-D problem with the potential independent of z, this expands to the differential equation

$$\frac{\partial}{\partial x}\varepsilon(x,y)\frac{\partial}{\partial x}\phi + \frac{\partial}{\partial y}\varepsilon(x,y)\frac{\partial}{\partial y}\phi = 0 \tag{4.68}$$

This is a generalized form of Laplace's equation.

The next step is to discretize this equation using the finite difference approach. For any region in which $\varepsilon(x,y)$ is constant, the permittivity can be factored out of the equation, and the stencil (4.46) can be applied. For regions in which $\varepsilon(x,y)$ changes with x or y, the stencil must be modified. We will consider the case of a planar dielectric interface, as shown in Figure 4.6. For the first term of (4.68), $\varepsilon(x,y)$ is evaluated on the boundary between the two regions. On the boundary, we will take the permittivity to be the average value, $(\varepsilon_1 + \varepsilon_2)/2$. Using the central difference rule for each of the x derivatives leads to

$$\frac{\partial}{\partial x}\varepsilon(x,y)\frac{\partial}{\partial x}\phi \simeq \frac{\frac{1}{2}(\varepsilon_1+\varepsilon_2)\frac{\phi_{i+1,j}-\phi_{i,j}}{h} - \frac{1}{2}(\varepsilon_1+\varepsilon_2)\frac{\phi_{i,j}-\phi_{i-1,j}}{h}}{h} \tag{4.69}$$

For the second term of (4.68), the finite difference approximation is

$$\frac{\partial}{\partial y}\varepsilon(x,y)\frac{\partial}{\partial y}\phi \simeq \frac{\varepsilon_1\frac{\phi_{i,j+1}-\phi_{i,j}}{h} - \varepsilon_2\frac{\phi_{i,j}-\phi_{i,j-1}}{h}}{h} \tag{4.70}$$

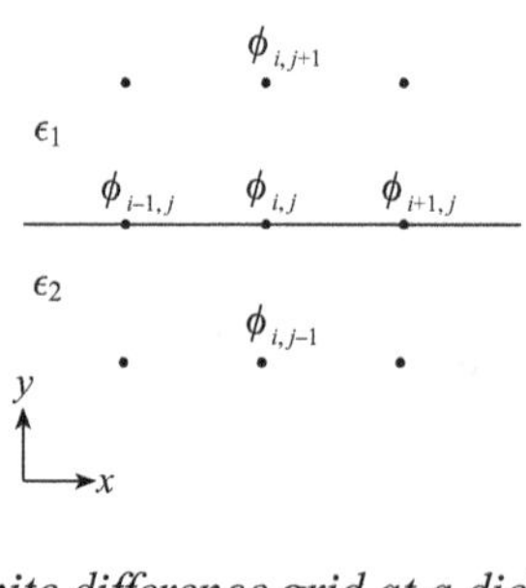

Figure 4.6 Finite difference grid at a dielectric interface

Combining the two terms and solving for $\phi_{i,j}$ produces the stencil

$$\phi_{i,j} = \frac{\varepsilon_1}{2(\varepsilon_1 + \varepsilon_2)}\phi_{i,j+1} + \frac{\varepsilon_2}{2(\varepsilon_1 + \varepsilon_2)}\phi_{i,j-1} + \frac{1}{4}\phi_{i+1,j} + \frac{1}{4}\phi_{i-1,j} \tag{4.71}$$

It can be seen that this stencil reduces to (4.46) if $\varepsilon_1 = \varepsilon_2$. The stencil must be adjusted if the dielectric interface is vertical instead of horizontal.

This modified stencil for dielectric interfaces allows the waveguide mode analysis algorithm of Section 4.3.2 to be applied more accurately to cases with dielectric materials within the cross section of the waveguide. The most important transmission lines that include dielectric layers are microstrip and coplanar waveguide (CPW), which are commonly used in radio frequency and microwave circuit designs. Deeper treatments of numerical methods for these types of transmission lines are found in other references, including [7].

4.4 2-D Finite Difference Time-Domain (FDTD) Method

So far, we have considered numerical methods for physical problems governed by the one-dimensional wave equation and the two-dimensional Laplacian operator. The next step in complexity is a problem with two spatial dimensions and time-dependent fields. The appropriate boundary value problem consists of Maxwell's equations as the governing system of partial differential equations together with a boundary condition on the behavior of the fields at the boundary of the simulation domain. We will assume that the fields and sources have a translational symmetry, so that field solutions are constant along one of the spatial dimensions and the invariant dimension can be omitted from the algorithm used to solve the boundary value problem.

We have already considered the FDTD method for 1-D problems in Section 4.2. For problems with more than one spatial dimension, the FDTD algorithm is more complex. One approach to solving time-domain EM problems with two spatial dimensions would be to discretize the 2-D wave equation, which is similar to (4.8) but with another second derivative term for the y coordinate. This works well for 2-D problems, but for 3-D problems the wave equation has certain mathematical solutions that are not solutions to Maxwell's equations (e.g., longitudinal waves). To avoid this difficulty, a more common approach in CEM is to discretize first-order Maxwell's equations directly instead. A method for doing this was first presented by Yee in 1966 [8].

4.4.1 2-D EM Problems

As discussed previously in Section 2.13.2, for 2-D EM problems in which sources and materials have a translational symmetry in one dimension, which is typically taken to be the z direction, the radiated electromagnetic field quantities will be independent of the invariant coordinate direction. The invariant axis is usually taken to be the z direction. All z derivative terms in Maxwell's equations are zero and the problem effectively has only two spatial dimensions.

In the physics literature, general field problems without symmetry are referred to as 3+1, or three spatial dimensions and one time dimension. The wave equation (4.8) represents a 1+1 problem, or a configuration with two translational symmetry directions. The Helmholtz BVPs in (4.51) and (4.52) might be thought of as 2+0, since time does not appear as an independent variable (the fields do vary in time, but the time dependence is sinusoidal and the phasor representations of the fields are constant). The time-dependent problems with a single translation symmetry considered in this section are referred to as 2+1. Table 4.1 lists a number of common partial differential equations for the analysis of waves and fields along with the number of spacetime dimensions.

For some problems, the materials are 2-D, but the sources and fields vary in z. These are sometimes called 2.5-D problems (see also Section 6.7). A waveguide and an infinite cylinder illuminated with a plane wave at an oblique incidence angle are examples of problems of this type. Numerical methods for 2-D problems can often be modified for 2.5-D analysis.

For a given set of materials and boundary conditions, there are two independent 2-D problems, depending on the vector direction of the sources. If the sources in the problem are currents flowing in the z direction, then the magnetic field can have only x and y components. Because the magnetic field is orthogonal to the z direction, this is called a transverse magnetic (TM) polarized problem. Sometimes the symmetry direction is indicated explicitly using a superscript: TM^z. If the current flows in the x–y plane, the electric field intensity vector is also in the x–y plane, and the problem is

Table 4.1 Partial differential equations of various dimensions

Name	PDE	Dimension
1-D wave equation	$\frac{\partial^2 u}{\partial x^2} = \frac{1}{c^2}\frac{\partial^2 u}{\partial t^2}$	1+1
2-D Laplace equation	$\frac{\partial^2 u}{\partial x^2} + \frac{\partial^2 u}{\partial y^2} = 0$	2+0
2-D Laplace eigenvalue problem	$\frac{\partial^2 u}{\partial x^2} + \frac{\partial^2 u}{\partial y^2} = \lambda u$	2+0
2-D Helmholtz equation	$\frac{\partial^2 u}{\partial x^2} + \frac{\partial^2 u}{\partial y^2} + k^2 u = 0$	2+0
2-D wave equation	$\frac{\partial^2 u}{\partial x^2} + \frac{\partial^2 u}{\partial y^2} = \frac{1}{c^2}\frac{\partial^2 u}{\partial t^2}$	2+1
3-D Laplace equation	$\frac{\partial^2 u}{\partial x^2} + \frac{\partial^2 u}{\partial y^2} + \frac{\partial^2 u}{\partial z^2} = 0$	3+0
3-D Laplace eigenvalue problem	$\frac{\partial^2 u}{\partial x^2} + \frac{\partial^2 u}{\partial y^2} + \frac{\partial^2 u}{\partial z^2} = \lambda u$	3+0
3-D Helmholtz equation	$\frac{\partial^2 u}{\partial x^2} + \frac{\partial^2 u}{\partial y^2} + \frac{\partial^2 u}{\partial z^2} + k^2 u = 0$	3+0
3-D wave equation	$\frac{\partial^2 u}{\partial x^2} + \frac{\partial^2 u}{\partial y^2} + \frac{\partial^2 u}{\partial z^2} = \frac{1}{c^2}\frac{\partial^2 u}{\partial t^2}$	3+1
Maxwell's equations	$\nabla \times \overline{E} = -\mu\frac{\partial}{\partial t}\overline{H},\ \nabla \times \overline{H} = \varepsilon\frac{\partial}{\partial t}\overline{E} + \overline{J}$	3+1

transverse electric or TE^z polarized. By linearity, 2-D problems with arbitrary source direction can be solved by decomposing the problem into TM and TE problems and analyzing each separately.

4.4.2 Yee Cell and 2-D FDTD Method for TM Polarized Fields

Maxwell's equations can be discretized using a construction known as a Yee cell [8]. For the TM^z polarization, the Yee cell is a graphical construction that defines three different rectangular grids, one for each of the nonzero field components E_z, H_x, and H_y. We will consider the samples of E_z to be the primary grid. The magnetic field component H_x is sampled on a grid staggered by one half grid step in the y direction, and H_y is sampled on a grid staggered by one half grid step in the x direction, as shown in Figure 4.7. The E and H grids are staggered because the central difference approximations for the first derivatives in the curl operator lead to derivatives evaluated at points that lie between the sample locations used in the difference formula.

A similar pair of staggered grids is required for the time dimension, because the central difference formula for the time derivatives in Maxwell's equations also leads to derivatives evaluated at a time point that lies between the samples used in the difference formula. The various grids defined by the Yee cell may appear complicated at first glance, but on further reflection it can be seen that the grids are at precisely the locations needed for central difference approximations to the spatial and temporal derivatives in Maxwell's equations to be consistent and mesh together into a seamless set of field update equations.

To represent all of the grids, we use the notation $(x_i, y_j, t_n) = (i\Delta x, j\Delta y, n\Delta t)$ where i, j, and n can be either integers or half-integers. In implementing the method, arrays used to store E_z, H_x, and H_y must have integer indices, so we interpret the i, j element of the array corresponding to E_z as the grid point (x_i, y_j), whereas for the magnetic field arrays the i, j element corresponds to the point $(x_i, y_{j+1/2})$ for H_x and $(x_{i+1/2}, y_j)$ for H_y.

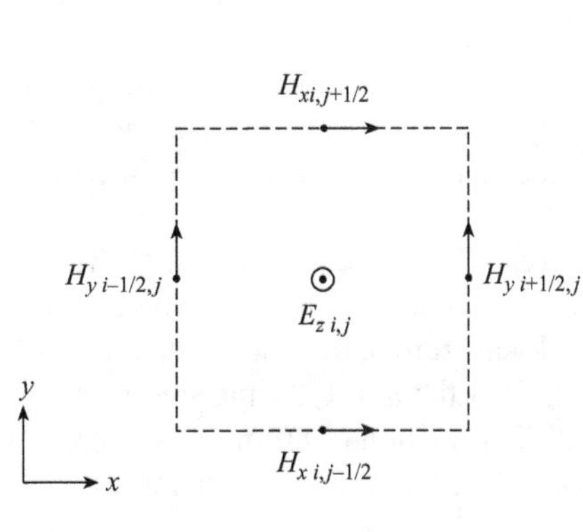

Figure 4.7 FDTD Yee cell for the 2-D TM^z polarization

For a TM^z problem, $E_x = E_y = H_z = 0$, and Ampère's and Faraday's law reduce to the PDEs

$$\frac{\partial E_z}{\partial y} = -\mu \frac{\partial H_x}{\partial t} \tag{4.72}$$

$$\frac{\partial E_z}{\partial x} = \mu \frac{\partial H_y}{\partial t} \tag{4.73}$$

$$\frac{\partial H_y}{\partial x} - \frac{\partial H_x}{\partial y} = \varepsilon \frac{\partial E_z}{\partial t} \tag{4.74}$$

Differencing these equations based on the grid points defined by the Yee cell gives

$$\frac{E^n_{z\,i,j+1} - E^n_{z\,i,j}}{\Delta y} = -\mu \frac{H^{n+1/2}_{x\,i,j+1/2} - H^{n-1/2}_{x\,i,j+1/2}}{\Delta t} \tag{4.75}$$

$$\frac{E^n_{z\,i+1,j} - E^n_{z\,i,j}}{\Delta x} = \mu \frac{H^{n+1/2}_{y\,i+1/2,j} - H^{n-1/2}_{y\,i+1/2,j}}{\Delta t} \tag{4.76}$$

$$\frac{H^{n+1/2}_{y\,i+1/2,j} - H^{n+1/2}_{y\,i-1/2,j}}{\Delta x} - \frac{H^{n+1/2}_{x\,i,j+1/2} - H^{n+1/2}_{x\,i,j-1/2}}{\Delta y} = \varepsilon \frac{E^{n+1}_{z\,i,j} - E^n_{z\,i,j}}{\Delta t} \tag{4.77}$$

To implement these difference equations in the FDTD algorithm, we must solve the difference equations for the forward time sample. This leads to the FDTD update equations

$$H^{n+1/2}_{x\,i,j+1/2} = H^{n-1/2}_{x\,i,j+1/2} - \frac{\Delta t}{\mu \Delta y}\left[E^n_{z\,i,j+1} - E^n_{z\,i,j}\right] \tag{4.78}$$

$$H^{n+1/2}_{y\,i+1/2,j} = H^{n-1/2}_{y\,i+1/2,j} + \frac{\Delta t}{\mu \Delta x}\left[E^n_{z\,i+1,j} - E^n_{z\,i,j}\right] \tag{4.79}$$

$$E^{n+1}_{z\,i,j} = E^n_{z\,i,j} + \frac{\Delta t}{\varepsilon \Delta x}\left[H^{n+1/2}_{y\,i+1/2,j} - H^{n+1/2}_{y\,i-1/2,j}\right] - \frac{\Delta t}{\varepsilon \Delta y}\left[H^{n+1/2}_{x\,i,j+1/2} - H^{n+1/2}_{x\,i,j-1/2}\right] \tag{4.80}$$

The first two equations come from Faraday's law and give the magnetic field at a future time step in terms of past values of the electric field. The third equation is Ampère's law and gives the electric field at a future time step in terms of past values of the magnetic field.

To solve the system of difference equations for the field samples equations, we use a "leapfrog" scheme. Beginning with initial values for the samples of E_z, H_x, and H_y (which are typically all zeros), the update equations (4.78) and (4.79) can be used to compute H_x and H_y at the $n + 1/2$ time step from E_z at time step n and H_x and H_y at time step $n - 1/2$. Equation (4.80) from Ampère's law can then be used to compute E_z at time step $n + 1$. Alternating between these two steps allows the solution to propagate forward in time.

While the update equations form the core of the FDTD algorithms, they are straightforward to implement in code. The most difficult parts of the solver from a programing point of view are the preprocessing and postprocessing steps that are needed to create the grid, define the sources, apply boundary conditions, and compute from simulated field values the desired result for a particular type of EM problem.

4.4.3 Dielectric and Conductive Materials

With the FDTD algorithm, conductive materials can be modeled by rederiving the update equations from Maxwell's equations with a nonzero value for the conductivity σ. This requires the addition of the term $\overline{\mathscr{J}} = \sigma\overline{\mathscr{E}}$ term to Ampère's law. Equation (4.74) becomes

$$\frac{\partial H_y}{\partial x} - \frac{\partial H_x}{\partial y} = \varepsilon\frac{\partial E_z}{\partial t} + \sigma E_z \tag{4.81}$$

Differencing this equation using the Yee cell leads to

$$\frac{H^{n+1/2}_{y\,i+1/2,j} - H^{n+1/2}_{y\,i-1/2,j}}{\Delta x} - \frac{H^{n+1/2}_{x\,i,j+1/2} - H^{n+1/2}_{x\,i,j-1/2}}{\Delta y} = \varepsilon\frac{E^{n+1}_{z\,i,j} - E^{n}_{z\,i,j}}{\Delta t} + \sigma\frac{E^{n+1}_{z\,i,j} + E^{n}_{z\,i,j}}{2} \tag{4.82}$$

For the last term on the right, we have used the averaging technique introduced in (4.18) to ensure that all terms in the difference equation are evaluated at the same point. Solving for $E^{n+1}_{z\,i,j}$ leads to the update equation

2-D FDTD Update Equation for E_z

$$\begin{aligned} E^{n+1}_{z\,i,j} = {} & \frac{\frac{\varepsilon}{\Delta t} - \frac{\sigma}{2}}{\frac{\varepsilon}{\Delta t} + \frac{\sigma}{2}} E^{n}_{z\,i,j} + \frac{1}{\Delta x\left(\frac{\varepsilon}{\Delta t} + \frac{\sigma}{2}\right)}\left[H^{n+1/2}_{y\,i+1/2,j} - H^{n+1/2}_{y\,i-1/2,j}\right] \\ & - \frac{1}{\Delta y\left(\frac{\varepsilon}{\Delta t} + \frac{\sigma}{2}\right)}\left[H^{n+1/2}_{x\,i,j+1/2} - H^{n+1/2}_{x\,i,j-1/2}\right] \end{aligned} \tag{4.83}$$

At dielectric or magnetic media interfaces, ε or μ vary abruptly with position. The FDTD method is reasonably accurate if the material parameters in the update equations are taken to be the values of the constitutive coefficients at the point where the difference equation is evaluated. Accuracy can be improved slightly by modifying the difference equations for Yee cells lying on the dielectric interface to incorporate the variation of the material coefficients within the Yee cell, as was done for the FD algorithm in Section 4.3.5. The integral forms of Maxwell's equations can be used to obtain the modified difference equations.

For good conductors, the relationship $\sigma/\omega \gg \varepsilon$ holds, where ω is the angular frequency. The conductivity of copper, for example, is $\sigma = 5.8 \times 10^7$ S/m. It is common in electromagnetic theory to model good conductors as perfect electric conductors. This typically results in a considerable simplification of the mathematical treatment.

PEC objects can be easily modeled using the Dirichlet boundary condition (see also the scattered field formulation of Section 4.5.3).

4.4.4 Anisotropic Materials

For some applications, the materials are anisotropic, and the constitutive relations become matrix relationships of the form

$$\begin{bmatrix} D_x \\ D_y \\ D_z \end{bmatrix} = \begin{bmatrix} \varepsilon_{11} & \varepsilon_{12} & \varepsilon_{13} \\ \varepsilon_{21} & \varepsilon_{22} & \varepsilon_{23} \\ \varepsilon_{31} & \varepsilon_{32} & \varepsilon_{33} \end{bmatrix} \begin{bmatrix} E_x \\ E_y \\ E_z \end{bmatrix} \tag{4.84}$$

$$\begin{bmatrix} B_x \\ B_y \\ B_z \end{bmatrix} = \begin{bmatrix} \mu_{11} & \mu_{12} & \mu_{13} \\ \mu_{21} & \mu_{22} & \mu_{23} \\ \mu_{31} & \mu_{32} & \mu_{33} \end{bmatrix} \begin{bmatrix} H_x \\ H_y \\ H_z \end{bmatrix} \tag{4.85}$$

In an anisotropic medium, the phase velocity of a wave depends on the polarization. It is straightforward, though tedious, to derive FDTD update equations for an anisotropic medium with constitutive relations given by (4.84) and (4.85). To simplify the treatment, we will assume that the coordinate system is aligned in such a way that the constitutive relationships are diagonal:

$$\begin{bmatrix} D_x \\ D_y \\ D_z \end{bmatrix} = \begin{bmatrix} \varepsilon_1 & 0 & 0 \\ 0 & \varepsilon_2 & 0 \\ 0 & 0 & \varepsilon_3 \end{bmatrix} \begin{bmatrix} E_x \\ E_y \\ E_z \end{bmatrix} \tag{4.86}$$

$$\begin{bmatrix} B_x \\ B_y \\ B_z \end{bmatrix} = \begin{bmatrix} \mu_1 & 0 & 0 \\ 0 & \mu_2 & 0 \\ 0 & 0 & \mu_3 \end{bmatrix} \begin{bmatrix} H_x \\ H_y \\ H_z \end{bmatrix} \tag{4.87}$$

In this case, the previously derived update equations need only to be adjusted so that μ_1 appears in (4.78), μ_2 appears in (4.79), and ε_3 appears in (4.80).

4.4.5 Stability Criterion

By deriving a numerical dispersion relation for the 2-D FDTD algorithm, it can be shown that the stability criterion for the spatial and temporal grid step sizes is

$$c\Delta t \leq \left[\frac{1}{\Delta x^2} + \frac{1}{\Delta y^2}\right]^{-1/2} \tag{4.88}$$

If $\Delta x = \Delta y$, then

2-D FDTD Stability Criterion

$$c\Delta t \leq \frac{1}{\sqrt{2}}\Delta x \tag{4.89}$$

This condition is slightly more stringent than the 1-D stability criterion. In higher dimensions, the factor on the right-hand side of the inequality becomes $1/\sqrt{n}$, where n is the number of spatial dimensions.

4.4.6 Boundary Conditions for the 2-D FDTD Method

The boundary conditions discussed in Section 4.2.4 can be extended to the 2-D and 3-D FDTD methods. To apply boundary conditions, one must determine which field components lie on the boundary. There are different ways to tile a computational domain with the Yee cell shown in Figure 4.7. One possibility is shown in Figure 4.8. If the number of Yee cells in the x and y directions are N_x and N_y, respectively, then the size of the array of samples of E_z is N_x by N_y, the array representing H_x is N_x by $N_y + 1$, and the array representing H_y is $N_x + 1$ by N_y.

Other ways to handle the boundary include arranging the Yee cells so that E_z points lie on all four sides of the boundary, to facilitate a Dirichlet boundary condition for E_z. Or, one can place the boundary of the computational domain along the E_z points on just the top and right sides so that the three field arrays have the same dimensions.

4.4.6.1 First-Order Mur Boundary Condition

The simple 1-D first-order Mur condition (4.19) can be used directly for the tangential field component on each of the four sides of a square computational domain. If the computational domain is tiled with Yee cells according to Figure 4.8, then the tangential field components are H_x on the top and bottom and H_y on the right and left sides. For the bottom side of the computational domain, the Mur ABC can be implemented using

```
Hx(1:NX, 1) = Hx1(1:NX,2) + a*(Hx(1:NX,2)-Hx1(1:NX,1));
```

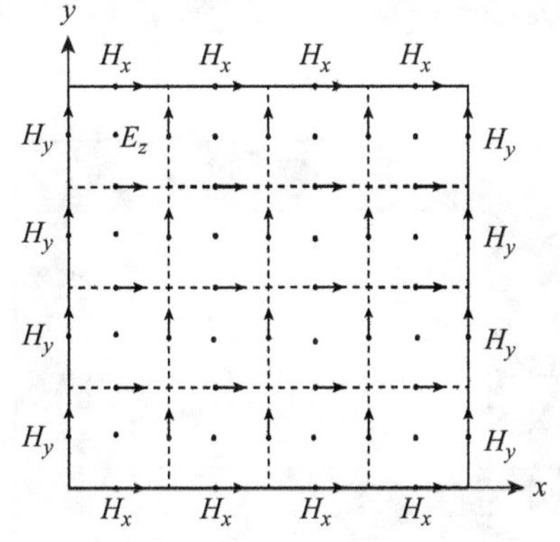

Figure 4.8 One possibility for tiling a square computational domain with Yee cells

where a is the coefficient $(r-1)/(r+1)$ in (4.19). The boundary condition is applied after the field components are updated. `Hx1` is the field array at the current time step and `Hx` is the updated field array at the $(n+1)$th time step. The other three boundary conditions are implemented in a similar way by changing the indices and field components appropriately.

4.4.6.2 Perfectly Matched Layer

The first-order Mur boundary condition we have used in this chapter is easy to implement, but waves arriving at the boundary at incidence angles other than normal undergo significant reflection, which contributes to modeling error. A perfectly matched layer that surrounds the simulation domain as shown in Figure 4.9 is one approach to achieving a lower reflection for waves over a large range of incidence angles. There are several ways to derive perfectly matched layers, including coordinate stretching, the split-field approach, and the uniaxial perfectly matched layer (UPML). We will choose the latter approach, since it is theoretically elegant, computationally efficient, and has a closer relationship to physically realizable materials than some other types of ABCs. We will consider the UPML for the TM^z polarization. The complete development for 3-D problems can be found in [1].

The UPML is based on the theoretical observation that the reflection coefficient at the interface between a lossy, anisotropic medium, and free space can be zero with certain choices for the material parameters. For this treatment, it is convenient to use the frequency domain or phasor representation. We will consider a medium with constitutive relations specified in such a way that Faraday's and Ampère's laws have the form

$$\nabla \times \overline{E} = -j\omega\mu\overline{\overline{s}}\overline{H} \tag{4.90}$$

$$\nabla \times \overline{H} = j\omega\varepsilon\overline{\overline{s}}\overline{E} \tag{4.91}$$

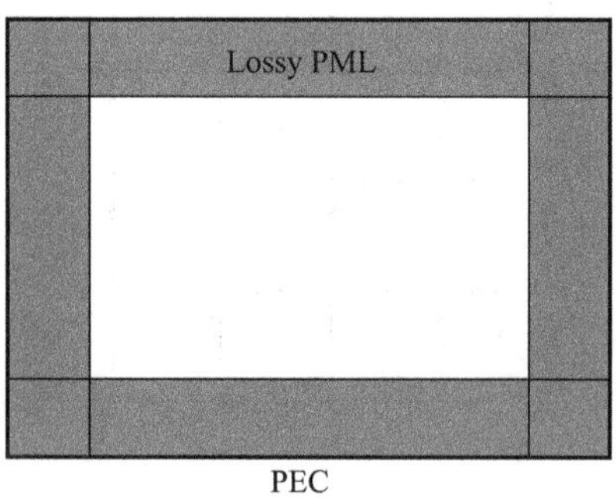

Figure 4.9 Lossy perfectly matched layer with PEC outer boundary

which represents a lossy, anisotropic medium. If this medium occupies the half-space $x < 0$, the half-space $x > 0$ is filled with a homogenous, isotropic medium (e.g., free space), and the matrix $\bar{\bar{s}}$ is

$$\bar{\bar{s}} = \begin{bmatrix} 1/s_x & 0 & 0 \\ 0 & s_x & 0 \\ 0 & 0 & s_x \end{bmatrix} \tag{4.92}$$

where

$$s_x = \kappa_x + \frac{\sigma_x}{j\omega\varepsilon} \tag{4.93}$$

then the reflection coefficient at the interface is identically zero. This is easy to see for a plane wave arriving at a normal incidence angle. In a homogeneous, isotropic medium, the characteristic impedance is $\eta_1 = \sqrt{\mu/\varepsilon}$. For a wave propagating in the x direction in the UPML medium, the electric field is in either the y or z directions, so the effective permittivity is εs_x, the effective permeability is μs_x, and the characteristic impedance is also $\eta_2 = \sqrt{\mu/\varepsilon}$. Since the characteristic impedances are identical in both media, there is no reflection at the interface. With a more involved derivation, it can be shown that the reflection coefficient is zero for any angle of incidence.

More generally, reflection-free boundary conditions in all three coordinate directions can be combined into a unified formulation with

$$\bar{\bar{s}} = \begin{bmatrix} s_y s_z/s_x & 0 & 0 \\ 0 & s_x s_z/s_y & 0 \\ 0 & 0 & s_x s_y/s_z \end{bmatrix} \tag{4.94}$$

where the parameters are chosen from

$$s_x = \kappa_x + \frac{\sigma_x}{j\omega\varepsilon} \tag{4.95a}$$

$$s_y = \kappa_y + \frac{\sigma_y}{j\omega\varepsilon} \tag{4.95b}$$

$$s_z = \kappa_z + \frac{\sigma_z}{j\omega\varepsilon} \tag{4.95c}$$

The parameters κ_x, κ_y, and κ_z can be set to one or adjusted if desired to improve the performance of the UPML. Outside the UPML, if we use the values $s_x = s_y = s_z = 1$, the material parameters reduce to those of free space. This is particularly convenient for the implementation of the FDTD algorithm, since the same update equations can be used for the interior of the simulation domain and the UPML regions at the boundaries.

The imaginary terms in (4.95a)–(4.95c) include conductivity parameters that allow the material to have loss, so that waves that pass into the medium decay in amplitude as they propagate. After a relatively small distance in the UPML, waves are attenuated to a small amplitude, and the layer can be terminated with a PEC boundary condition with a negligible increase in the reflection coefficient.

4.4.6.3 UPML for the TM^z Polarization

It remains to discretize Maxwell's equations for the UPML medium so that they can be implemented in an FDTD code. For the TM^z polarization, the relevant equations are

$$\frac{\partial E_z}{\partial y} = -j\omega s_y \mu \frac{s_z}{s_x} H_x \tag{4.96a}$$

$$\frac{\partial E_z}{\partial x} = j\omega s_x \mu \frac{s_z}{s_y} H_y \tag{4.96b}$$

$$\frac{\partial H_y}{\partial x} - \frac{\partial H_x}{\partial y} = j\omega s_x \varepsilon \frac{s_y}{s_z} E_z \tag{4.96c}$$

Before finite difference stencils can be applied, these equations must be transformed into the time domain. According to the properties of the Fourier transform, each factor of $j\omega$ becomes a time derivative. Since s_x, s_y, and s_z appear in the denominators of the equations and include this factor, however, the inverse Fourier transform is difficult to evaluate. To avoid this problem, we can define the auxiliary field variables

$$B_x = \mu \frac{s_z}{s_x} H_x \tag{4.97a}$$

$$B_y = \mu \frac{s_z}{s_y} H_y \tag{4.97b}$$

$$D_z = \varepsilon \frac{s_y}{s_z} E_z \tag{4.97c}$$

Within the UPML, these quantities are not strictly equal to the usual magnetic and electric flux densities, but outside the UPML, $s_x = s_y = s_z = 1$, and these expressions reduce to the standard constitutive relations for an isotropic medium. Using these variables in (4.96a)–(4.96c) and rearranging (4.97a)–(4.97c) leads to

$$\frac{\partial E_z}{\partial y} = -j\omega \kappa_y B_x - \frac{\sigma_y}{\varepsilon} B_x \tag{4.98a}$$

$$\frac{\partial E_z}{\partial x} = j\omega \kappa_x B_y + \frac{\sigma_x}{\varepsilon} B_y \tag{4.98b}$$

$$\frac{\partial H_y}{\partial x} - \frac{\partial H_x}{\partial y} = j\omega \kappa_x D_z + \frac{\sigma_x}{\varepsilon} D_z \tag{4.98c}$$

$$j\omega \varepsilon \kappa_x B_x + \sigma_x B_x = j\omega \varepsilon \mu \kappa_z H_x + \mu \sigma_z H_x \tag{4.98d}$$

$$j\omega \varepsilon \kappa_y B_y + \sigma_y B_y = j\omega \varepsilon \mu \kappa_z H_y + \mu \sigma_z H_y \tag{4.98e}$$

$$j\omega \varepsilon \kappa_z D_z + \sigma_z D_z = j\omega \varepsilon^2 \kappa_y E_z + \varepsilon \sigma_y E_z \tag{4.98f}$$

Since all factors of $j\omega$ are in the numerator, these equations can be readily transformed into the time domain.

In the time domain, the equations that must be discretized to implement the UPML for the TM^z polarization are

$$\frac{\partial E_z}{\partial y} = -\kappa_y \frac{\partial B_x}{\partial t} - \frac{\sigma_y}{\varepsilon} B_x \tag{4.99a}$$

$$\frac{\partial E_z}{\partial x} = \kappa_x \frac{\partial B_y}{\partial t} + \frac{\sigma_x}{\varepsilon} B_y \tag{4.99b}$$

$$\frac{\partial H_y}{\partial x} - \frac{\partial H_x}{\partial y} = \kappa_x \frac{\partial D_z}{\partial t} + \frac{\sigma_x}{\varepsilon} D_z \tag{4.99c}$$

$$\varepsilon\kappa_x \frac{\partial B_x}{\partial t} + \sigma_x B_x = \varepsilon\mu\kappa_z \frac{\partial H_x}{\partial t} + \mu\sigma_z H_x \tag{4.99d}$$

$$\varepsilon\kappa_y \frac{\partial B_y}{\partial t} + \sigma_y B_y = \varepsilon\mu\kappa_z \frac{\partial H_y}{\partial t} + \mu\sigma_z H_y \tag{4.99e}$$

$$\varepsilon\kappa_z \frac{\partial D_z}{\partial t} + \sigma_z D_z = \varepsilon^2\kappa_y \frac{\partial E_z}{\partial t} + \varepsilon\sigma_y E_z \tag{4.99f}$$

Equation (4.99c) is nearly identical in form to (4.81), so the update equation for (4.99c) can be obtained by modifying the coefficients and field variables in (4.83). Update equations for (4.99a) and (4.99b) can be obtained using a derivation similar to that of (4.83).

The update equations for (4.99a)–(4.99c) are

$$B^{n+1/2}_{x\,i,j+1/2} = -\frac{2\varepsilon\Delta t}{(2\varepsilon\kappa_y + \sigma_y\Delta t)\Delta y}\left(E^n_{z\,i,j+1} - E^n_{z\,i,j}\right) + \frac{2\varepsilon\kappa_y - \sigma_y\Delta t}{2\varepsilon\kappa_y + \sigma_y\Delta t} B^{n-1/2}_{x\,i,j+1/2} \tag{4.100a}$$

$$B^{n+1/2}_{y\,i+1/2,j} = \frac{2\varepsilon\Delta t}{(2\varepsilon\kappa_x + \sigma_x\Delta t)\Delta x}\left(E^n_{z\,i+1,j} - E^n_{z\,i,j}\right) + \frac{2\varepsilon\kappa_x - \sigma_x\Delta t}{2\varepsilon\kappa_x + \sigma_x\Delta t} B^{n-1/2}_{x\,i,j+1/2} \tag{4.100b}$$

$$\begin{aligned} D^{n+1}_{z\,i,j} &= \frac{2\varepsilon\Delta t}{(2\varepsilon\kappa_x + \sigma_x\Delta t)\Delta x}\left(H^{n+1/2}_{y\,i+1/2,j} - H^{n+1/2}_{y\,i-1/2,j}\right) \\ &\quad - \frac{2\varepsilon\Delta t}{(2\varepsilon\kappa_x + \sigma_x\Delta t)\Delta y}\left(H^{n+1/2}_{x\,i,j+1/2} - H^{n+1/2}_{x\,i,j-1/2}\right) + \frac{2\varepsilon\kappa_x - \sigma_x\Delta t}{2\varepsilon\kappa_x + \sigma_x\Delta t} D^n_{z\,i,j} \end{aligned} \tag{4.100c}$$

In (4.100a), the material parameters are evaluated at the grid point $i, j + 1/2$; in (4.100b), the parameters are evaluated at the grid point $i + 1/2, j$; and in (4.100c); the parameters are evaluated at the grid point i, j.

It remains to find update equations for (4.99d)–(4.99f). Applying finite difference stencils to (4.99d)–(4.99f) and solving for the future time sample leads to the update equations

$$H^{n+1/2}_{x\,i,j+1/2} = \frac{2\varepsilon\kappa_z - \sigma_z\Delta t}{2\varepsilon\kappa_z + \sigma_z\Delta t} H^{n-1/2}_{x\,i,j+1/2} + \frac{1}{\mu}\frac{2\varepsilon\kappa_x + \sigma_x\Delta t}{2\varepsilon\kappa_z + \sigma_z\Delta t} B^{n+1/2}_{x\,i,j+1/2} + \frac{1}{\mu}\frac{-2\varepsilon\kappa_x + \sigma_x\Delta t}{2\varepsilon\kappa_z + \sigma_z\Delta t} B^{n-1/2}_{x\,i,j+1/2} \tag{4.101a}$$

$$H^{n+1/2}_{y\,i+1/2,j} = \frac{2\varepsilon\kappa_z - \sigma_z\Delta t}{2\varepsilon\kappa_z + \sigma_z\Delta t} H^{n-1/2}_{y\,i+1/2,j} + \frac{1}{\mu}\frac{2\varepsilon\kappa_y + \sigma_y\Delta t}{2\varepsilon\kappa_z + \sigma_z\Delta t} B^{n+1/2}_{y\,i+1/2,j} + \frac{1}{\mu}\frac{-2\varepsilon\kappa_y + \sigma_y\Delta t}{2\varepsilon\kappa_z + \sigma_z\Delta t} B^{n-1/2}_{y\,i,j} \tag{4.101b}$$

$$E^{n+1}_{z\,i,j} = \frac{2\varepsilon\kappa_y - \sigma_y\Delta t}{2\varepsilon\kappa_y + \sigma_y\Delta t} E^{n}_{z\,i,j} + \frac{1}{\varepsilon}\frac{2\varepsilon\kappa_z + \sigma_z\Delta t}{2\varepsilon\kappa_y + \sigma_y\Delta t} D^{n+1}_{z\,i,j} + \frac{1}{\varepsilon}\frac{-2\varepsilon\kappa_z + \sigma_z\Delta t}{2\varepsilon\kappa_y + \sigma_y\Delta t} D^{n}_{z\,i,j} \tag{4.101c}$$

In these equations, the material parameters are evaluated at the same grid point as the field quantity on the left-hand side. Within the loop over time in the FDTD algorithms, the update equations (4.100a)–(4.100c) are used to find $B^{n+1/2}_{x\,i,j+1/2}$, $B^{n+1/2}_{y\,i+1/2,j}$, and $D^{n+1}_{z\,i,j}$. The update equations (4.101a)–(4.101c) are then used to find $H^{n+1/2}_{x\,i,j+1/2}$, $H^{n+1/2}_{y\,i+1/2,j}$, and $E^{n+1}_{z\,i,j}$.

Finally, we need to specify values for the coefficients s_x, s_y, and s_z. By adjusting the values of these coefficients, we can model free space in the interior region of the simulation domain as well as an absorbing boundary condition on the sides of the domain. For a rectangular simulation domain, we must consider three cases:

Interior region of the simulation domain: $s_x = s_y = s_z = 1$.

Boundaries at $x = x_{\min}$ *and* $x = x_{\max}$:

$$s_x = \kappa_x + \frac{\sigma_x}{j\omega\varepsilon}$$
$$s_y = 1$$
$$s_z = 1$$

Boundaries at $y = x_{\min}$ *and* $y = x_{\max}$:

$$s_x = 1$$
$$s_y = \kappa_y + \frac{\sigma_y}{j\omega\varepsilon}$$
$$s_z = 1$$

Overlapping UPML regions at corners:

$$s_x = \kappa_x + \frac{\sigma_x}{j\omega\varepsilon}$$

$$s_y = \kappa_y + \frac{\sigma_y}{j\omega\varepsilon}$$

$$s_z = 1$$

The remaining parameters, κ_x, κ_y, and κ_z, can be set to unity or adjusted to improve the numerical performance of the UPML.

With the continuous equations (4.99a)–(4.99f), the conductivity parameter of the perfectly matched layer can be set to any value, and the reflection coefficient remains identically zero. When the partial differential equations are discretized, the reflection at the PML interface is no longer zero due to discretization error, particularly if the conductivity parameter jumps from zero in the interior of the simulation domain to a large, nonzero value in the PML. Reflections can be minimized by gradually changing the conductivity parameter from zero at the PML interface to a finite value at the outer edge of the PML. Various functions can be employed to grade the conductivity. We will choose polynomial grading, for which

$$\sigma_x(x) = \left(\frac{x}{d}\right)^m \sigma_{x,\max} \tag{4.102}$$

where x is the distance to the interface between the UPML region and the interior of the simulation domain, and d is the thickness of the UPML. σ_y is graded similarly with a polynomial in y. Optimal values for m and $\sigma_{x,\max}$ can be found empirically or analytically. In practice, $m = 2$ and $\sigma_{x,\max} = 1$ work well.

4.4.7 Preprocessing

Preprocessing for the FDTD algorithm involves grid generation, specification of the radiation or scattering problem parameters, and generation of material constitutive parameter arrays. For the 2-D FDTD method, grid generation is similar to the 2-D FD method described in Section 4.3. Parameters of the radiation or scattering problem include source type, location, and pulse duration. For plane wave incident fields, the frequency and incidence angle must be specified.

Materials in the simulation domain are described with arrays of constitutive parameters. If there are a small number of materials, to reduce the memory requirement, the array may consist of integers that indicate a material type. If storage is not an issue, then materials can be represented with arrays of the material parameters ε_r, μ_r, and σ at each grid point.

A convenient approach to defining material parameters is to create arrays of coordinates of each grid point and to use logical operations on the grid point coordinates to determine the material parameter each grid point. If `X` and `Y` are N_x-by-N_y arrays of x and y coordinates of the grid points, the following code fragment can be used to fill an array of permittivity values for a dielectric circular cylinder:

Circular Dielectric Cylinder Model

```
epsr = ones(Nx,Ny);    % Relative permittivity
x_ctr = 0;             % x coordinate cylinder center
y_ctr = 0;             % y coordinate cylinder center
a = 0.5;               % Radius of circular cylinder
epsr_cyl = 2;          % Relative permittivity

% Find indices of grid points inside cylinder
idx = find((X - x_ctr).^2 + (Y - y_ctr).^2 <= a^2);

% Set permittivity at grid points inside cylinder
epsr(idx) = epsr_cyl;
```

The array `epsr` is used in the FDTD update equation (4.83) to define the permittivity at each grid point. Arrays `sigm` and `mur` for the conductivity and relative permeability can be created similarly. This approach can be extended to more complex polygonal geometries using the `inpolygon` function.

4.5 FDTD Modeling for Scattering Problems

Electromagnetics problems can be classified into two groups: (1) radiation problems, which consist of a source near a material structure; and (2) scattering problems, which involve a distance source launching a wave that is incident on a material structure or scatterer. An antenna with a source at the input port is a radiation problem, and radar scattering from an aircraft is a radiation problem. Mathematically, the two kinds of problems are closely related, since both involve finding unknown fields given a known source and material distribution, but the location of the source and the physical interpretation of the solution are different.

For scattering problems, the goal is to find the reflected or scattered fields given a source that illuminates an object, or scatterer. We define the incident field as the field radiated by the source if the scatterer were not present. The incident field is typically easy to find or given in analytical form, since it usually is the field radiated by a simple type of source in free space. The scatterer causes a perturbation to the fields, and if we can find that perturbation the problem is solved. The total field in a scattering problem consists of the sum of the incident field and the perturbation caused by the scatterer:

$$\overline{E}^{\text{tot}} = \overline{E}^{\text{i}} + \overline{E}^{\text{s}} \tag{4.103a}$$

$$\overline{H}^{\text{tot}} = \overline{H}^{\text{i}} + \overline{H}^{\text{s}} \tag{4.103b}$$

where the incident fields $\overline{E}^{\text{i}}$ and $\overline{H}^{\text{i}}$ are the fields that would be measured if the scatterer were not present. We now consider the various aspects of scattering problems in detail.

4.5.1 Incident Field

The incident field in a scattering problem is most often a plane wave. Because the source in a scattering problem is typically far from the scatterer, the illuminating field radiated by the source can be approximated near the scatterer as a plane wave. Moreover, any incident field can be decomposed into a combination of plane waves arriving from different directions, so that in phasor form

$$\overline{E}^{\text{i}}(\overline{r}) = \int d\hat{k}\, \overline{E}_0^{\text{i}}(\overline{k})\, e^{-j\overline{k}\cdot\overline{r}} \tag{4.104}$$

where $\overline{E}_0^{\text{i}}(\overline{k})$ is the phasor amplitude of each plane wave component of the incident field, and the integral is over the direction of pr opagation. For these reasons, plane wave scattering is a fundamental problem in electromagnetic theory. Less common incident field types include the fields radiated by line currents and dipoles near enough to the scatterer that the fields cannot be approximated by a plane wave.

For a single incident plane wave, the phasor electric field intensity vector can be expressed using the vector notation of (2.57), so that

$$\overline{E}^{\text{i}}(\overline{r}) = \overline{E}_0^{\text{i}} e^{-j\overline{k}^{\text{i}}\cdot\overline{r}} \tag{4.105}$$

The wave vector is

$$\overline{k}^{\text{i}} = -k(\sin\theta\cos\phi\,\hat{x} + \sin\theta\sin\phi\,\hat{y} + \cos\theta\,\hat{z}) \tag{4.106}$$

The angles θ and ϕ represent the spherical angle of arrival or direction of arrival (DOA) of the incident wave.

Assuming that the constant vector $\overline{E}_0$ is real, the time-dependent electric field is

$$\overline{E}^{\text{i}}(\overline{r}, t) = \overline{E}_0^{\text{i}} \cos(\overline{k}^{\text{i}} \cdot \overline{r} - \omega t) \tag{4.107}$$

If the plane wave is TM^z polarized and propagating in the $+x$ direction, the plane wave simplifies to

$$E_z^{\text{i}}(x, t) = \cos(k_0 x - \omega t) \tag{4.108}$$

where k_0 is the wave number, and ω is the angular frequency of the wave. Since the cosine function jumps abruptly to a nonzero value if the FDTD algorithm begins at time $t = 0$, it is common to reduce signal transients by changing the phase of the incident wave so that the cosine becomes a sine. The plane wave can be represented in MATLAB notation as

```
Ez_inc = sin(k0*x - omega*t);
```

More generally, for a wave propagating in the x–y plane, the time-varying electric field is given by (2.63).

4.5.2 Scattered Field

The scattered field is defined to be the perturbation to the incident field caused by the scatterer. It should be kept in mind that the measurable field that exists around the scatterer is the total field, and the incident and scattered fields represent physically motivated components that make up the total field.

The physical interpretation of the scattered field is quite intuitive, except in the shadow of the scatterer where the total field is small. In a shadow region, the scattered field is actually large, rather than small in magnitude. This may seem strange, but it is a consequence of the definition of the scattered field according to ((4.103a) and (4.103b)). Because the total field is the sum of the incident and scattered fields, to have a small total field the scattered field must be nearly equal in magnitude and opposite in sign relative to the incident field. Since the scattered field is the perturbation caused by the scatterer, and the shadow represents a large perturbation of the incident field, it is reasonable that the scattered field would be large in the shadow region.

4.5.3 Scattered Field Formulation

Because the incident field in a scattering problem is given and the scattered field is unknown, it is common when implementing numerical methods to reformulate problems so that the scattered field is modeled instead of the total field. Since Maxwell's equations are linear, both the incident and scattered fields satisfy Maxwell's equations, but with different material parameters and sources. The incident field satisfies Maxwell's equations with values for the permittivity, permeability, and conductivity parameters equal to that in free space, and the source is the excitation that radiates the incident field. In the case of the total field, the material parameters that enter into Maxwell's equations are simply those of the given scatterer. If we formulate Maxwell's equations using (4.103a) and (4.103b) in terms of the scattered fields, the material structures in the problem become equivalent sources that radiate the scattered fields. One advantage of the scattered field formulation is that the source that radiates the incident field does not need to be included in the model, since the effect of the incident field is incorporated in the equivalent sources.

4.5.3.1 PEC Scatterer

We will consider the scattered field formulation first for the case of a PEC scatterer. In this case, we can find the equivalent source using the electric field boundary condition at the surface of the scatterer, which is

$$\hat{n} \times (\overline{E}^{\mathrm{i}} + \overline{E}^{\mathrm{s}}) = 0 \tag{4.109}$$

If we model only the scattered fields in a numerical simulations, then the boundary condition at the scatterer surface becomes

$$\hat{n} \times \overline{E}^{\mathrm{s}} = -\hat{n} \times \overline{E}^{\mathrm{i}} = \overline{M}_s \tag{4.110}$$

This fictitious magnetic surface current can be viewed as a source impressed on the surface of the PEC object that radiates the scattered field in the presence of the scatterer. In the FDTD algorithm, this is implemented as a hard source.

For the TM^z polarization, the tangential component at the surface of any 2-D scatterer of the electric field is in the z direction. The equivalent source relationship at the surface of the scatterer simplifies to

Equivalent Source for PEC Scatterers

$$E_z^s(\overline{\rho}) = E_z^i(\overline{\rho})\big|_{\overline{\rho} \text{ on } S} \tag{4.111}$$

Since sources inside a PEC object do not radiate, it is possible to set all tangential electric field grid sample values on and inside the scatterer equal to the equivalent source rather than to try to identify which grid points actually lie on the scatterer surface. Thus, (4.111) can be implemented conveniently using a list of the indices of all grid points on and inside the scatterer.

4.5.3.2 Dielectric Scatterer

For a magnetic or dielectric scatterer, the equivalent source is implemented as a soft source. We use the scattered field decomposition (4.103a) and (4.103b) in Maxwell's equations to obtain

$$\nabla \times (\overline{E}^i + \overline{E}^s) = -\mu \frac{\partial}{\partial t}(\overline{H}^i + \overline{H}^s) \tag{4.112a}$$

$$\nabla \times (\overline{H}^i + \overline{H}^s) = \varepsilon \frac{\partial}{\partial t}(\overline{E}^i + \overline{E}^s) + \sigma(\overline{E}^i + \overline{E}^s) \tag{4.112b}$$

By definition, the incident field satisfies Maxwell's equations in free space ($\varepsilon = \varepsilon_0$, $\mu = \mu_0, \sigma = 0$), so that

$$\nabla \times \overline{E}^i = -\mu_0 \frac{\partial \overline{H}^i}{\partial t} \tag{4.113a}$$

$$\nabla \times \overline{H}^i = \varepsilon_0 \frac{\partial \overline{E}^i}{\partial t} \tag{4.113b}$$

Subtracting ((4.113a) and (4.113b)) from ((4.112a) and (4.112b)) leads to

Equivalent Sources for Magnetic and Dielectric Scatterers

$$\nabla \times \overline{E}^s = -\mu \frac{\partial \overline{H}^s}{\partial t} - \underbrace{(\mu - \mu_0)\frac{\partial \overline{H}^i}{\partial t}}_{\text{Magnetic source } \overline{M}} \tag{4.114a}$$

$$\nabla \times \overline{H}^s = \varepsilon \frac{\partial \overline{E}^s}{\partial t} + \sigma \overline{E}^s + \underbrace{(\varepsilon - \varepsilon_0)\frac{\partial \overline{E}^i}{\partial t} + \sigma \overline{E}^i}_{\text{Electric source } \overline{J}} \tag{4.114b}$$

where the equivalent sources are given in terms of known quantities. These fictitious volume currents can be viewed as impressed sources that radiate the scattered field in the presence of the dielectric scatterer. In effect, we have transformed the scattering problem into a radiation problem. With the scattered field formulation implemented in the FDTD algorithm, all material objects have soft current sources impressed on them. As the FDTD algorithm steps forward in time, these impressed sources radiate the scattered fields $\overline{E}^s$ and $\overline{H}^s$.

4.5.4 *Scattering Amplitudes, Scattering Width, and Radar Cross Section*

With radiation and scattering problems, we are often more interested in the fields far away from the radiating object or scatterer than in the near fields. For time-harmonic problems, far from a scatterer all fields oscillate harmonically as a function of distance and decay as $\rho^{-1/2}$ for 2-D problems or r^{-1} for 3-D problems. In characterizing scattered fields, it is convenient to factor out the asymptotic ρ or r dependence of the field to obtain a function of angle only. We also normalize by the intensity of the incident field. The resulting quantity is called a scattering amplitude and is defined for 2-D and 3-D problems according to

$$S(\phi^{s};\phi^{i}) = \lim_{\rho\to\infty}\sqrt{\frac{\pi k\rho}{2j}}e^{jk\rho}\frac{E^{s}(\rho,\phi^{s})}{E^{i}} \quad \text{(2-D)} \tag{4.115}$$

$$S(\theta^{s},\phi^{s};\theta^{i},\phi^{i}) = \lim_{r\to\infty} kre^{jkr}\frac{E^{s}(r,\theta^{s},\phi^{s})}{E^{i}} \quad \text{(3-D)} \tag{4.116}$$

where θ^i and ϕ^i are the spherical angles of the wavevector of the incident plane wave, and the scattered field is evaluated at the point (ρ,ϕ^s) in 2-D or (r,θ^s,ϕ^s) in 3-D. E^i is the phasor amplitude of the incident plane wave, typically evaluated at the origin. The constants are chosen so that the scattering amplitude is dimensionless. The electric field can be replaced by the magnetic field in these expressions.

Since scattering amplitudes are defined for time-harmonic fields, all the field quantities in the scattering amplitude definitions are represented as phasors. For time-domain algorithms such as FDTD, to use (4.115) and (4.116) the time-domain fields must be transformed to phasors.

The scattering amplitude depends on the properties of the scatterer, the angle of arrival of the incident field, and the scattering direction at which E^s is evaluated. The field quantities in the previous definitions represent components of the incident and scattered field vectors, which means that the scattering amplitude also depends on the polarization of the incident and scattered fields.

Scattered fields are also often characterized by the scattering width for 2-D problems or radar cross section (RCS) for 3-D problems. The definitions are

2-D Scattering Width and 2-D Scattering Cross Section (RCS)

$$\sigma_{2\text{-D}}(\phi^s;\phi^i) = \lim_{\rho\to\infty} 2\pi\rho\frac{|E^s(\rho,\phi^s)|^2}{|E^i|^2} = \frac{4}{k}|S|^2 \tag{4.117}$$

$$\sigma_{3\text{-D}}(\theta^s,\phi^s;\theta^i,\phi^i) = \lim_{r\to\infty} 4\pi r^2\frac{|E^s(r,\theta^s,\phi^s)|^2}{|E^i|^2} = \frac{4\pi}{k^2}|S|^2 \tag{4.118}$$

The scattering width $\sigma_{2\text{-D}}$ has units of length, and the RCS $\sigma_{3\text{-D}}$ has units of area. Physically, the RCS is the area that receives the amount of incident power that when radiated isotropically leads to the same power density as the scattered field in a given direction. RCS and scattering width are shape and material dependent, so a physically large object such as a stealth aircraft may have a small RCS.

The scattering coefficient of a large object or surface is defined to be the RCS divided by the area of the object that is illuminated by the incident field. A flat conducting plate typically has a larger peak scattering coefficient than objects with curved surfaces, since curvature tends to reflect or diffract energy in many directions rather than in one primary direction.

The RCS of a scatterer is analogous to the gain pattern of an antenna. The expression for the RCS of a large, flat plate, for example, is similar in form to the expression for the effective receiving area of an aperture antenna of the same shape.

4.5.5 Bistatic and Monostatic Scattering

The scattering amplitude, scattering width, and RCS of a given object depend on the direction of the incident field as well as the scattering direction. In practice, these directions are determined by the relative positions of the transmitter, receiver, and scatterer.

In a bistatic system, the transmitter that launches an incident plane wave is at a different location from the receiver that measures the scattered field. For monostatic systems, the transmitter and receiver are in the same location. In this case, $\phi^s = \phi^i$ and $\theta^s = \theta^i$. This is called the backscattering direction since a wave in this direction propagates toward the transmitter or in the opposite direction as the incident field. For obvious reasons, a monostatic configuration is very common in applications such as radar.

Forward scattering is the scattering amplitude or RCS in the same direction as the incident field. Since the forward direction typically lies in the shadow of the object, the forward-scattering amplitude is often large in relation to the scattering amplitude in other scattering directions. Another scattering direction of importance is the specular direction, or the direction of maximum scattering for a large, flat object for which the incident wave reflects predominantly according to the equal angle or mirror law.

Bistatic scattering for a single incident plane wave direction and many scattering directions is typically easier to simulate than monostatic scattering for many incident directions, because the bistatic far fields can be found for all directions in postprocessing from the results of one simulation for the given incident field. In the monostatic case, the simulation must be rerun for each incident field direction.

4.6 Postprocessing the FDTD Solution

The direct result of an FDTD simulation consists of electric and magnetic fields as a function of position and time. Often, the desired final output of the simulation is not the fields but some quantity derived from the computed fields, such as impedances, far-field radiation patterns, scattering amplitudes, and radar cross sections. These are computed in postprocessing.

4.6.1 Frequency-Domain (Phasor) Fields

Most quantities computed in postprocessing are functions of frequency. To compute frequency-domain quantities defined in terms of the phasor representation of the fields, the time-dependent fields must be converted from the time domain to the frequency or phasor domain.

When frequency-domain quantities are needed, the source used in the simulation can be either a broadband pulse or a time-harmonic, single-frequency source. For a pulse excitation, the Fourier transform of the radiated or scattered fields can be used to find derived frequency-domain quantities over the bandwidth of the pulse. For a time-harmonic excitation, frequency-domain quantities are obtained only at the frequency of the source.

4.6.1.1 Two Equation, Two Unknown Method

The simplest approach to conversion from the time domain to the frequency domain is the two equation, two unknown (2E2U) method. At each point in the simulation domain, a time-harmonic field can be expressed in the form

$$E_z(\bar{r}, t) = A\cos(\omega t + \phi) \tag{4.119}$$

where ω is known and the amplitude A and the phase ϕ of the signal are unknown. At a given grid point $\bar{r}$, during the FDTD simulation we can store the values of the field at two different times:

$$\begin{aligned} v_1 &= A\cos(\omega t_1 + \phi) \\ v_2 &= A\cos(\omega t_2 + \phi) \end{aligned}$$

where $t_1 = n_1\Delta t$, and $t_2 = n_2\Delta t$. The two equations can be solved for the amplitude A and the phase ϕ to obtain

$$\begin{aligned} \phi &= \tan^{-1}\left[\frac{v_2\sin(\omega t_1) - v_1\sin(\omega t_2)}{v_1\cos(\omega t_2) - v_2\cos(\omega t_1)}\right] \\ A &= \left|\frac{v_1}{\sin(\omega t_1 + \phi)}\right| \end{aligned}$$

Typically, t_1 and t_2 are chosen to be near the end of a simulation, when transients have died down and the fields have reached a steady-state solution. The two time points can be chosen to be adjacent, so that n_1 and n_2 are the next-to-last and final time points in the FDTD simulation, respectively, but to avoid rounding errors for small values of the time step Δt, it can be better to choose n_1 and n_2 to be several time steps apart.

4.6.1.2 Fourier Transform Method

A more general approach to conversion from the time domain to the frequency domain can be derived from the definition (2.45) of the phasor representation. This formula can be inverted using the Fourier transform to find an expression for the phasor $\overline{E}(\overline{r})$ in terms of the time-varying electric field $\overline{\mathscr{E}}(\overline{r}, t)$.

For convenience, we will consider only one component of the electric field, E_z. If the field $E_z(\overline{r}, t)$ is time harmonic with frequency ω, then it is a sinusoidal function with period $T = 2\pi/\omega$. For a sinusoidal function, only two coefficients of the Fourier series representation are nonzero:

$$a_1 = \frac{1}{T}\int_T E_z(\overline{r}, t)e^{-j\omega t}\, dt$$

$$a_{-1} = \frac{1}{T}\int_T E_z(\overline{r}, t)e^{j\omega t}\, dt$$

Since the time-domain field is by definition real, it can be seen from these expressions that $a_{-1} = a_1^*$. Using the Fourier series representation of E_z, we have

$$E_z(\overline{r}, t) = a_1 e^{j\omega t} + a_1^* e^{-j\omega t} \tag{4.120}$$

By comparing this expression with (2.45), it can be seen that the phasor representation is $E_z(\overline{r}) = 2a_1$. Using the definition of the Fourier coefficient a_1, the electric field phasor can be expressed in terms of the time-varying field as

$$E_z(\overline{r}) = \frac{2}{T}\int_0^T E_z(\overline{r}, t)e^{-j\omega t}\, dt \tag{4.121}$$

This integral is essentially a Fourier transform, so that as expected the phasor form of the electric field is related to the Fourier transform of the time-domain field.

The integral in (4.121) can be approximated using the discrete Fourier transform (DFT), which can be rapidly computed using the fast Fourier transform (FFT) algorithm. If we approximate the integral using a Riemann sum or midpoint integration rule, we obtain

$$E_z(\overline{r}) \simeq \frac{2}{T}\sum_1^N e^{-j\omega t_n} E_z(\overline{r}, t_n)\, \Delta t \tag{4.122}$$

where $t_n = (n-1)\Delta t$, and $\Delta t = T/N$. The FFT is defined by

$$\text{FFT}\{E_z(\overline{r}, t_n)\}_m = \sum_{n=1}^N e^{-j2\pi(m-1)(n-1)/N} E_z(\overline{r}, t_n) \tag{4.123}$$

Comparing these two expressions shows that the frequency points corresponding to the results of the FFT are

$$\omega_m = \frac{2\pi(m-1)}{N\Delta t} = (m-1)\frac{2\pi}{T} \tag{4.124}$$

Since $\omega = 2\pi/T$ is the frequency of the time-harmonic wave, we can see that $\omega_2 = \omega$. The second sample of the FFT therefore contains the phasor representation of the time-domain fields, so that

Time-domain to Phasor Transformation

$$E_z(\bar{r}, \omega) \simeq \frac{2}{N}\text{FFT}\{E_z(\bar{r}, t_n)\}_2 \tag{4.125}$$

where the subscript 2 indicates that the second element of the vector returned by the FFT operation should be selected. The FFT is computed for the N time samples of the field at the point $\bar{r}$ produced by the FDTD algorithm for one full period of the time-harmonic field.

If the field time dependence is broadband rather than time-harmonic, this method can be generalized by using all the values returned by the FFT algorithm to analyze the frequency-domain properties of a system at multiple frequencies using a single simulation run.

4.6.2 Near Field to Far Field Transformation

It is generally impractical to simulate a large domain that includes the far field of a radiating object or scatterer. Using a Green's function and a radiation integral, far fields can be computed from near fields in postprocessing. One way to do this is to surround all material objects and sources in the simulation domain with a closed surface and find equivalent sources on the surface that when impressed in free space radiate the same fields as the scatterer outside the contour or surface. This allows the scatterer and all conductors, dielectrics, and other materials to be replaced by a current source in free space. The free-space radiation integral then can be used to compute the far fields radiated by the source. The closed surface is sometimes called a Huygens surface, because Huygens's principle can be used to determine the equivalent sources on the surface and find the fields radiated by the sources outside the surface.

For 2-D problems, the Huygens surface is an infinite cylinder. Since the simulation domain is a 2-D slice in 3-D space, we need to consider only the intersection of the infinite cylinder with the 2-D slice, which is a one-dimensional closed contour. Even though the contour is one-dimensional, we will still refer to it as a Huygens surface. For convenience when working with rectangular grids, the Huygens surface is often chosen to be a box as shown in Figure 4.10. Once the simulation has reached steady state, electric and magnetic fields are stored along the surface. The far fields can then be computed by relating the electric and magnetic fields on the Huygens

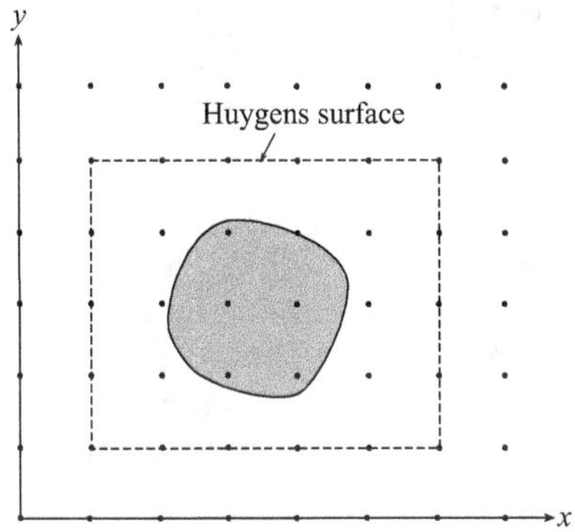

Figure 4.10 FDTD grid with a Huygens surface enclosing a scatterer. Tangential fields on the surface can be transformed into equivalent sources in free space, allowing the radiation integral to be used to obtain far fields

surface to equivalent current sources in free space and using the far-field radiation integral.

Far fields are typically computed in the frequency domain, so the time-dependent fields are converted to phasor form before the near-to-far transformation is applied. Accordingly, all field quantities in this section are represented as phasors.

The equivalence principle of electromagnetic field theory covered in Section 2.13.7 can be used to relate the fields on the surface to equivalent sources that if impressed in free space would radiate the same fields outside the surface. In the equivalent problem, suppose that the fields inside the surface are arbitrary fields $\overline{E}_1$ and $\overline{H}_1$ that are different from the actual fields $\overline{E}$ and $\overline{H}$ radiated or scattered by the structure inside the surface. By the electric and magnetic field boundary conditions, the sources

$$\overline{M}_s = -\hat{n} \times (\overline{E} - \overline{E}_1) \tag{4.126}$$

$$\overline{J}_s = \hat{n} \times (\overline{H} - \overline{H}_1) \tag{4.127}$$

must lie on the surface. If we choose $\overline{E}_1$ and $\overline{H}_1$ to be zero, then

$$\overline{M}_s = -\hat{n} \times \overline{E} \tag{4.128}$$

$$\overline{J}_s = \hat{n} \times \overline{H} \tag{4.129}$$

This pair of surface currents when impressed in free space radiates the same fields outside the Huygens surface as the original materials and sources that were inside the surface. We can therefore use the free-space radiation integral to compute the fields radiated by these sources. The sources (4.128) and (4.129) with $\overline{E}_1 = \overline{H}_1 = 0$ are obtained using Love's equivalence principle (see Section 2.13.7).

Each sample of the equivalent currents on the grid corresponds to a time-harmonic line source that is oriented in the z direction. To find the field radiated by the equivalent currents, we need to sum over the fields radiated by each line source. The line current intersects the x–y plane at one point, which we can represent using a delta function $\delta(x-x')\delta(y-y')$, where the source is located at (x', y'). Because the source is given by a delta function in the x–y plane, we refer to the field radiated by a line source as the 2-D Green's function.

4.6.2.1 2-D Green's Function

We will now derive the 2-D Green's function by computing the field radiated by a line source carrying a time-harmonic electric current of amplitude I in the $+z$ direction. Because the source distribution is invariant in the z direction, it is convenient to use the cylindrical coordinate system. We will assume that the line current is along the z-axis. The approach we will follow is to begin with a general series expansion for the fields radiated by an arbitrary source and to determine the unknown coefficients in the series such that it reduces to the field radiated by the line source.

Since the current associated with the line source is infinitely long and flows in the z direction, only the z component of the electric field will be nonzero. The general solution for the z component of the electric field in the cylindrical coordinate system can be expressed in terms of Bessel functions as

$$E_z(\rho,\phi,z) = \sum_{m=0}^{\infty}\left[A_m H_m^{(1)}(k_\rho\rho) + B_m H_m^{(2)}(k_\rho\rho)\right]\left[C_m e^{jm\phi} + D_m e^{-jm\phi}\right] \times \left[E_m e^{jk_z z} + F_m e^{-jk_z z}\right] \quad (4.130)$$

where $H_m^{(1)}(x)$ is the Hankel function of the first kind and $H_m^{(2)}(x)$ is the Hankel function of the second kind as defined in Section 2.11, and k_p and k_z are constrained by the dispersion relation (2.74).

In (4.130), the coefficients A_m, B_m, C_m, D_m, and E_m are determined by the value of $E_z(\rho,\phi,z)$. For a given radiating source, if these coefficients can be determined, then (4.130) provides a solution for E_z.

We will now use the symmetry of the source to show that many of the coefficients in the series (4.130) are zero. Because the line source is rotationally symmetric, the fields do not vary in the ϕ or z directions, so $k_z = 0$ and $C_m = D_m = 0$ for $m \neq 0$. Using the dispersion relation (2.74), we find that $k_\rho = k$. Since $H_m^{(1)}(x) \sim e^{jx}/\sqrt{x}$ for large x, the first term in (4.130) represents an incoming wave that propagates from $\rho = \infty$ toward the line source at $z = 0$. By the principle of causality, the radiated fields must represent an outgoing wave propagating in the $+\rho$ direction from the source outward to $\rho = \infty$, so we must have $A_0 = 0$. These arguments simplify the expression for E_z considerably, since only a single term in the series remains:

$$E_z = B_0 H_0^{(2)}(k\rho) \quad (4.131)$$

To find the constant B_0, we use Faraday's law to find the magnetic field,

$$\begin{aligned}\overline{H} &= -\frac{1}{j\omega\mu}\nabla \times \hat{z}E_z(\rho) \\ &= \frac{1}{j\omega\mu}\hat{\phi}\frac{\partial E_z}{\partial \rho} \\ &= \frac{1}{j\eta}\hat{\phi}B_0 H_0^{(2)'}(k\rho) \end{aligned} \tag{4.132}$$

where $\eta = \sqrt{\mu/\varepsilon}$. For small x, $H_0^{(2)'}(x) \simeq -2j/(\pi x)$. Using Ampère's law in integral form for a very small loop around the wire

$$\begin{aligned} I &= \oint \overline{H} \cdot d\bar{l} \\ &\simeq \int_0^{2\pi} \rho d\phi \frac{1}{j\eta} B_0 \frac{-2j}{\pi k \rho} \\ &= -\frac{4}{k\eta} B_0 \end{aligned} \tag{4.133}$$

so that $B_0 = -Ik\eta/4$. This leads to the result

$$E_z(\rho) = -I\frac{k\eta}{4}H_0^{(2)}(k\rho) \tag{4.134}$$

for the electric field radiated by a line source. With $I = 1$, this is the Green's function for 2-D electric current sources in the TM^z polarization.

4.6.2.2 2-D Radiation Integral

The 2-D Green's function, or the field radiated by a line source, allows one to find the fields radiated by an arbitrary z-directed source distribution that is infinite in the z direction. If we confine our attention to the x–y plane, the line source can be viewed as a 2-D delta function. An arbitrary z-directed source distribution that is infinite in the z direction can be viewed as a superposition of line sources. Such a source can be written in the form

$$\overline{J} = J_z(x,y)\hat{z} \tag{4.135}$$

The electric field of the source distribution can be found by convolving the Green's function (4.134) with the current distribution $J_z(x,y)$. This leads to the radiation integral

$$\overline{E}(\overline{\rho}) = -\frac{k\eta}{4}\hat{z}\int d\overline{\rho}'\, H_0^{(2)}(k|\overline{\rho} - \overline{\rho}'|)J_z(\overline{\rho}') \tag{4.136}$$

where $\overline{\rho}' = x'\hat{x} + y'\hat{y}$ is a shorthand notation for the source point (x', y'), and the differential $d\overline{\rho}$ represents $\rho d\rho d\phi$ or $dx\,dy$. The convolution integral is over the primed coordinates ρ' and ϕ', which are integrated out in (4.136), leaving only the unprimed coordinates in $\overline{\rho} = x\hat{x} + y\hat{y}$, which represents the observation point (x, y). The source point $\overline{\rho}'$ is integrated over the region where the source is nonzero, and $\overline{\rho}$ is the point where the resultant electric field is evaluated.

For the near-to-far transformation, both electric and magnetic equivalent current sources are required. Magnetic sources have not been observed in nature, but they are convenient in mathematical modeling of electromagnetic fields. We need to add an additional term to the radiation integral (4.136) to allow for equivalent magnetic currents. The derivation of a radiation integral for a magnetic current distribution that is infinite in the z direction and flowing in the $\hat{x}$ or $\hat{y}$ directions is similar to the derivation of (4.136). If we combine the radiation integrals for electric and magnetic currents, we obtain

$$\overline{E}(\overline{\rho}) = -\frac{k\eta}{4}\hat{z}\int d\overline{\rho}'\, H_0^{(2)}(k|\overline{\rho}-\overline{\rho}'|)J_z(\overline{\rho}') + \nabla\times\frac{j}{4}\int d\overline{\rho}'\, H_0^{(2)}(k|\overline{\rho}-\overline{\rho}'|)\overline{M}(\overline{\rho}') \tag{4.137}$$

If the electric and magnetic currents are surface sources obtained using (4.128) and (4.129), the integration over $\overline{\rho}'$ reduces to an integral over the Huygens surface. In this case, (4.137) becomes a 2-D mathematical expression of Huygens's principle, which states that an arbitrary source can be replaced by a distribution of point sources on a surface enclosing the original source without changing the fields outside the surface. This is the basis for the near-to-far transformation that is commonly used as a postprocessing step with the 2-D FDTD algorithm.

4.6.2.3 Far-Field Radiation Integral

Since we are interested in the fields far from the scatterer, we can simplify the radiation integral by expanding the integrand for $r \gg r'$ and retaining the leading-order term in the far-field limit. This is done using the far-field approximation

$$|\overline{r}-\overline{r}'| \simeq r - \hat{r}\cdot\overline{r}', \quad r \gg r' \tag{4.138}$$

When restricted to the x–y plane, we write this as

$$|\overline{\rho}-\overline{\rho}'| \simeq \rho - \hat{\rho}\cdot\overline{\rho}' \tag{4.139}$$

We will also use the asymptotic expansion

$$H_0^{(2)}(x) \simeq \sqrt{\frac{2j}{\pi x}}e^{-jx}, \quad x\to\infty \tag{4.140}$$

for the second-kind Hankel function. Using these approximations in (4.137) leads to

$$\overline{E}(\overline{\rho}) = \frac{k}{4}\sqrt{\frac{2j}{\pi k\rho}}e^{-jk\rho}\int d\overline{\rho}'\, e^{jk\hat{\rho}\cdot\overline{\rho}'}\left[-\eta J_z(\overline{\rho}')\hat{z} + \hat{\rho}\times\overline{M}(\overline{\rho}')\right] \tag{4.141}$$

For the TM^z polarization, $\overline{M}$ has no z component, so that $\hat{\rho}\times\overline{M} = \hat{\rho}\times(M_\rho\hat{\rho} + M_\phi\hat{\phi}) = M_\phi\hat{z}$, and we can express the integral in the form

$$E_z(\overline{\rho}) = \frac{k}{4}\sqrt{\frac{2j}{\pi k\rho}}e^{-jk\rho}\int d\overline{\rho}'\, e^{jk\hat{\rho}\cdot\overline{\rho}'}\left[-\eta J_z(\overline{\rho}') + M_\phi(\overline{\rho}')\right] \tag{4.142}$$

Alternately, we can write the integrand in terms of the $\hat{x}$ and $\hat{y}$ components of $\overline{M}$. Since $\hat{\rho} = \cos\phi\hat{x} + \sin\phi\hat{y}$,

$$\begin{aligned} M_\phi \hat{z} &= \hat{\rho} \times \overline{M} \\ &= (\cos\phi\hat{x} + \sin\phi\hat{y}) \times (M_x\hat{x} + M_y\hat{y}) \\ &= (M_y \cos\phi - M_x \sin\phi)\hat{z} \end{aligned}$$

from which we can identify $M_\phi = M_y \cos\phi - M_x \sin\phi$.

Both terms in the radiation integral (4.137) or the far-field limit (4.141) represent electric fields in the z direction, so all of these result apply to the TM^z polarization. A radiation integral for the TE^z polarization can be derived by considering $\hat{x}$- and $\hat{y}$-directed electric currents and z-directed magnetic currents. To avoid a lengthy derivation, the electromagnetic duality principle can be used to transform the radiation integral for the TM^z polarization to the TE^z polarization.

4.6.2.4 Implementation of the Near-to-Far Transformation

The far-field radiation integral can be used to transform near fields from the FDTD algorithm to far fields. If the rectangular Huygens surface in Figure 4.10 is used, then on the left-hand side of the contour, for example, $\hat{n} = -\hat{x}$, and the surface currents are

$$\begin{aligned} \overline{M}_s &= -(-\hat{x}) \times \hat{z}E_z \\ &= -E_z\hat{y} \\ \overline{J}_s &= -\hat{x} \times (H_x\hat{x} + H_y\hat{y}) \\ &= -H_y\hat{z} \end{aligned}$$

In terms of the tangential fields, the radiation integral for the left side of the Huygens surface is

Far Field Radiation Integral for Huygens Surface Left Side

$$E_z = \frac{k}{4}\sqrt{\frac{2j}{\pi k \rho}}e^{-jk\rho} \int_{y_1}^{y_2} dy' \, e^{jk\hat{\rho}\cdot\overline{\rho}'} \left[\eta H_y(x_1, y') - E_z(x_1, y') \cos\phi\right] \qquad (4.143)$$

where the lower left corner of the Huygens surface is at (x_1, y_1) and the upper left corner is at (x_1, y_2). The exponent in the integrand can be expanded by recalling that $\overline{\rho}$ is the point where the far field is evaluated, so that $\hat{\rho}$ is a unit vector in the direction of the far-field point. From this definition, we have that

$$\begin{aligned} \hat{\rho} \cdot \overline{\rho}' &= (\cos\phi^s\hat{x} + \sin\phi^s\hat{y}) \cdot (x'\hat{x} + y'\hat{y}) \\ &= x' \cos\phi^s + y' \sin\phi^s \end{aligned}$$

On the left side of the Huygens surface, this becomes $\hat{\rho} \cdot \overline{\rho}' = x_1 \cos \phi^s + y' \sin \phi^s$. Expressions for the other three sides can be obtained in a similar way.

The integral in (4.143) must be evaluated numerically using a quadrature rule. Since the fields in the integrand are available only at the FDTD grid points, it is natural to use the midpoint quadrature rule with integration points at the FDTD grid points. If the Huygens surface is chosen to pass through E_z grid points, it will lie halfway between the required magnetic field points on all four sides. For maximum accuracy, the magnetic field samples on either side of the contour should be averaged to obtain values on the contour. The magnetic field samples are also at staggered time steps relative to the electric field, so consecutive time samples of the magnetic field should be averaged as well.

The steps required to implement the near-to-far transformation are as follows:

Implementing the Near-to-Far Transformation

1. Near the end of the loop over time in the FDTD algorithm, when the simulation has run long enough that transients have decayed or propagated away from the Huygens surface, store the electric and magnetic field samples lying on the Huygens surface for one period of the time-harmonic field. Use temporal and spatial averaging as previously described so that samples of the electric and magnetic fields are evaluated at the same point in space and time.
2. In postprocessing, transform the time-domain fields to the phasor domain using the FFT algorithm.
3. Within a loop over the far-field angle ϕ^s, compute the far-field radiation integral and add the resulting values for each of the sides of the Huygens surface.
4. From the far electric field, calculate the scattering amplitude, scattering width, or other desired final values.

Integrals such as (4.143) are typically evaluated using the midpoint rule and implemented in code as a sum over a product of the factors in the integrand. Care must be taken to ensure that in each term of the sum, the point $\overline{\rho}'$ in the complex exponential $e^{jk\hat{\rho}\cdot\overline{\rho}'}$ is the same point at which the fields $H_x(\overline{\rho}')$, $H_y(\overline{\rho}')$, and $E_z(\overline{\rho}')$ are sampled.

4.6.3 Other Types of Postprocessing

Many other quantities can be computed from the electromagnetic fields simulated using the FDTD method. In studies of interactions of EM fields with biological tissues, the absorption rate of energy by conducting tissues is of great importance. This can be calculated by integrating near fields to obtain the power dissipated by induced currents over the material region. For antenna analysis, the far fields calculated using a near-to-far transformation can be used to obtain the directivity or other antenna parameters.

4.6.3.1 Antenna Parameters

The near-to-far transformation can be used for both radiation and scattering problems. For scattering problems, far fields are used to compute scattering amplitudes, scattering widths, and RCS. In the case of radiation problems, far fields are used to compute antenna parameters such as the radiation pattern, directivity, or radiation resistance.

In terms of the phasor far electric field intensity vector, the time-average radiated power flux density is

$$S(\bar{r}) = \frac{|\bar{E}(\bar{r})|^2}{2\eta} \tag{4.144}$$

The antenna radiation pattern is

$$f(\theta, \phi) = \lim_{r \to \infty} \frac{S(\bar{r})}{S_{\max}(r)} \tag{4.145}$$

where $S_{\max}(r)$ is the maximum value of the magnitude of the power flux density over all angles θ and ϕ at the radial distance r.

The time-average total radiated power is the integral of the power flux density over a sphere:

$$P_{\text{rad}} = \int_0^{2\pi} \int_0^{\pi} S(r, \theta, \phi) r^2 \sin\theta \, d\theta \, d\phi \tag{4.146}$$

The maximum antenna directivity is

$$D = \frac{S_{\max}}{P_{\text{rad}}/(4\pi r^2)} \tag{4.147}$$

To evaluate the directivity in postprocessing, the integrals in (4.146) must be computed using a numerical integration rule. If the far-field angles are evenly spaced in θ and ϕ, the integral can be approximated by the double summation

$$P_{\text{rad}} \simeq \sum_{m=1}^{N_\theta} \sum_{n=1}^{N_\phi} S(\theta_m, \phi_n) \sin\theta_n \, \Delta\theta \, \Delta\phi \tag{4.148}$$

where $\Delta\theta$ and $\Delta\phi$ are the spacings between the points θ_n, $n = 1, 2, \ldots, N_\theta$, and ϕ_m, $m = 1, 2, \ldots, N_\phi$, respectively.

For more accurate results, the Gauss–Legendre quadrature approach discussed in Section 5.7 and used in Problem 5.6 can be used to evaluate the integral over a sphere. Since antenna problems are typically not well approximated as two dimensional, this type of postprocessing is used with the 3-D FDTD algorithm.

4.6.3.2 Impedances

For radiating structures such as antenna, we are often interested in the input impedance at a terminal junction. At a gap between conductors where a current source of strength I is introduced, we can integrate to find the voltage across the gap using

$$V(t) = -\int_P \overline{E} \cdot d\bar{l} \tag{4.149}$$

where P is a path across the gap. Alternately, if the voltage V at a source introduced into a conductor is given, the current flowing through the conductor is

$$I(t) = \oint_C \overline{H} \cdot d\bar{l} \tag{4.150}$$

where C is a closed contour around the conductor. We then convert from the time domain to the phasor domain and obtain the impedance at frequency ω using

$$Z(\omega) = \frac{V(\omega)}{I(\omega)} \tag{4.151}$$

One difficulty associated with the computation of impedances is that antenna terminals often involve small metallic features, and it is difficult for a grid with finite step size to resolve the fields near fine features of the radiating structure. If accurate impedances are desired, care must be taken to ensure that the fields near the terminals of an antenna are well modeled.

4.7 Code Verification

As discussed in Section 1.3, there are several approaches for verifying that a numerical algorithm has been implemented properly and assessing the accuracy of the results. The most common approach is to test the algorithm using a simple geometry for which the scattering or radiation solution is available in closed form. The circular cylinder (2-D) and sphere (3-D) are common test cases, because analytical series solutions for scattered fields can be derived for these geometries.

4.7.1 Scattering by a Circular Cylinder

We derive here the exact solution for a TM^z-polarized plane wave incident on a PEC circular cylinder. Suppose that the incident field is a plane wave traveling in the $-x$ direction, so that in phasor form,

$$E_z^{\mathrm{i}}(x) = E_0 e^{jkx} \tag{4.152}$$

where E_0 is a constant. Because of the symmetry of the circular cylinder, the solution for any other angle of incidence can be found by rotating the scattered field obtained for this particular plane wave.

The general solution for the scattered field E_z^{s} in the cylindrical coordinate system is given by (4.130). Since the fields scattered by the cylinder are outgoing only, the coefficients A_m of the first kind Hankel function are zero. Because of the symmetry of the scatterer and the incident field, the scattered fields are independent of z, so that $k_z = 0$ and $k_\rho = k$. This simplifies the solution for the scattered field to

$$E_z^{\mathrm{s}} = \sum_{n=-\infty}^{\infty} B_n H_n^{(2)}(k\rho)e^{jn\phi} \tag{4.153}$$

All we need to do now is to find the values of the coefficients B_n. This can be done by using the cylindrical wave expansion of a plane wave or the Jacobi–Anger expansion

$$e^{jkx} = \sum_{n=-\infty}^{\infty} j^n J_n(k\rho)e^{jn\phi} \tag{4.154}$$

where $J_n(x)$ is the Bessel function described in Section 2.11. The cylindrical wave expansion allows one to write the incident field as

$$E_z^{\mathrm{i}}(\rho,\phi) = E_0 \sum_{n=-\infty}^{\infty} j^n J_n(k\rho)e^{jn\phi} \tag{4.155}$$

The unknown coefficients B_n can be found by combining the incident and scattered fields to form the total field and applying the boundary condition for the electric field at the surface of the cylinder. The coefficients B_n must be such that the tangential component of the total field vanishes at the cylinder surface.

The electric field boundary condition at the surface of the PEC cylinder is

$$\hat{\rho} \times (\overline{E}^{\mathrm{s}} + \overline{E}^{\mathrm{i}})\Big|_{\rho=a} = 0 \tag{4.156}$$

Because both the incident and scattered electric fields are in the z direction, the boundary condition implies that $E_z^{\mathrm{i}} = -E_z^{\mathrm{s}}$ at $\rho = a$. We can see that both (4.153) and (4.155) are Fourier series in ϕ. Since the two functions $E_z^{\mathrm{i}}(a,\phi)$ and $-E^{\mathrm{s}}(a,\phi)$ are equal, the coefficients of each term in the two Fourier series must be equal. This leads to the solution

$$B_n = -E_0 \frac{j^n J_n(ka)}{H_n^{(2)}(ka)} \tag{4.157}$$

Substituting this result into the scattered field gives

$$E_z^{\mathrm{s}}(\rho,\phi) = -E_0 \sum_{n=-\infty}^{\infty} \frac{j^n J_n(ka)}{H_n^{(2)}(ka)} H_n^{(2)}(k\rho)e^{jn\phi} \tag{4.158}$$

which is an exact solution for the scattering of a plane wave by a circular PEC cylinder. This is referred to as the Mie series solution for the circular cylinder.

The scattering amplitude corresponding to $\overline{E}^{\mathrm{s}}$ can be found using the asymptotic expansion

$$H_n^{(2)}(x) \sim j^n \sqrt{\frac{2j}{\pi x}}e^{-jx}, \quad x \to \infty \tag{4.159}$$

of the second-kind Hankel function of order n. The scattered far field is

$$E_z^s(\rho, \phi) \simeq -E_0 \sqrt{\frac{2j}{\pi k \rho}} e^{-jk\rho} \sum_{n=-\infty}^{\infty} \frac{(-1)^n J_n(ka)}{H_n^{(2)}(ka)} e^{jn\phi}, \quad \rho \to \infty \tag{4.160}$$

By comparing this expression with the definition (4.115), the scattering amplitude can be identified as

Circular PEC Cylinder Scattering Amplitude

$$S(\phi) = -\sum_{n=-\infty}^{\infty} \frac{(-1)^n J_n(ka)}{H_n^{(2)}(ka)} e^{jn\phi} \tag{4.161}$$

where ϕ is the scattering angle, which we have denoted earlier as ϕ^s. The plane wave (4.152) was incident from the $-x$ direction, so that the incidence angle is $\phi^i = 0$. For an arbitrary incidence angle, $\phi = \phi^s - \phi^i$.

The terms of this series are small for $|n| > ka$, since the Hankel function begins to grow exponentially when the order n is larger than the argument ka. For numerical implementation, the infinite series can be truncated at some value N that is modestly larger than ka, so that the sum is evaluated only over a finite number of terms with $|n| \leq N$.

4.8 3-D FDTD Method

Now that we have developed the Yee cell discretization for 2-D problems, it is straightforward to extend the FDTD algorithm to 3-D. To derive the 3-D FDTD update equations, we expand Maxwell's equations into component form to obtain

$$\nabla \times \overline{E} = -\mu \frac{\partial \overline{H}}{\partial t} \qquad \nabla \times \overline{H} = \varepsilon \frac{\partial \overline{E}}{\partial t}$$

$$\Downarrow \qquad\qquad \Downarrow$$

$$\begin{aligned}
&\hat{x}: \frac{\partial E_y}{\partial z} - \frac{\partial E_z}{\partial y} = \mu \frac{\partial H_x}{\partial t} \qquad \frac{\partial H_z}{\partial y} - \frac{\partial H_y}{\partial z} = \varepsilon \frac{\partial E_x}{\partial t} \\
&\hat{y}: \frac{\partial E_z}{\partial x} - \frac{\partial E_x}{\partial z} = \mu \frac{\partial H_y}{\partial t} \qquad \frac{\partial H_x}{\partial z} - \frac{\partial H_z}{\partial x} = \varepsilon \frac{\partial E_y}{\partial t} \\
&\hat{z}: \frac{\partial E_x}{\partial y} - \frac{\partial E_y}{\partial x} = \mu \frac{\partial H_z}{\partial t} \qquad \frac{\partial H_y}{\partial x} - \frac{\partial H_x}{\partial y} = \varepsilon \frac{\partial E_z}{\partial t}
\end{aligned} \tag{4.162}$$

This system of six coupled partial differential equations can be discretized using the staggered grids defined by the 3-D Yee cell shown in Figure 4.11. The three spatial grids and the time grid are defined by $\overline{r}_{ijk} = (i\Delta x, j\Delta y, k\Delta z)$ and $t_n = n\Delta t$ with integer and half-integer values for the indices i, j, k, and n.

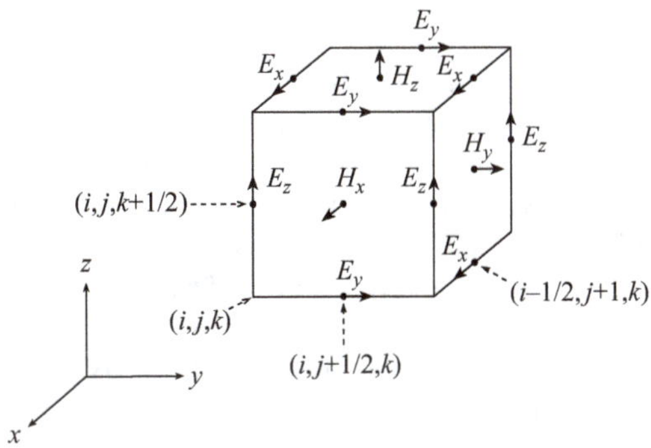

Figure 4.11 Yee cell that defines three staggered rectangular spatial grids for discretization of Ampère's and Faraday's laws

If we apply the central difference stencil to the $\hat{x}$ component of Faraday's law at the point $(\overline{r}_{i,j+1/2,k+1/2}, t_n)$, we obtain the discretization

$$\frac{E^n_{y\,i,j+1/2,k+1} - E^n_{y\,i,j+1/2,k}}{\Delta z} - \frac{E^n_{z\,i,j+1,k+1/2} - E^n_{z\,i,j,k+1/2}}{\Delta y} = \mu \frac{H^{n+1/2}_{x\,i,j+1/2,k+1/2} - H^{n-1/2}_{x\,i,j+1/2,k+1/2}}{\Delta t} \tag{4.163}$$

In the first term on the left-hand side, the z derivative is reflected in the difference between the indices $k+1$ and k on E_y. In the second term, the indices $j+1$ and j arise from the y derivative. On the right-hand side, the time indices $n+1/2$ and $n-1/2$ perform the t derivative. Because the central difference formula leads to a derivative evaluated at the center of the interval between two samples, it can be seen that all three terms are evaluated at the same point in space and time.

This difference equation can be solved for $H^{n+1/2}_{x\,i,j+1/2,k+1/2}$ to obtain an update equation for H_x at a future time step in terms of the values of E_y, E_z, and H_x at past time steps:

3-D FDTD Update Equation for H_x

$$H^{n+1/2}_{x\,i,j+1/2,k+1/2} = H^{n-1/2}_{x\,i,j+1/2,k+1/2} + \frac{\Delta t}{\mu}\left(\frac{E^n_{y\,i,j+1/2,k+1} - E^n_{y\,i,j+1/2,k}}{\Delta z} - \frac{E^n_{z\,i,j+1,k+1/2} - E^n_{z\,i,j,k+1/2}}{\Delta y}\right) \tag{4.164}$$

Differencing the remaining five equations in the same way provides three equations that can be solved to obtain time update equations for $\overline{E}^{n+1}$ and three that can be solved for $\overline{H}^{n+1/2}$. These update equations are the core of the 3-D FDTD method.

4.8.1 PEC Cavity

Because of the complexity of the 3-D FDTD method, we will begin by developing the algorithm with simple boundary conditions and postprocessing. The simplest boundary condition is vanishing tangential electric fields at a flat PEC surface. An example problem with this type of boundary condition is the analysis of a PEC resonant cavity. Using the FDTD method, the resonance frequencies of the cavity can be computed in postprocessing from the fields excited by a source inside the cavity.

In the case of a cubical cavity, it is convenient to set up the Yee cells for the field grids inside the PEC cube as shown in Figure 4.12. On all six of the outer faces of the cube, the boundary field samples are tangential electric field and normal magnetic field components. From the PEC boundary conditions, these field components must vanish at the cavity wall, so this arrangement makes the required boundary condition very simple to implement.

While the half-integer notation for the Yee cell grid points is convenient for mathematical expressions like (4.163), to implement the algorithm in code the half-integer grid points must be mapped to integer array indices. From Figure 4.12, it can be seen that the arrays for the six field components will have different sizes. Let N_x, N_y, and N_z represent the number of Yee cells in the PEC cube on each side. In the figure, $N_x = N_y = N_z = 4$. The array elements represent samples of the fields at points located at spatial coordinates according to

$$\begin{aligned}
\texttt{Ex(i,j,k)} &= E_x[(i-\tfrac{1}{2})\Delta x, (j-1)\Delta y, (k-1)\Delta z] \\
\texttt{Ey(i,j,k)} &= E_y[(i-1)\Delta x, (j-\tfrac{1}{2})\Delta y, (k-1)\Delta z] \\
\texttt{Ez(i,j,k)} &= E_z[(i-1)\Delta x, (j-1)\Delta y, (k-\tfrac{1}{2})\Delta z] \\
\texttt{Hx(i,j,k)} &= H_x[(i-1)\Delta x, (j-\tfrac{1}{2})\Delta y, (k-\tfrac{1}{2})\Delta z] \\
\texttt{Hy(i,j,k)} &= H_y[(i-\tfrac{1}{2})\Delta x, (j-1)\Delta y, (k-\tfrac{1}{2})\Delta z] \\
\texttt{Hz(i,j,k)} &= H_z[(i-\tfrac{1}{2})\Delta x, (j-\tfrac{1}{2})\Delta y, (k-1)\Delta z]
\end{aligned}$$

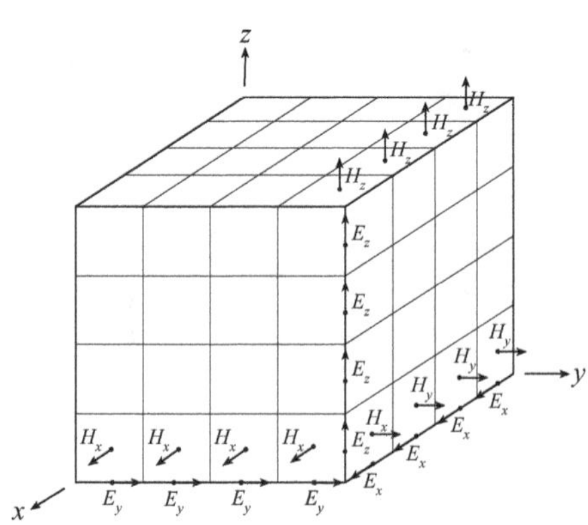

Figure 4.12 Yee cells for the interior of a cubical PEC cavity

On the left-hand sides of these expressions, i, j, and k are integer array indices, rather than the half-integers used in (4.163). The ranges

`Ex(i,j,k)`	$1 \leq i \leq N_x,$	$1 \leq j \leq N_y + 1,$	$1 \leq k \leq N_z + 1$
`Ey(i,j,k)`	$1 \leq i \leq N_x + 1,$	$1 \leq j \leq N_y,$	$1 \leq k \leq N_z + 1$
`Ez(i,j,k)`	$1 \leq i \leq N_x + 1,$	$1 \leq j \leq N_y + 1,$	$1 \leq k \leq N_z$
`Hx(i,j,k)`	$1 \leq i \leq N_x + 1,$	$1 \leq j \leq N_y,$	$1 \leq k \leq N_z$
`Hy(i,j,k)`	$1 \leq i \leq N_x,$	$1 \leq j \leq N_y + 1,$	$1 \leq k \leq N_z$
`Hz(i,j,k)`	$1 \leq i \leq N_x,$	$1 \leq j \leq N_y,$	$1 \leq k \leq N_z + 1$

define the sizes of the arrays. At the PEC surface, the normal magnetic field and the tangential electric field must vanish. Since all the grid points on the outer boundary in Figure 4.12 are either tangential electric or normal magnetic fields, the boundary samples vanish, and only the interior grid points need to be updated in the FDTD algorithm.

If a time-harmonic source is placed inside the cavity, the excited fields will be small except at resonant frequencies of the cavity. At a resonance frequency, a resonant mode is excited by the source, unless the source happens to be located at a null of the resonant mode or is orthogonal to the mode. Since the quality factor of a PEC cavity is infinite, energy radiated by the source is not absorbed, and the magnitudes of the electromagnetic fields inside the cavity grows without bound. If a broadband (pulsed) source is placed in the cavity, the cavity will "ring" at the resonance frequencies of the structure. This can be used to determine the resonances numerically using the FDTD algorithm by computing the Fourier transform of the field inside the cavity.

4.8.2 Stability Criterion

For the 3-D FDTD algorithm, the stability criterion is

$$c\Delta t \leq \left[\frac{1}{\Delta x^2} + \frac{1}{\Delta y^2} + \frac{1}{\Delta z^2}\right]^{-1/2} \tag{4.165}$$

If $\Delta x = \Delta y = \Delta z$, then the condition simplifies to

3-D FDTD Stability Criterion

$$c\Delta t \leq \frac{1}{\sqrt{3}}\Delta x \tag{4.166}$$

4.9 Summary

We have surveyed applications of the finite difference approach, including Laplace eigenvalue solutions for waveguide modal analysis, the finite difference time domain method for wave propagation in 2-D and 3-D, boundary conditions for finite difference algorithms, and postprocessing. In Chapter 8, we will consider the finite element method, which is closely related to the finite difference method. The accuracy of finite difference algorithms is controlled by the grid spacing. As the grid spacing decreases, the numerical solution generally becomes more accurate, but the computation time increases.

Problems

4.1 Derive the update equation (4.16) from the 1-D wave equation.

4.2 Implement the 1-D FDTD algorithm with zero initial conditions and a Gaussian pulse launched from the left side of the computational domain.

(a) Apply the Dirichlet and Neumann boundary conditions at the other side of the domain. Does the pulse reflect completely in both cases? What is different about the reflection for the two boundary conditions?

(b) Implement the Mur ABC at the right-hand boundary of the simulation domain. Verify that a propagating pulse is not reflected at the boundary.

(c) Change the phase velocity or the grid step size so that $c\Delta t/\Delta x$ is less than one. Launch a square pulse into a free-space medium from one side of the domain. What happens to the pulse as it propagates? Explain your answer using the concept of dispersion. What happens if $c\Delta t/\Delta x > 1$?

4.3 Modify the 1-D FDTD code so that it models a single interface between free space ($\varepsilon = \varepsilon_0$) and a medium having permittivity $\varepsilon_1 = \varepsilon_{1r}\varepsilon_0$, where ε_{1r} is the relative permittivity. Use the relationship $c = 1/\sqrt{\mu\varepsilon}$ to determine the phase velocity that appears in the wave equation for the two regions. Launch a Gaussian pulse of unit amplitude toward the interface. After the pulse reaches the interface, the amplitude of the transmitted pulse will be the plane wave transmission coefficient for normal incidence (why does the pulse reflect with the same amplitude as a plane wave?). Embed the FDTD code in a loop to compute the transmission coefficient over a range of values for ε_{1r}, such as `1:.1:4` in MATLAB notation. Plot the numerical results on top of the exact solution as a function of ε_{1r}. How accurate are the numerical results? Can you change any parameters of the FDTD algorithm to increase the accuracy?

4.4 Starting from Maxwell's equations, derive the 1-D FDTD update equation (4.25) for a 1-D problem with a time-harmonic, infinite plane current source term.

4.5 Derive equation (4.32). Model the delta function in the plane current source as a rec function with width Δx and height $1/\Delta x$.

4.6 Modify the 1-D FDTD code of Problem 4.2 so that there is a time-harmonic current source in free space a small distance from the left boundary. The source should launch plane waves in both directions with an electric field of unit amplitude. Use a "turn-on" function so that the amplitude of the source is continuous at $t = 0$.

(a) Set the grid parameters so that $c\Delta t/\Delta x < 1$ (to mimic a 2-D or 3-D FDTD code, for which there is no "magic" dispersion-free choice of grid parameters).

After the plane wave propagates to the boundaries, stop the simulation, and plot the computed electric field at a fixed time as a function of position. Overlay the exact solution for the field radiated by the plane current (see Problem 2.7).

(b) Plot the magnitude of the difference between the numerical and exact solutions. Does the error grow with distance from the source? Compute the numerical value for k from the 1-D FDTD dispersion relation, and plot the expected error on top of the actual error. Do they match? How far could the plane wave propagate before the numerical and exact solutions are completely out of phase? What parameters in the code could be changed to decrease the error?

(c) Repeat for $c\Delta t/\Delta x = 1$.

4.7 Derive a finite element stencil for the second-order derivative operator

$$\frac{\partial^2}{\partial x \partial y}$$

4.8 Write a finite difference code to compute the cutoff frequencies of the first 20 TM modes of a rectangular waveguide with dimensions 1 cm × 2 cm. The matrix has a block structure that can be found by creating the matrix discretization of the Laplacian by hand on paper for small grids and identifying the pattern. Write a MATLAB script to create a matrix with the same structure for a grid of arbitrary size. The MATLAB function `diag` is useful in generating the matrix blocks. Compare numerical results with the analytical cutoff frequencies. To find matrix eigenvalues, one can use the function `eig`, or `eigs` if the matrix is large. How rapidly do the FD results approach the exact results as the number of grid points increases? Is the accuracy the same for all the modes? Why or why not?

4.9 (a) Find the cutoff frequencies for the first five TM modes of an equilateral triangle waveguide of side 2 cm. Make sure the answer is in the ballpark by estimating the lowest TM cutoff frequency using a circular waveguide as a lower bound. Hints: Create a function `G = numgrid_poly(p,n)`, similar to the built-in MATLAB function `numgrid`, which creates a numbered grid for an arbitrary polygon with vertices defined by an array `p` with rows of the form `[x1 y1 x2 y2]`, where the first pair of coordinates represents the first vertex of an edge and the second pair represents the second vertex of the edge. The numgrid is constructed from a subset of an `n` × `n` rectangular grid. Use the MATLAB function `inpolygon` to find interior grid points. The numbered grid can be used with the function `A = delsq(G)` to find the matrix discretization of the Laplacian. (b) Use the code to study the mode cutoff frequencies for waveguides with one or two other cross-sectional shapes. How do the cutoff frequencies compare?

4.10 Modify the finite difference algorithm of Problem 4.8 to include a dielectric layer within the rectangular waveguide of thickness 0.5 cm by 2 cm with relative permittivity $\varepsilon_r = 2$. How does the dielectric layer affect the waveguide mode cutoff frequencies?

4.11 For the rectangular waveguide mode solver in Problem 4.8, write a code to visualize the waveguide mode transverse field patterns. The transverse field pattern is given by the eigenfunction in (4.56) and is a part of the electric field $E_{z,mn}(x,y,z) = u_{mn}(x,y)e^{\pm k_{z,mn}z}$ associated with the mode. Using the `eig` function with two output arguments, compute the eigenvectors of the discretized Laplacian operator. The `eig` function does not always sort the eigenvalues in any particular order, so use the `sort` command to find the lowest order mode. Use the `reshape` function to convert the corresponding eigenvector into an N_x-by-N_y array, where N_x and N_y represent the size of the grid used to discretize the Laplacian for the rectangular waveguide. (a) Plot the lowest order mode pattern using the `imagesc` command. Does the mode pattern satisfy the Dirichlet boundary condition? (b) Plot other mode patterns of low order and high order. Can you connect the spatial variation of the mode pattern to the accuracy of the calculated cutoff frequency?

4.12 Derive the 2-D FDTD-TE update equations based on the Yee cell for TE^z polarized fields.

4.13 Implement the 2-D FDTD-TM scheme based on the Yee cell. Use the first-order Mur absorbing boundary condition for the tangential field components at the boundaries. To test the code, place a hard time-harmonic 2-D point source (3-D line source) in the middle of the grid. A cylindrical wave should propagate outward.

4.14 Use the MATLAB function `inpolygon` to develop code to introduce dielectric or PEC polygons into the simulation domain. Create several obstacles illuminated by a 2-D hard point source. Allow the fields to propagation across the domain and interact with the obstacles. Plot the power in the domain in dB as a 2-D density plot using the `imagesc` function. This code represents a 2-D propagation model that could be used in planning the locations of cellular base station antennas. Can you identify shadowing and diffraction effects?

4.15 Using the code from Problem 4.14, change the source so that it consists of four 2-D point sources spaced a half wavelength apart. Run the simulation to verify that the array directs energy in the broadside direction. Adjust the relative phasing of the sources to steer the beam at a 45° angle away from broadside. Plot the power in the simulation domain for both cases.

4.16 Model a parallel plate waveguide by introducing two parallel PEC plates into the 2-D FDTD simulation domain. The PEC plates will be lines in two dimensions with a zero Dirichlet boundary condition for E_z. Excite the waveguide using a time-harmonic hard point source between the two plates. (a) Set the source frequency to be lower than the fundamental mode cutoff frequency for the parallel plate waveguide. Can you see the exponential decay of the evanescent fields in the waveguide? (b) Set the source frequency to be between the fundamental mode cutoff frequency and the cutoff frequency of the next-order mode. Do the fields in the waveguide converge to the expected half-sinusoidal mode pattern as the higher order waveguide modes decay away from the source? This approach can be used to generate modal source distributions, a method sometimes referred to as boot-strapping.

4.17 Reconfigure the 2-D FDTD code to use the scattered field formulation for dielectrics and conductive materials. Set the incident field to be a time-harmonic plane wave traveling in the $+x$ direction. Use the `imagesc` command to plot the electric field scattered by a circular cylinder of radius 0.5 λ for both a PEC cylinder and a dielectric cylinder with $\varepsilon_r = 2$ (see Section 3.1.2.1 for a tip on using the `imagesc` command properly to visualize fields in the x–y plane).

4.18 Modify the 2-D FDTD-TM code to store tangential fields on a rectangular Huygens surface enclosing the scatterer for one full time cycle of a plane wave incident field, after waiting for transients to die down. In postprocessing, convert the time-domain fields to phasors using the FFT algorithm, and evaluate the far-field radiation integral for equivalent electric and magnetic currents on the Huygens surface to transform the near fields to far fields. Test the code by computing the bistatic scattering width of a circular PEC cylinder of one-half wavelength radius over a few hundred scattering angles from 0° to 180°. Compare with the exact solution for the circular PEC cylinder. What does the peak of the bistatic scattering width correspond to physically? Hint: The first-order Mur boundary condition is an imperfect ABC, so reflections from the boundary can lead to an inaccurate numerical result. To reduce error due to reflections, make the simulation domain larger than the scatterer, place the Huygens surface close to the scatterer, and terminate the simulation before boundary reflections return to the Huygens surface.

4.19 Derive update equations for the 3-D FDTD algorithm. (a) Solve the difference equation (4.163) for $H^{n+1/2}_{x\,i,j+1/2,k+1/2}$ to obtain an update equation for H_x. (b) Using the Yee cell approach, derive similar update equations for H_y, H_z, E_x, E_y, and E_z.

4.20 Implement the 3-D FDTD method for a cubical cavity resonator that is 10 cm on a side. Use the PEC boundary condition on the cavity walls. To make the boundary condition easy to implement, set up the grids so that the outermost face of each Yee cell on all six sides of the cube has normal magnetic fields and tangential electric fields (this will lead to different size arrays for the six field components). To excite the cavity, impress a pulse source at some point in the domain not on a plane of symmetry and away from the cavity walls. To avoid a DC component in the field solution, it is convenient to use the derivative of a Gaussian as the pulse shape, so that

$$E_z(i,j,k) = E_z(i,j,k) + [t/(4t_0) - 1]e^{-(t-4t_0)^2/t_0^2} \tag{4.167}$$

where $t_0 = 80\,\text{ps}$. (What is the bandwidth of this pulse?) The interior of the PEC cube forms a resonant cavity that will "ring" at its resonance frequencies, while other frequency components excited by the source will die out as the source decays to zero. As the simulation runs, store one component of the fields at another point in the domain, also not on a plane of symmetry. Run the simulation for about $N_t\Delta t = 20\,\text{ns}$. (What determines the required simulation time?) In postprocessing, compute the FFT of the stored field. Plot the magnitude of the FFT versus frequency in GHz. Compare the result with the exact resonance frequencies of the PEC cavity by marking the exact resonances as vertical lines on the plot. The simulation should take only a minute or so with a reasonably fast processor and with the grid step sizes chosen so that the computational domain is about 25 × 25 × 25 grid cells. Fields in the cavity can be conveniently visualized using the MATLAB `slice` command.

4.21 For the 2-D FDTD-TM code, add the UPML boundary condition. Test the boundary condition by introducing a hard pulsed source in the interior of the simulation domain, storing the field at some location in the domain over a time long enough for the source to decay to zero and reflections to travel back and forth across the domain several times, and plotting on a dB scale the power in the field versus time. Reflections should be smaller by 30 dB or more than the initial pulse.

4.22 Use the UPML boundary condition for the PEC circular cylinder scattering problem of Problem 4.18. Does the accuracy of the computed scattering width increase with the improved boundary condition? Why or why not?

4.23 Implement the first-order Mur boundary condition with the 3-D FDTD algorithm from Problem 4.20. Apply the ABC equations to the tangential components of the electric field on each of the six sides of the cubical computational domain. Test the ABC by modeling a hard source at the center of the computational domain and ensuring that the radiated field is absorbed on all sides of the computational domain.

4.24 Using the FDTD code from Problem 4.23, model a dipole antenna as two thin conducting cylinders $\lambda/4$ in length, parallel to the z-axis, and separated by a small feed gap. Choose the gap width to be equal to one Yee cell in length. Excite the antenna with a time-harmonic hard electric source along one of the edges of the Yee cell, so that the source extends from one dipole arm to the other, across the feed gap. Plot the z component of the radiated electric field on a slice through the antenna in the x–z plane.

References

[1] A. Taflove and S. C. Hagness, *Computational Electrodynamics: The Finite-Difference Time-Domain Method*, 2nd edn. Norwood, MA: Artech House, 2000.

[2] D. M. Sullivan, *Electromagnetic Simulation Using the FDTD Method*. New York, NY: IEEE Press, 2000.

[3] A. Elsherbeni and V. Demir, *The Finite Difference Time Domain Method for Electromagnetics: With MATLAB Simulations*. London: SciTech, 2008.

[4] C. Furse, D. H. Roper, D. N. Buechler, D. A. Christensen, and C. H. Durney, "The problem and treatment of DC offsets in FDTD simulations," IEEE Trans. Antennas Propag., vol. 48, no. 8, pp. 1198–1201, 2000.

[5] M. Kac, "Can one hear the shape of a drum?," Am. Math. Mon., vol. 73, no. 4, pp. 1–23, 1966.

[6] C. Gordon, D. L. Webb, and S. Wolpert, "One cannot hear the shape of a drum," Bulletin of the American Mathematical Society, vol. 27, no. 1, pp. 134–138, 1992.

[7] M. N. O. Sadiku, *Numerical Techniques in Electromagnetics With MATLAB*. Boca Raton, FL: CRC Press, 2009.

[8] K. S. Yee, "Numerical solution of initial boundary-value problems involving Maxwell's equations in isotropic media," IEEE Trans. Antennas Propag., vol. 14, pp. 302–307, 1966.

Chapter 5
Numerical Integration

In computational science and engineering, evaluation of integrals numerically is a fundamental problem with many applications. Computational electromagnetics codes often use integration routines that are evaluated thousands of times to fill a large matrix, so efficient methods for numerical integration are very important. From the point of view of numerical analysis, the theory of numerical integration is mathematically rich, with interesting theoretical connections to other areas such as interpolation, orthogonal functions, and matrix theory. Davis and Rabinowitz [1] is a standard reference, but many other papers and books are available on the topic.

In this chapter, we will explore some of the basic techniques used for numerical integration. The goal of numerical integration to compute an accurate approximation to an integral

$$I = \int_a^b f(x)\, dx \tag{5.1}$$

with as few evaluations of the integrand $f(x)$ as possible. Numerical integration is also called numerical quadrature, which is a reference to the idea of estimating the area under a curve by adding the areas of quadrilaterals. The approximate value of the integral has the general form

Numerical Integration in General Form

$$\hat{I} = \sum_{n=1}^{N} w_n f(x_n) \tag{5.2}$$

where w_n are integration weights, and x_n are variously referred to as integration points, nodes, or abscissas. The weights and nodes are chosen to make the approximate value $\hat{I}$ of the integral as accurate as possible, so that $\hat{I} \simeq I$. To minimize the computation time required to evaluate the integration rule, the number of nodes N should be as small as possible.

5.1 Types of Integration Rules

An algorithm for choosing weights w_n and nodes x_n as a function of the desired number of nodes N is called a quadrature rule or integration rule. There is no way to choose nodes and weights so that (5.2) is accurate for all possible integrands $f(x)$. Consequently, any given integration rule can be accurate only for a restricted class of functions, and there are many different methods for specifying the nodes and weights for various types of integrands.

For smooth functions, or functions that are well approximated by low-order polynomials, the following integration rules are typically used:

Integration Rules for Smooth Functions

Newton–Cotes rules: For this type of quadrature rule, the x_n are evenly spaced, and the w_n are chosen so that (5.2) is *exact* if $f(x)$ is a low-order polynomial. The order of the integration rule is the degree of the highest order polynomial that is integrated exactly by the quadrature rule.

Extended Newton–Cotes rules: The region of integration is divided into many subintervals and a Newton–Cotes rule is used to evaluate the integral on each subinterval.

Romberg integration: This is a method for piecing together low-order Newton–Cotes quadrature rules to produce higher order integration rules. Both Newton–Cotes and Romberg integration improve the accuracy of the basic polynomial integration rules.

These methods are inaccurate if the integrand is singular or the region of integration is unbounded. For singular integrals, other integration rules can be used:

Integration Rules for Smooth and Singular Functions

Gaussian quadrature: In this method, both weights and the locations of the nodes are treated as unknowns, so the integration points are not evenly spaced. The integrand is also factored into the product of a fixed weight function $w(x)$ and a low-order polynomial $f(x)$. The weight function $w(x)$ need not be smooth, which allows singular integrals to be accurately evaluated. For a given number of integration points, Gaussian quadrature is typically more accurate than a Newton–Cotes quadrature rule.

Transformations: Variable substitutions can be used to transform singular integrands and improper regions of integration so that Newton–Cotes rules can be used.

Singularity subtraction: This approach is often used in computational electromagnetics (CEMs) and consists of subtracting a known singularity from the integrand that can be integrated in closed form, leaving a smoother integral to evaluate numerically.

5.2 Newton–Cotes Quadrature Rules

The most basic integration rule is variously known as a Riemann sum evaluated at interval midpoints, Euler's rule, or the midpoint rule. This integration rule was introduced briefly in Section 3.3. The region of integration $[a, b]$ is divided into N equal intervals of width $h = (b - a)/N$. The nodes are the center points of the intervals, so that

$$x_n = a + (n - 1/2)h, \quad n = 1, 2, \ldots, N \tag{5.3}$$

and the weights w_n are all equal to h. If the integrand of (5.1) is a polynomial of zero or first order, the quadrature rule gives the exact value of the integral. More generally, if the integrand $f(x)$ is a smooth, slowly varying function, the quadrature rule will be accurate. If $f(x)$ varies rapidly with respect to the integration step size h, the result will be inaccurate.

An illustration of the midpoint rule is shown in Figure 5.1. Graphically, error is caused by the difference between the areas of the rectangles on each interval and the area under the curve. The more rapidly the integrand changes with x, the larger the difference between the rectangle areas and the area under the curve.

5.2.1 Error Analysis of the Midpoint Rule

We can analyze the accuracy of the midpoint rule quantitatively by integrating a complex exponential and determining the error as a function of its spatial frequency. If $f(x) = e^{jkx}$, the exact value of the integral on the interval $[-a/2, a/2]$ is

$$I = \int_{-a/2}^{a/2} e^{jkx}\, dx = \frac{\sin(ka/2)}{k/2} \tag{5.4}$$

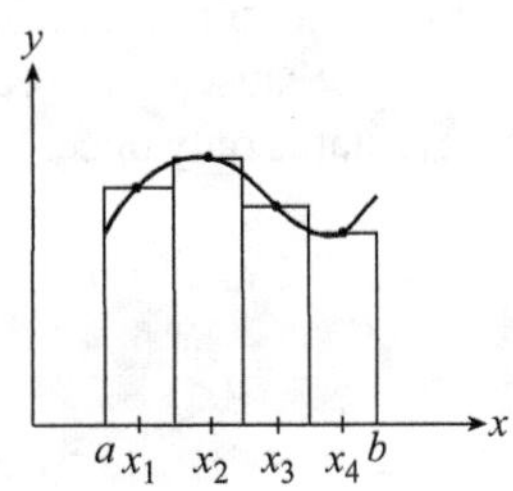

Figure 5.1 Illustration of the midpoint integration rule. Error is caused by the discrepancy between the areas of the rectangles in each interval and the area under the curve

The quadrature rule gives

$$\begin{aligned}
\hat{I} &= h\sum_{1}^{N} e^{jk(-a/2+(n-1/2)h)} \\
&= h\frac{e^{jk(-a/2+h/2)} - e^{jk(-a/2+(N+1/2)h)}}{1-e^{jkh}} \\
&= h\frac{\sin(ka/2)}{\sin(kh/2)} \qquad (5.5)
\end{aligned}$$

This function is sometimes called the Dirichlet function or periodic sinc function. The relative quadrature error is

$$\frac{I-\hat{I}}{I} = 1 - \frac{kh/2}{\sin(kh/2)} \qquad (5.6)$$

This result has an interesting parallel with the error in computing the first derivative using the central difference approximation. If $\hat{f}'(x)$ denotes the central difference approximation to the derivative of $f(x)$ at x, the relative error is

$$\frac{f'(x)-\hat{f}'(x)}{f'(x)} = 1 - \frac{\sin(kh/2)}{kh/2} \qquad (5.7)$$

The sinc function in this expression appears in (5.6) inverted. For small kh, the integration rule overestimates the exact value of the integral, whereas the central difference formula underestimates the derivative.

By expanding (5.6) for small kh, we find that the magnitude of the relative quadrature error is approximately

$$\left|\frac{I-\hat{I}}{I}\right| \simeq \frac{k^2h^2}{24} \qquad (5.8)$$

This shows that the integration rule is accurate to second order in h, so that doubling the number of integration points reduces the integration error by a factor of four. If we were to approximate an integral using N nodes, and the result were not good enough, to get another digit of accuracy, we would need to use roughly $\sqrt{10}N$ integration points.

Even though the accuracy of the midpoint rule is not especially good, for many applications the midpoint rule is ideal because it is so easy to implement, particularly when a given integral needs to be evaluated only once. With MATLAB®, for example, a single line of code suffices:

Midpoint Rule

```
I = sum(f((a+dx/2):dx:(b-dx/2)))*dx;
```

where `f(x)` is the integrand, `a` and `b` define the region of integration, and `dx = (b-a)/N` is the integration step size in terms of the number of integration points `N`.

If the integrand can be rapidly calculated, the midpoint rule requires negligible computation time with thousands or even millions of integration points. This makes the midpoint rule the algorithm of choice for many "one-off" problems where a quick value for an integral is needed.

For other applications, high accuracy is critical or the integrand must be evaluated many times, and a quadrature rule that provides an accurate result with fewer integration points is required. As we will see in the following sections, quadrature rules are available that are far more accurate for a given number of integration points than the midpoint rule.

5.2.1.1 Code Example: Midpoint Rule Error as a Function of Integrand Frequency

We have already seen in previous chapters that assessing and understanding the accuracy of an algorithm is an important aspect of numerical analysis. The following code example illustrates how the integration error obtained with the midpoint rule depends on the properties of the integrand.

Midpoint Rule Error

```
% Midpoint rule integration error

% Define integrand
f = @(k,x) exp(j*k*x);

% Define region of integration
a = 0;
b = 1;

% Number of integration points and step size
N = 10;
h = (b-a)/N;

% Integration points
x = (a+h/2):h:(b-h/2);

list = 0:2:40;
for n = 1:length(list),
     k = list(n);

     % Reference value of integral
     I0 = integral(@(x) f(x,k),a,b);

     % Approximate value of integral using midpoint rule
     I = sum(f(k,x))*h;
```

```
        % Compute relative error
        err(n) = abs(I - I0)/abs(I0);

        err_theory(n) = abs(1-1./sinc(k*h/(2*pi)));
    end % loop over n1

    figure(1)
    plot(list,err*100,'.')
    hold on
    plot(list,err_theory*100,'-')
    hold off
    xlabel('k')
    ylabel('Relative Error (%)')
```

The MATLAB function `integral` is used to compute a reference value for the integral. Since the integral used in this code can be evaluated symbolically in closed form, the exact value of the integrand could also be computed directly. The `integral` function allows the code to be applied to other integrals that cannot be evaluated in closed form.

The predicted integration error (5.7) and the actual integration error obtained using this code are shown in Figure 5.2 for the integrand of (5.4). The interval of

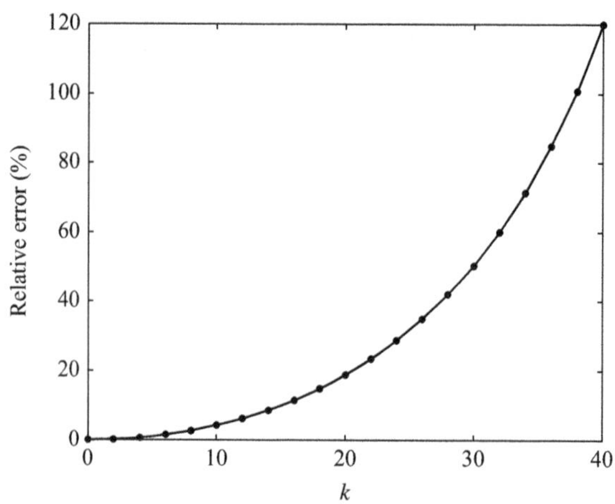

Figure 5.2 Error obtained with the midpoint integration rule applied to the integrand e^{jkx} *as a function of the spatial frequency k of the complex exponential. Actual integration error is indicated by dots, and the solid line is the absolute value of the theoretically predicted error* (5.7)

integration is [0, 1]. Ten integration points are used, and the integration step size is $h = 0.1$. The error curves represent the absolute value of the error expressed as a percentage:

$$\text{Relative error (\%)} = 100\left|\frac{I - \hat{I}}{I}\right| \tag{5.9}$$

For small values of k, the integrand is relatively smooth on the interval of integration, and the midpoint rule yields an accurate result. As k becomes large, the integrand oscillates more rapidly, and the midpoint rule is no longer inaccurate.

5.2.2 Higher Order Newton–Cotes Rules

With the midpoint rule, the node points are evenly spaced, and the weights are equal. By allowing the weights used in the integration rule to be unequal and choosing their values judiciously, it is possible to get better accuracy for smooth integrands with fewer integration points than required by the midpoint rule. The basic idea is to solve for the values of the weights that make the integration rule exact over a specific set of functions.

Suppose we choose the weights such that the quadrature rule integrates N given functions f_k, $n = 1, 2, \ldots N$ exactly. This leads to a system of N linear equations,

$$\sum_{1}^{N} w_n f_k(x_n) = I_k, \quad k = 1, 2, \ldots, N \tag{5.10}$$

where

$$I_k = \int_a^b f_k(x)\, dx \tag{5.11}$$

This linear system can be solved for the weights w_n. A natural choice for the functions $f_k(x)$ are the monomials $1, x, x^2, \ldots, x^{N-1}$, so that $f_k(x) = x^{k-1}$. In this case, the linear system in matrix form is

$$\begin{bmatrix} x_1^0 & x_2^0 & \cdots & x_N^0 \\ x_1^1 & x_2^1 & \cdots & x_N^1 \\ & \vdots & & \\ x_1^{N-1} & x_2^{N-1} & \cdots & x_N^{N-1} \end{bmatrix} \begin{bmatrix} w_1 \\ w_2 \\ \vdots \\ w_N \end{bmatrix} = \begin{bmatrix} I_1 \\ I_2 \\ \vdots \\ I_N \end{bmatrix} \tag{5.12}$$

This type of matrix is known as Vandermonde. After substituting values for the node points x_n, the linear system can be inverted to find the weights.

For the first few orders, we obtain the following:

One interval of the Euler rule or Riemann sum: $N = 1$, $a = 0$, $b = h$. x_1 is arbitrary, and solving the linear system leads to $w_1 = h$. If x_1 is chosen to be the center of the interval, this becomes the midpoint rule. The approximate value of the integral is

$$\hat{I} = hf(x_1) \tag{5.13}$$

Trapezoidal rule: $N = 2$, $a = 0$, $b = h$. Choosing $x_1 = 0$, $x_2 = h$ and solving the linear system yield the weights $w_1 = w_2 = h/2$. The approximate value of the integral is

$$\hat{I} = h/2f(x_1) + h/2f(x_2) \tag{5.14}$$

Simpson's rule: $N = 3$, $a = 0$, $b = 2h$. With the nodes $x_1 = 0$, $x_2 = h$, $x_3 = 2h$, the linear system leads to $w_1 = h/3$, $w_2 = 4h/3$, $w_3 = h/3$. The approximate value of the integral is

$$\hat{I} = h/3f(x_1) + 4h/3f(x_2) + h/3f(x_3) \tag{5.15}$$

We can analyze the accuracy of these integration rules by again considering the integrand e^{jkx}. Expanding the complex exponential in a Taylor series leads to

$$I = \int_a^b dx \left[1 + jkx + \frac{(jkx)^2}{2!} + \frac{(jkx)^3}{3!} + \frac{(jkx)^4}{4!} + \cdots\right] \tag{5.16}$$

By construction, an Nth-order rule will integrate at least the first N terms of this series exactly. The error can be approximated by the first term that does not integrate to zero, which is

$$\text{Error} \simeq \frac{|jk|^N}{N!} \sum_1^N w_n x_n^N \simeq \frac{(kh)^N}{N!} \frac{b^{N+1} - a^{N+1}}{N+1} \tag{5.17}$$

Since the dependence of the error term on h is of the form h^N, the integration error decreases roughly by a factor of 2^N if the integration step size h is halved and the number of integration points doubled. We refer to this as order N convergence.

By this analysis, the Riemann sum ($N = 1$ case) is first-order accurate, but if the nodes are chosen to be the midpoints of each interval, the first order error term vanishes, and the error becomes second order. Consequently, the order of the midpoint rule is better than might be expected.

For the higher order quadrature rules, to gain another digit of accuracy in the computed value of an integral, we need to increase the number of integration points only by one or two instead of a factor of more than three if we were to add more points to the midpoint rule. This illustrates the remarkable improvement in accuracy that can be realized with higher order algorithms. Since the error decreases as h^N with N in the exponent, the convergence rate with the order N for these rules is exponential. Unfortunately, this exponential convergence comes at a price.

5.2.3 Extended Newton–Cotes Rules

While the Newton–Cotes quadrature rules offer significantly greater accuracy than the midpoint rule, they are numerically unstable as the order increases. The matrix in

(5.12) becomes ill-conditioned and difficult to invert for large N. Even for moderate orders, the integration rules are nonrobust, because the accuracy of the integral may be very good for some integrands but poor for those that deviate from a polynomial behavior. Because of this, the Newton–Cotes rules are rarely used directly beyond third or fourth order.

Another way to improve the accuracy of an integration rule is to piece together low-order rules to obtain an extended rule that is more robust for large numbers of integration points. The extended rules have the same order of accuracy as the single-step rules in the sense of the power-law decay rate or log-log slope of the error with respect to the integration step size h, but the absolute accuracy of the extended rules is much better, since the region of integration is divided into smaller subintervals.

5.2.3.1 Extended Midpoint Rule

The midpoint rule is simple enough to extend to multiple integration points that it is typically introduced directly in its extended form, as we have done in Sections 3.3 and 5.2. As will be seen in Chapter 6, there are occasional cases where a single step of the midpoint rule is used, but the extended form with many integration points across the interval of integration is most common, since an integration rule with a single point is not accurate enough for many applications.

5.2.3.2 Extended Trapezoidal Rule

The extended trapezoidal rule is

$$\begin{aligned}\hat{I} &= h\left[f(x_1)/2 + \underbrace{f(x_2)/2 + f(x_2)/2}_{} + \cdots + \underbrace{f(x_{N-1})/2 + f(x_{N-1})/2}_{} + f(x_N)/2\right] \\ &= h\left[f(x_1)/2 + f(x_2) + f(x_3) + \cdots + f(x_{N-1}) + f(x_N)/2\right] \qquad (5.18)\end{aligned}$$

where $x_1, x_2, \ldots, x_N$ are evenly spaced points in the interval $[a, b]$, with $x_1 = a$ and $x_N = b$. The braces in the upper equation mark pairs of function evaluations from adjacent applications of the trapezoidal rule. At interior integration points, adjacent function evaluations are at the same point and can be combined.

The extended trapezoidal rule can be implemented in MATLAB by introducing a vector of weighting coefficients

```
w = [1/2, ones(1,N-2), 1/2];
```

where N is the desired number of integration points. The code in Section 3.3.1 can be modified to evaluate the extended trapezoidal rule by changing the summation to

```
I = sum(w.*f(x))*h
```

This approach can also be used to implement the extended Simpson's rule, with the appropriate adjustment to the coefficients stored in the vector `w`. The trapezoidal rule is implemented in the `trapz` function, but without the integration step size h.

5.2.3.3 Extended Simpson's Rule

The extended Simpson's rule is

$$\hat{I} = h\,[f(x_1)/3 + 4f(x_2)/3 + 2f(x_3)/3 + 4f(x_4)/3 + \cdots \\ + 2f(x_{N-2})/3 + 4f(x_{N-1})/3 + f(x_N)/3] \tag{5.19}$$

The required vector of weighting coefficients w can be generated by looping over integration points and using logic to select from the values 1/3, 4/3, and 2/3 depending on the loop index.

5.2.3.4 Refining Extended Rules

An important property of the extended rules is that they can be refined to more integration points without throwing away previous function evaluations. If the approximate value for N points is $\hat{I}_N$, the value for an extended trapezoidal rule of $2N$ points is equal to $\hat{I}_N/2$ plus h multiplied by the sum of function evaluations halfway between all the nodes of the N point integration. This procedure can be repeated until the difference between successive refinements of the integration, $|\hat{I}_{2N} - \hat{I}_N|$, is smaller than a given tolerance.

5.2.4 Romberg Integration

Another extension of the Newton–Cotes rules arises from the interesting observation that a linear combination of the trapezoidal rule for N and $2N$ points gives the same result as Simpson's rule:

$$\hat{I}_{2N}^{\text{Simpson}} = \frac{4}{3}I_{2N}^{\text{trap}} - \frac{1}{3}I_N^{\text{trap}} \tag{5.20}$$

This relationship can be systematically generalized to obtain increasingly high-order integration rules from combinations of lower order integrations. This provides an efficient way to increase the accuracy of a numerical integral with fewer function evaluations than simpler schemes like the midpoint rule.

5.3 Gaussian Quadrature

The polynomial quadrature rules we have studied previously have fixed, evenly spaced nodes, and variable weights. Since the nodes are fixed, the weights represent N degrees of freedom that must be chosen to specify the integration rule. Because there are N degrees of freedom, the integration rule can integrate the N polynomials of order zero through $N - 1$ exactly. By allowing the nodes to vary as well as the weights, the number of degrees of freedom associated with the integration rule doubles and an integration rule with variable weights and nodes can be designed to integrate polynomials of order up to $2N - 1$ exactly. This represents a substantial increase in accuracy. The framework for designing weights and nodes simultaneously is called Gaussian quadrature.

Another feature of Gaussian quadrature is that the integrand can include a factor, referred to as a weight function and typically written as $w(x)$, that is not constrained to be a polynomial. This substantially increases the power of Gaussian quadrature rules compared with the Newton–Cotes rules treated in the previous section, since the weight function may be singular. The two key generatizations introduced by Gaussian quadrature integration rules are as follows:

1. Both weights and nodes are variables. This allows an N-node Gaussian quadrature rule to integrate polynomials of order up to $2N - 1$ exactly, which increases accuracy significantly for smooth functions.
2. The integrand is factored, and the quadrature rule has the form

$$\int_a^b f(x)w(x)\,dx \simeq \sum_1^N w_k f(x_k) \tag{5.21}$$

where $w(x)$ is a weight function. Only $f(x)$ needs to be polynomial or close to polynomial for accurate integration. This allows the full integrand, including the weight function, to be nonsmooth or singular. The weights and nodes of the rule depend on the particular choice of weight function $w(x)$. When implementing a Gaussian quadrature rule, it is important to remember that the weight function is a part of the integrand on the left-hand side of (5.21) but is not included in the summation on the right-hand side when the quadrature rule is evaluated.

The most common weight functions are as follows:

Common Weight Functions for Gaussian Quadrature

$w(x) = 1$	Gauss–Legendre
$w(x) = e^{-x}, 0 \leq x < \infty$	Gauss–Laguerre
$w(x) = e^{-x^2}, -\infty < x < \infty$	Gauss–Hermite
$w(x) = \frac{1}{\sqrt{1-x^2}}, -1 \leq x \leq 1$	Gauss–Chebyshev

The names associated with these weight functions are also attached to families of orthogonal polynomials. As will be seen shortly, there is a close connection between Gaussian quadrature rules and orthogonal polynomials. Quadrature rules associated with common weight functions such as these are sometimes referred to as classical and integration rules for other weight functions are nonclassical Gaussian quadrature rules. For common weight functions, numerical values for the nodes and weights of Gaussian quadrature rules are often tabulated in mathematical handbooks and other references [2].

The improved accuracy of Gaussian quadrature rules and the ability to include a weight function in the integrand are valuable for many applications, but these benefits do come at a modest cost. The nodes vary in location for different orders of the quadrature rule, so that function evaluations cannot be reused as the order increases.

This is generally not a serious problem, because high accuracy can often be obtained using relatively few function evaluations. Another issue is that the weights and nodes of a Gaussian quadrature rule cannot be found simply by solving a linear system as in the case of the Newton–Cotes rules considered earlier. For common weight functions, the weights and nodes have been tabulated and are readily available in printed or online references, but for uncommon weight functions, finding the nodes and weights requires a fairly sophisticated algorithm.

A Gaussian quadrature rule is defined by the set of equations

$$\begin{aligned}\sum_{k=1}^{N} w_k f_1(x_k) &= \int_a^b f_1(x)w(x)\,dx \\ \sum_{k=1}^{N} w_k f_2(x_k) &= \int_a^b f_2(x)w(x)\,dx \\ &\;\vdots \\ \sum_{k=1}^{N} w_k f_{2N}(x_k) &= \int_a^b f_{2N}(x)w(x)\,dx\end{aligned} \tag{5.22}$$

where $f_n(x) = x^{n-1}$ is a monomial of order $n-1$. There are N weights and N nodes, for a total of $2N$ unknowns, so $2N$ equations are required to uniquely specify the weights and nodes of the Gaussian quadrature rule. Unlike (5.12), these equations are nonlinear, because the unknown nodes appear inside the arguments of the functions $f_n(x)$.

In principle, one could solve the nonlinear equations (5.22) for any given set of $2N$ functions f_n using a numerical algorithm. Because of the difficulty of solving nonlinear equations, this has not been done until relatively recently. Ma, Rokhlin, and Wandzura have successfully implemented a nonlinear solution method for (5.22) in the general case of functions f_n that are arbitrary up to a fairly nonrestrictive consistency condition [3]. This represents a brute-force, direct method for determining the weights and nodes for a given weight function $w(x)$ and set of functions for which the rule should be exact. For the case of polynomial functions, a more elegant solution method was developed long before computers were available by the great mathematician Gauss.

5.3.1 Orthogonal Polynomials and Gaussian Quadrature

Gauss's elegant solution method for the nonlinear equations (5.22) relies on the theory of orthogonal polynomials. The notion of orthogonality requires an inner product. Using the weight function $w(x)$, we can define an inner product by

$$\langle f, g\rangle = \int_a^b f(x)g(w)w(x)\,dx \tag{5.23}$$

This is referred to as a weighted L_2 inner product. This integral does not satisfy the properties required for it to be an inner product for all weight functions. A necessary condition is that $w(x)$ is nonnegative on the open interval (a, b). If the functions f and g are complex, a complex conjugate is included on f in the integrand of the defining integral. Here, the function spaces involved in the derivation are real, so we omit the complex conjugate in the inner product.

A set of orthonormal polynomials $p_n(x)$ satisfies

$$\langle p_m, p_n \rangle = \begin{cases} 1 & m = n \\ 0 & \text{otherwise} \end{cases} \tag{5.24}$$

By convention, some families of polynomials are scaled in such a way that $\langle p_m, p_m \rangle$ is not necessarily equal to one, in which case the polynomials are orthogonal but not orthonormal (Legendre polynomials being a common example). Orthogonal polynomials for a given weight function $w(x)$ can be constructed by Gram-Schmidt orthogonalization of the monomials $1, x, x^2, \ldots$. In this way, it can be shown that a family of orthonormal polynomials exists for each weight function and interval $[a, b]$. Families of orthogonal polynomials have been tabulated and studied for common weight functions.

The goal in designing a Gaussian quadrature rule is to find nodes and weights such that

$$\int_a^b f(x)w(x)\,dx = \sum_{k=1}^{N} w_k f(x_k) + \varepsilon \tag{5.25}$$

is exact ($\varepsilon = 0$) if $f(x)$ is a polynomial of degree $2N - 1$ or less. Let $f(x)$ be a polynomial of degree at most $2N - 1$. Dividing $f(x)$ by the orthogonal polynomial $p_N(x)$ gives

$$f(x) = p_N(x)Q_{N-1}(x) + R(x) \tag{5.26}$$

where Q_{N-1} is a polynomial of order $N - 1$, and R is a remainder polynomial of degree $N - 1$. If we multiply by the weight function and integrate both sides over the interval $[a, b]$, we get

$$\int_a^b f(x)w(x)\,dx = \int_a^b p_N(x)Q_{N-1}(x)w(x)\,dx + \int_a^b R(x)w(x)\,dx \tag{5.27}$$

The first term on the right-hand side is zero, since Q_{N-1} can be expressed as a linear combination of the orthogonal polynomials $p_0, p_1, \ldots, p_{N-1}$ and so must be orthogonal to p_N. This leaves us with

$$\int_a^b f(x)w(x)\,dx = \int_a^b R(x)w(x)\,dx \tag{5.28}$$

Now, we let the nodes x_k be the N roots of $p_N(x)$. With this choice of the nodes, from (5.26) we have that $R(x_k) = f(x_k)$.

We now introduce another special set of polynomials, the Nth-order Lagrange polynomials

$$L_k(x) = \prod_{l=1, l \neq k}^{N} \frac{x - x_l}{x_k - x_l} \tag{5.29}$$

These functions are equal to one at $x = x_k$ and zero at all the other points x_l for $l \neq k$. Because of this property, they are called interpolatory functions. Since $R(x)$ is of order $N - 1$, we can write it as a sum of Lagrange polynomials. Using the previously obtained result, $R(x_k) = f(x_k)$ leads to

$$R(x) = \sum_{k=1}^{N} f(x_k) L_k(x) \tag{5.30}$$

Using this expression in (5.28) results in

$$\int_a^b f(x) w(x)\, dx = \sum_{k=1}^{N} f(x_k) \underbrace{\left[\int_a^b L_k(x) w(x)\, dx \right]}_{\text{weights } w_k} \tag{5.31}$$

By comparing the right-hand side with (5.25), we can identify the term in square brackets as the weights w_k.

To simplify the computation of the weights, we can express the Lagrange polynomials in terms of the orthogonal polynomials. An interpolatory Lagrange polynomial can be constructed from any polynomial using

$$L_k(x) = \frac{p_N(x)}{(x - x_k) p'_N(x_k)} \tag{5.32}$$

where in this case $p_N(x)$ is a Legendre polynomial. It is easy to see from the way we chose the nodes that this polynomial is zero at all of the nodes except for x_k, so that the polynomial on the right-hand side has the same $N - 1$ zeros as L_k. By L'Hôpital's rule, the polynomial is equal to one at $x = x_k$. Because a polynomial of order $N - 1$ is completely determined by N specified values, the function on the right must be equal to L_k.

Combining these results provides the following prescription for the weights and nodes of the Gaussian quadrature rule:

1. The functions $p_n(x)$ are orthonormal polynomials with respect to the weight function $w(x)$ on the interval $[a, b]$.
2. Nodes are given by the zeros of the Nth-order orthogonal polynomial: $p_N(x_k) = 0$, $k = 1, 2, \ldots, N$.

3. The weights are

$$w_k = \int_a^b \frac{p_N(x)}{(x - x_k)p'_N(x_k)} w(x)\, dx \tag{5.33}$$

It can be shown that the weights can also be expressed in the equivalent but more convenient form

$$w_k = -\frac{A_N}{A_{N-1}} \frac{\langle p_{N-1}, p_{N-1} \rangle}{p_{N-1}(x_k)p'_N(x_k)} \tag{5.34}$$

where A_n is the coefficient of x^n in $p_n(x)$.

5.3.2 Gauss–Legendre Quadrature ($w(x) = 1$)

The simplest weight function is $w(x) = 1$. By convention, the region of integration is taken to be $[a, b] = [-1, 1]$. The orthogonal polynomials for this weight function are the Legendre polynomials

$$\begin{aligned} P_0(x) &= 1 \\ P_1(x) &= x \\ P_2(x) &= \frac{1}{2}(3x^2 - 1) \\ &\vdots \\ P_n(x) &= \frac{1}{2^n n!} \frac{d^n}{dx^n}(x^2 - 1)^n \end{aligned}$$

The nodes can be found using a zero finding algorithm such as Newton's method or the Newton–Raphson method (see Section 3.6). A helpful fact in implementing a root finder for the Legendre polynomials is that the zeros of P_{n+1} are in between those of $P_n(x)$, so the zeros of one value of n can be used as starting guesses for the zeros of P_{n+1}. Alternately, the approximation

$$x_k \simeq \cos\left[\frac{\pi(k - 1/4)}{(n + 1/2)}\right] \tag{5.35}$$

can be used to initialize Newton's method.

The weights as given by (5.34) can be simplified using the following properties of the Legendre polynomials:

$$\int_{-1}^{1} |P_n(x)|^2\, dx = \frac{2}{2n + 1} \tag{5.36}$$

$$(x^2 - 1)P'_n(x) = nxP_n(x) - nP_{n-1}(x) \tag{5.37}$$

Table 5.1 Gauss–Legendre quadrature weights and nodes

N	x_n	w_n
2	±0.577350	1.000000
3	0	0.888889
	±0.774597	0.555556
4	±0.339981	0.652145
	±0.861136	0.347855
5	0	0.568889
	±0.538469	0.478629
	±0.906180	0.236927

$$\frac{A_n}{A_{n-1}} = 2 - 1/n \tag{5.38}$$

The first relationship defines the normalization of the polynomials with respect to the inner product with weight function $w(x) = 1$, the second is a recursion relation between polynomials of different orders and the derivative of a Legendre polynomial, and the third is the ratio of the leading polynomial coefficients in successive polynomials. With these identities, the weights can be simplified to

$$w_k = \frac{2}{(1 - x_k^2)[P_N'(x_k)]^2} \tag{5.39}$$

which is convenient for numerical evaluation. To compute the derivative of $P_n(x)$, a central difference approximation could be used, but the recursion relation (5.37) provides an exact way to evaluate the derivative in closed form. The weights and nodes for Gauss–Legendre quadrature for the first few orders are shown in Table 5.1.

The integration points for the Gauss–Legendre.

5.4 Nonclassical Gaussian Quadrature Rules

Classical Gaussian quadrature rules involve common weight functions for which the orthogonal polynomials are known in closed form. In principle, a Gaussian quadrature rule can be obtained for any integrable weight function. Straightforward approaches for doing this where the orthogonal polynomials are not readily available were plagued by numerical instability, until an insight by Sack and Donovan [4] solved the problem. Their approach was to use Legendre polynomials as a basis for the space of polynomials instead of monomials. Ma *et al.*'s brute-force nonlinear solver can also be used to find Gaussian quadrature rules for nonclassical weight functions.

Another method for computing Gaussian quadrature rules for nonclassical weight functions is based on an interesting connection with linear algebra. We first use an auxiliary quadrature rule, such as the midpoint rule, to cast the integral (5.21) into matrix form. The auxiliary N point rule is

$$I_w = \int_a^b f(x)w(x)\,dx \simeq \sum_1^N f(y_n)w(y_n)h \tag{5.40}$$

where N is very large (perhaps 10^5 or so). If we define a diagonal matrix

$$A = \begin{bmatrix} y_1 & & & \\ & y_2 & & \\ & & \ddots & \\ & & & y_N \end{bmatrix} \tag{5.41}$$

and vector with elements $b_n = [hw(y_n)]^{1/2}$, we can express the auxiliary quadrature rule as

$$I_w \simeq b^T f(A) b \tag{5.42}$$

Next, we will find a reduced-order representation of A with the form

$$\begin{bmatrix} A \\ (N \times N) \end{bmatrix} \simeq \begin{bmatrix} C \\ (N \times K) \end{bmatrix} \underbrace{\begin{bmatrix} \Lambda \\ (K \times K) \end{bmatrix}}_{\text{small}} \begin{bmatrix} C^T & (K \times N) \end{bmatrix} \tag{5.43}$$

where Λ is a diagonal matrix with dimension K much smaller than N. The matrix C will have the property that $C^T C = I_{K \times K}$. This factorization will be computed using the Lanczos algorithm from linear algebra. It will turn out that the fact that an N by N matrix can be approximated by a much smaller K by K matrix implies that the N-point auxiliary integration rule can be replaced by a K-point Gaussian quadrature rule.

5.4.1 Lanczos Algorithm

In Chapter 7, we will study the methods for solving the linear system $Ax = b$ and for finding the eigenvectors and eigenvalues of the matrix A. If A is very large, direct methods for solving the linear system or finding the eigenvalues and eigenvectors require too much computation time to be practical, so faster iterative methods that find approximate solutions have been developed. One such iterative method is the Lanczos algorithm. We will use the Lanczos algorithm here to generate the matrix factorization (5.43) needed to compute weights and nodes for nonclassical Gaussian quadrature rules.

In pseudo-code, the Lanczos algorithm is as follows:

$$\beta_0 = 0,\ q_0 = 0,\ q_1 = b/\|b\|$$

for $n = 1, 2, 3, \ldots, K$

$$v = Aq_n - \beta_{n-1}q_{n-1}$$

$$\alpha_n = q_n^T v$$

$$v = v - \alpha_n q_n$$

$$\beta_n = \|v\|$$

$$q_{n+1} = v/\beta_n$$

end loop

Scalar quantities are denoted by Greek letters (α_n and β_n), and Roman letters denote vectors (b, q_n, and v).

At the Kth step, the coefficients produced by the iteration can be used to generate the K by K tridiagonal Jacobi matrix

$$T_K = \begin{bmatrix} \alpha_1 & \beta_1 & & & \\ \beta_1 & \alpha_2 & \beta_2 & & \\ & \beta_2 & \alpha_3 & \ddots & \\ & & \ddots & \ddots & \beta_{K-1} \\ & & & \beta_{K-1} & \alpha_K \end{bmatrix} \tag{5.44}$$

and an N by K matrix with columns given by the vectors q_n,

$$Q_K = \left[\, q_1\ q_2\ \cdots\ q_K \,\right] \tag{5.45}$$

These two matrices provide the approximate tridiagonal matrix factorization

$$A \simeq Q_K T_K Q_K^T \tag{5.46}$$

Typically, K is small compared with N, and the matrix T_K is much smaller than A. For this reason, this factorization is sometimes referred to as a reduced-order representation of A, since it allows the properties of a large matrix to be approximated using a small matrix.

There are several applications for the factorization (5.46). The eigenvalues of T_K approximate K of the eigenvalues of A, so the Lanczos algorithm provides a computationally efficient method for estimating the spectrum of a large matrix. The factorization can be used to find an approximation to the solution to the linear system $Ax = b$, in which case the algorithm is referred to as the conjugate gradient method (see Section 7.5).

Here we will use the factorization in a different way to create a reduced-order representation of the form of (5.43). We will first diagonalize the tridiagonal matrix, so that

$$T_K = V \Lambda V^T \tag{5.47}$$

where Λ has the eigenvalues of T_K along its main diagonal, and the columns of V are the eigenvectors of T. The matrix C in (5.43) is then given by $Q_K V$. The reduced order representation (5.43) is

$$A \simeq \underbrace{Q_K V}_{C} \Lambda \underbrace{V^T Q_K^T}_{C^T} \tag{5.48}$$

We will also make use of two important properties of the approximate matrix factorization: the columns of Q_K are mutually orthogonal, and the first column q_1 is $b/\|b\|$.

5.4.2 Weights and Nodes

Using the reduced-order representation of A, the auxiliary quadrature rule can be written as

$$I_w \simeq b^T f(C\Lambda C^T)b \tag{5.49}$$

If we expand the function f as a Taylor series, this becomes

$$\begin{aligned} I_w &\simeq b^T \left[a_0 I + a_1 C\Lambda C^T + a_2 (C\Lambda C^T)^2 + \cdots\right] b \\ &= b^T C \left[a_0 I + a_1 \Lambda + a_2 \Lambda^2 + \cdots\right] C^T b \\ &= b^T C f(\Lambda) C^T b \end{aligned} \tag{5.50}$$

From the properties of the Lanczos algorithm,

$$\begin{aligned} b^T C &= b^T Q_K V \\ &= \begin{bmatrix} 1 & 0 & \cdots & 0 \end{bmatrix} V \|b\| \\ &= \begin{bmatrix} V_{11} & V_{12} & \cdots & V_{1K} \end{bmatrix} \|b\| \end{aligned} \tag{5.51}$$

Combining these results leads to

$$\begin{aligned} I_w &\simeq \begin{bmatrix} V_{11} & V_{12} & \cdots & V_{1K} \end{bmatrix} \begin{bmatrix} f(\lambda_1) & & & \\ & f(\lambda_2) & & \\ & & \ddots & \\ & & & f(\lambda_K) \end{bmatrix} \begin{bmatrix} V_{11} \\ V_{12} \\ \vdots \\ V_{1K} \end{bmatrix} \|b\|^2 \\ &= \sum_{k=1}^{K} \underbrace{\left[V_{1k}^2 \|b\|^2\right]}_{\text{weights } w_k} f(x_k) \end{aligned} \tag{5.52}$$

where the nodes x_k are the eigenvalues of T_K. What we have accomplished is to transform the N-point integration rule in (5.40) into a K-point integration rule.

How do we know the integration rule in (5.52) is a Gaussian quadrature rule? For the rule to be Gaussian, it must exactly integrate (5.40) if f is a polynomial of degree

$2K - 1$ or less. This can be proved from the fact that the approximate factorization (5.43) is exact in the sense that

$$Ax = C\Lambda C^T x \tag{5.53}$$

as long as x is in the subspace span$\{b, Ab, A^2b, \ldots, A^{K-1}b\}$.

One might also ask the question, "Why not simply use the original N-point quadrature rule to compute I_w?" Once we have done the work of finding the weights and nodes (x_k, w_k) of the Gaussian quadrature rule, we can get the same accuracy for integrals of the form of (5.1) with many orders of magnitude fewer evaluations of $f(x)$. For example, N may be on the order of 10^5, whereas K can be around five for similar accuracy.

5.5 Implementation

Implementing this method for finding Gaussian quadrature rules requires the following algorithmic steps:

Algorithm Steps for Generating Nonclassical Gaussian Quadrature Rules

1. Choose an auxiliary quadrature rule (y_n, u_n), $n = 1, 2 \ldots, N$.
2. Set $A = \text{diag}(y_1, y_2, \ldots, y_N)$, and $b_n = [u_n w(y_n)]^{1/2}$.
3. Run the Lanczos algorithm for K steps using the matrix A and the vector b.
4. Find the eigenvalues and eigenvectors of T_K. The nodes are the eigenvalues and the weights are the squares of the first components of each of the eigenvectors scaled by $\|b\|^2$.

Since the matrix A is diagonal, the zero off-diagonal elements need not be stored. A can be a vector of the off-diagonal elements, and the matrix-vector multiply Aq_n in the Lanczos algorithm can be implemented efficiently using the MATLAB `.*` operator.

To obtain accurate weights and nodes for the Gaussian quadrature rule, the auxiliary quadrature rule must be very accurate, which means that N must be large. For Gauss–Legendre quadrature, $N = 10^5$ gives about seven digits of accuracy for the weights and nodes. As the order K of the Gaussian quadrature rule increases, N must also grow to ensure that the weights and nodes are accurate. The efficiency of the method can be improved by choosing a more accurate auxiliary quadrature rule than the midpoint rule, but the execution time is fast enough that for moderate values of K the midpoint rule is adequate.

5.6 Other Methods for Singular Integrands

In computational electromagnetics, integrands with singularities are often encountered. Gaussian quadrature has been used on occasion [5], but other methods are more widely used. Singularity subtraction has been the most common approach due to its conceptual simplicity. In recent years, there has been a move to integration methods based on transformations, which typically yield more accurate results.

5.6.1 Singularity Subtraction

Some integrands, particularly those involving Green's functions, have a type of singularity that can be approximated in analytical form. Consider the 2-D Green's function, which is a scaled Hankel function. This function has the limiting behavior

$$H_0^{(2)}(x) \simeq 1 - j\frac{2}{\pi}\left[\ln\left(\frac{x}{2}\right) + \gamma\right], \quad x \to 0 \tag{5.54}$$

where $\gamma \simeq 0.5772$ is the Euler–Mascheroni constant. Adding and subtracting the leading singularity leads to

$$H_0^{(2)}(x) = \left[H_0^{(2)}(x) + j\frac{2}{\pi}\ln\left(\frac{x}{2}\right)\right] - j\frac{2}{\pi}\ln\left(\frac{x}{2}\right) \tag{5.55}$$

The last term in this expression can be integrated exactly, and the term in square brackets is finite, making it easier to integrate with a quadrature rule. The integrand before and after singularity subtraction is shown in Figure 5.3.

If we integrate the Hankel function from 0 to 1, the exact value is approximately $0.9197 + j0.6371$. Using the midpoint rule with 10 points to integrate the Hankel function yields $0.9205 + j0.5942$. Due to the singularity of the Hankel function, there is significant error in the imaginary part. If we integrate only the term in square

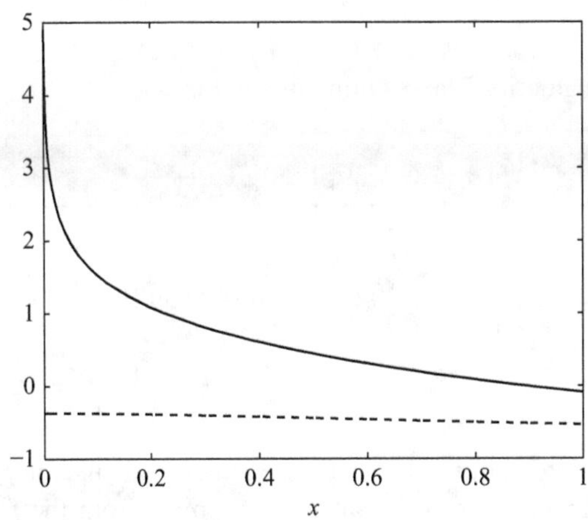

Figure 5.3 Singularity subtraction for $H_0^{(2)}(x)$. The solid line is the imaginary part of the Hankel function, and the dashed line is the imaginary part of the Hankel function with the leading singularity subtracted, as given by the first term in brackets in (5.55)

brackets in (5.55) with the quadrature rule and add the exact value of the integral of the analytical singularity, which is

$$-\int_0^1 \frac{2}{\pi} \ln\left(\frac{x}{2}\right) = j\frac{2}{\pi}[\ln(2) + 1] \tag{5.56}$$

we obtain $0.9205 + j0.6373$, which has a much more accurate imaginary part.

This approach often works well in practice, but singularity subtraction has a fundamental limitation. Although the integrand after subtraction of the leading singularity is finite, it may be singular in a mathematical sense, in that a derivative of the integrand may be infinite. A finite singularity can typically be integrated more accurately than one that becomes infinite, but the function is still nonpolynomial in nature and hence cannot be integrated as accurately as a smooth function. Implementing the analytical integration for polygonal domains of integration or more complex geometrical shapes can also be tedious.

5.6.2 Transformation Rules

Transformation rules can be used to map an unbounded integration domain to a bounded domain or a singular integrand to a smooth integrand. A standard quadrature rule can then be applied to the transformed integral.

If $f(x)$ has a power-law singularity at $x = a$ of the form

$$f(x) \sim (x-a)^{-\gamma}, \quad x \to a \tag{5.57}$$

then we can use the transformation

$$x(t) = t^{1/(1-\gamma)} + a \tag{5.58}$$

to eliminate the singularity. The resulting integral is

Transformation Rule for Power-law Singularities

$$\int_a^b f(x)\,dx = \frac{1}{1-\gamma} \int_0^{(b-a)^{1-\gamma}} t^{\gamma/(1-\gamma)} f\left[t^{1/(1-\gamma)} + a\right] dt \tag{5.59}$$

A simple midpoint integration rule can be used to evaluate the integral in t. By using the transformation rule to map the integration points from the t domain to the x domain, the evenly spaced and equally weighted integration rule in the t domain can be viewed as an unevenly spaced and unequally weighted integration rule in the x domain. Since the transformation "stretches" the singularity to make it finite, the integration points are concentrated near the singularity and are relatively sparse where the original integrand is smooth.

For unbounded integration domains, the transformation

$$\int_a^b f(x)\,dx = \int_{1/b}^{1/a} \frac{1}{t^2} f(1/t)\,dt \tag{5.60}$$

where $ab > 0$ can be used. If $a > 0$ and $b \to \infty$, the transformed range of integration is $0 \le x \le 1/a$, which is finite.

5.6.2.1 Code Example: Integration Using a Transformation Rule

This code compares integration of a singular function with and without the use of a transformation rule. The midpoint integration rule is applied to the original integrand, and the same rule is applied to the integral after transformation. The MATLAB function `integral` is used to obtain an accurate reference value for the integral.

Integration Using a Transformation Rule

```
% Define the integrand
f = @(x) x.^(-1/2);
a = 0;
b = 1;

% Reference value of integral
I0 = integral(f,a,b)

% Direct numerical integration with midpoint rule
dx = 0.3;
x = (a+dx/2):dx:(b-dx/2);
I1 = sum(f(x))*dx

% Numerical integration using the same midpoint rule
% but for the transformed variable
gamma = 1/2;
f2 = @(gamma,t) t.^(gamma/(1-gamma)).* ...
      ((t.^(1/(1-gamma))) .^(-gamma))/(1-gamma);
N = length(x);
t1 = 0;
t2 = (b-a)^(1-gamma);
dt = (t2-t1)/N;
t = (t1+dt/2):dt:(t2-dx/2);
I2 = sum(f2(gamma,t))*dt
```

The `...` command indicates a continuation, so the commands on the next line of the code are treated as part of the same line by the MATLAB compiler.

The exact value of the integral is 2, and the value with a 10-point midpoint rule is 1.5682, which means that the midpoint rule is highly inaccurate when applied to the singular integrand. Using a transformation rule to remove the singularity in the integrand and then applying the same 10-point midpoint rule to the transformed integrand leads to a much more accurate result. In fact, since the transformed integrand is constant, the midpoint rule after transformation yields the exact value of the integral.

5.7 Multidimensional Integrals

All the integrals we have considered so far have been one-dimensional. Numerical algorithms for 2-D and 3-D problems often require evaluation of integrals over multidimensional regions. For a 2-D integral of the form

$$I = \int_c^d \int_a^b f(x,y)\,dx\,dy \tag{5.61}$$

with a rectangular region of integration, a product integration rule can be used. A product rule is a simple combination of two one-dimensional integration rules. If the integration points on $a \le x \le b$ are x_n and the integration points for $c \le y \le d$ are y_n, the integration points for the 2-D region are of the form (x_m, y_n).

A simple product midpoint integration rule is illustrated in Figure 5.4. The x coordinates of the integration points are chosen according to the midpoint rule on the interval $1 \le x \le 2$ with spacing $\Delta x = 0.1$, and the y coordinates are chosen according

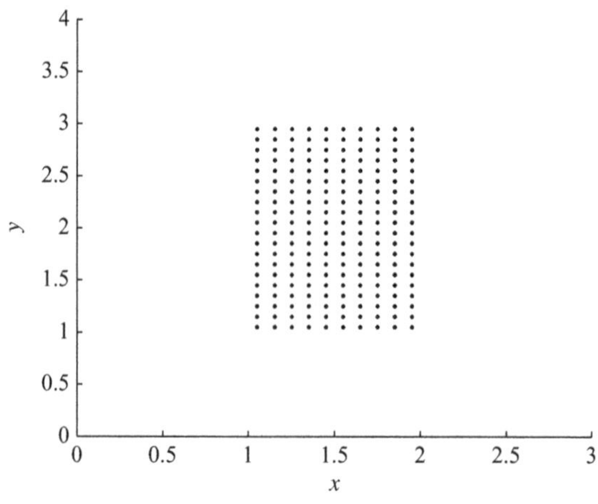

Figure 5.4 Two-dimensional product integration rule for the region $1 \le x \le 2$, $1 \le y \le 3$

to the midpoint rule on $1 \leq y \leq 3$ with spacing $\Delta y = 0.1$. This product rule can be applied to an integral of the form

$$I = \int_1^3 \int_1^2 f(x,y)\, dx\, dy \tag{5.62}$$

If the result is not accurate enough, the integration step sizes Δx and Δy can be decreased. Since the total number of integration points is the product of the number of points for the one-dimensional x and y integration rules, multidimensional integrals can be computationally costly to evaluate.

The x and y coordinates need not be chosen using the same integration rule. In electromagnetics, integrals over a sphere of the form

$$I = \int_0^{2\pi} \int_0^{\pi} f(\theta, \phi) \sin\theta \, d\theta d\phi \tag{5.63}$$

are often encountered. Making the substitution $u = \cos\theta$ changes the integral to

$$I = \int_0^{2\pi} \int_{-1}^{1} f(\cos^{-1} u, \phi)\, du\, d\phi \tag{5.64}$$

This integral can be efficiently computed with a product rule generated from Gauss–Legendre quadrature for u and the midpoint rule for ϕ.

For nonrectangular regions of integration, product integration rules cannot be used directly. One approach is to use a generalization of the Riemann sum by dividing the integration domain into N subregions with areas w_n and choosing one point in each subregion. These approaches are easy to implement but are not optimal in the sense that other sets of integration points and weights can yield a more accurate value for a given integral with fewer function evaluations. A method for integrating the 3-D Green's function over triangles is considered in Section 6.8.3.

While better quadrature rules than the product rules exist for multidimensional integrals, finding the required weights and nodes can be difficult. The theory of orthogonal polynomials used to develop Gaussian quadrature rules does not generalize in a straightforward way to arbitrary two-dimensional integration regions. There do exist many approaches for finding optimal multidimensional quadrature rules, some reaching extreme levels of mathematical sophistication, but these are beyond the scope of the present treatment. See [6] for an example of a numerical extension of Gaussian quadrature to higher dimensions.

5.7.1 Monte Carlo Integration

Numerical techniques that use randomness to estimate a quantity along with its statistical properties are broadly referred to as Monte Carlo methods. Monte Carlo methods are used to solve differential equations, model physical systems with fluids or interacting materials and structures, analyze complex business and finance problems, and have been applied in a wide range of other fields. These methods range in sophistication from simple estimators that can be easily implemented in a few lines of code to sophisticated mean field or Markov chain models used in physics and mathematics. Monte Carlo methods are used when a physical model is too complicated to solve using direct or deterministic techniques.

To implement a Monte Carlo method, a probability distribution function (PDF) is defined for inputs to a system or calculation, inputs are generated using a pseudorandom algorithm based on the PDF, a model or formula is evaluated with each set of inputs, and the statistics of the model outputs are combined to solve a problem.

The Monte Carlo approach can be used to evaluate integrals numerically. Using the fundamental theorem of expectation, an integral can be expressed as a statistical expectation. If x is a random variable with uniform PDF $f(x) = 1$ on the interval $[0, 1]$,

$$I = \int_0^1 g(x)f(x)dx = \mathrm{E}[g(x)] \tag{5.65}$$

where $\mathrm{E}[g(x)]$ is the expectation of the random variable $g(x)$. We can estimate the expectation using samples of the random variable x by computing the mean of $g(x)$ applied to each sample, so that

$$I \simeq \frac{1}{N}\sum_{n=1}^{N} g(x_n) \tag{5.66}$$

This summation is similar to the midpoint rule on the interval $[0, 1]$ with N sample points and integration weight $1/N$, except that the sample points are not evenly spaced. If the interval of integration is different from $[0, 1]$, the sample points must be scaled to be uniformly distributed on the interval of integration, and the Jacobian of the scaling transformation must be included in (5.66).

For simple 1-D integrals such as (5.65) the Monte Carlo method typically requires more samples for a given accuracy than the deterministic integration methods considered previously in this chapter. Improved performance can be achieved by using variants of the basic Monte Carlo integration method such as antithetic variates, importance sampling, or recursive stratified sampling.

5.8 MATLAB's Built-in Numerical Integration Functions

Many of the quadrature rules discussed in this chapter are available as MATLAB library functions. Several more efficient generalizations of these basic quadrature rules are also implemented. MATLAB integration routines include the following:

MATLAB Integration Functions

`trapz.m`: Evaluates the trapezoidal rule for a vector of function values, assuming a unit spacing between nodes.

`quadgk.m`: Adaptive Gauss–Kronrod quadrature. Additional integration points are added until the change between successive values of the integral are below a given tolerance. Gauss–Kronrod quadrature is a modification of Gaussian quadrature for which the set of nodes for a higher order rule includes points from a lower order rule. This is a type of nested quadrature rule and is more efficient for adaptive implementation because function evaluations from earlier steps can be reused as the order of the quadrature rule becomes larger.

`integral.m`: Newer implementation of adaptive Gauss–Kronrod quadrature. Allows vector-valued functions.

`integral2.m` and `integral3.m` generalize the 1-D integration method to double and triple integrals.

When using these built-in integration routines, an essential MATLAB feature is the function handle (see `help function_handle`). The `@` symbol returns a handle to a named MATLAB function, and the handle can be passed as an argument to an integration routine. This allows different functions to be integrated numerically without having to edit the integration function script.

Problems

5.1 Write a script to integrate a function using the extended Euler (left or right endpoints), midpoint, trapezoid, and Simpson's quadrature rules. Use the `@(x)` command to create a function handle that can be passed to each integration routine. Integrate the function $\cos(5x)$ over the interval $[0, 1]$. Plot the error for each rule as a function of the number of integration points. Use a `loglog` plot to show power-law curves as straight lines. With respect to the integration step size h, what is the order of convergence of the error for each rule?

5.2 Implement an algorithm to compute the weights and nodes of the Gauss–Legendre quadrature rule for arbitrary order. To find the nodes, use Newton's method to compute the zeros of a Legendre polynomial of arbitrary order. To find the weights, the derivative of the Legendre polynomial is required. This can be found using a recursion relation for the derivative of a Legendre polynomial in terms of a linear combination of two Legendre polynomials. Verify the algorithm by comparison to tabulated results.

5.3 To the plot from Problem 5.1, add an error curve for the Gauss–Legendre integration rule implemented in Problem 5.2. How does the convergence rate of the Gauss–Legendre quadrature rule with an increasing number of nodes compare to that of the Newton–Cotes rules?

5.4 For the integration rules in Problems 5.1 and 5.3, change the integrand to $\log|x - \pi/4|$ and plot the error versus the number of integration points for each rule. Explain the results.

5.5 Consider the integral

$$I = \int_0^1 x^2 \cos(10x^2)\, dx$$

(a) How many quadrature points are required to accurately evaluate this and other integrals with a similar form using the midpoint rule? Why? (b) Is there a quadrature rule that could accurately evaluate this integral using a few points or only one point? Why or why not?

5.6 A short linear antenna of length l can be modeled as a Hertzian dipole with radiated far field

$$\overline{E}(\overline{r}) \simeq jk\eta \frac{e^{-jkr}}{4\pi r}\hat{\theta} I_0 l \sin\theta \tag{5.67}$$

(a) Use the Gauss–Legendre quadrature method for integration over a sphere to compute the power radiated by the dipole by integrating the power flux density $S_r = |\overline{E}|^2/(2\eta)$. Compare the result to the analytical solution

$$P_{\text{rad}} = \eta \frac{(kI_0 l)^2}{12\pi} \tag{5.68}$$

(b) Compare the accuracy of the Gauss–Legendre method to the product midpoint rule over a sphere with the same number of integration points.

5.7 Implement the Lanczos method for computing weights and nodes of Gaussian quadrature rules for an arbitrary weight function.

(a) Check your code using $w(x) = 1$ and the weights and nodes for Gauss–Legendre quadrature from Problem 5.2. For Gauss–Legendre quadrature, the region of integration must be set to a positive interval such as $0 \leq x \leq 2$ so that the diagonal elements of the A matrix are positive.

(b) Change the weight function to $-\ln(x)$, $0 \leq x \leq 1$. Generate N node quadrature rules for a few values of N and integrate the function $-\ln(x)\sin(5x)$ from 0 to 1 using the quadrature rule. Compare to exact results (which can be obtained using the MATLAB symbolic toolbox `int` command or a table of integrals). How fast does the accuracy increase with N? How do these results compare with the results from Problem 5.4?

5.8 Evaluate the integral

$$I = \int_0^1 x^{-1/2}\, dx$$

using (a) the midpoint rule and (b) the transformation given by (5.59), with the midpoint rule in the transformed domain. Compare the accuracy of the two approaches.

5.9 Implement product integration rules to evaluate

$$I = \int_0^{\pi} \int_0^{\pi} \sin\left(x^2 + y^2\right) dx\, dy$$

based on (a) the midpoint rule and (b) the extended Simpson's rule. How many integration points are required to achieve four digits of accuracy?

5.10 Apply the Monte Carlo integration method to the integral in Problem 5.9. Compare the performance of the method to the product midpoint rule by plotting the error in the integral versus the number of sample points or integration points. Use a loglog scale for the plot.

5.11 Apply the Monte Carlo integration method to the N-dimensional integral

$$I = \int_0^1 dx_1 \int_0^1 dx_2 \cdots \int_0^1 dx_N \left(x_1^2 + x_2^2 + \cdots + x_N^2\right) \tag{5.69}$$

(a) Compare the result to the analytical value of the integral for increasing dimensions N. Plot the error as a function of the number of sample points for $N = 3$, $N = 4$, $N = 5$, and so on. (b) Compare the accuracy of the Monte Carlo integration method to the product midpoint rule for increasing dimensions. (c) Is there a dimension for which the Monte Carlo is more accurate than the midpoint rule for the same number of integration points?

References

[1] P. Davis and P. Rabinowitz, *Methods of Numerical Integration*. New York, NY: Academic Press, 1975.

[2] A. H. Stroud and D. Secrest, *Gaussian Quadrature Formulas*. Englewood Cliffs, NJ: Prentice-Hall, 1966.

[3] J. Ma, V. Rokhlin, and S. Wandzura, "Generalized Gaussian quadrature rules for systems of arbitrary functions," SIAM J. Numer. Anal., vol. 33, no. 3, pp. 971–996, 1996.

[4] R. A. Sack and A. F. Donovan, "An algorithm for Gaussian quadrature given modified moments," Numer. Math., vol. 18, pp. 465–478, 1971/72.

[5] S. Wandzura, "Accuracy in computation of matrix elements of singular kernels," in *11th Annual Review of Progress in Applied Computational Electromagnetics*, Naval Postgraduate School, vol. II, Monterey, CA, 1995, pp. 1170–1176.

[6] E. K. Ryu and S. P. Boyd, "Extensions of gauss quadrature via linear programming," Found. Comput. Math., vol. 15, no. 4, pp. 953–971, 2015.

Chapter 6
Integral Equations and the Method of Moments

The partial differential equations (PDEs) that are used to model the behavior of physical systems can often be transformed into integral equations. This provides an alternative basis for the development of numerical solution algorithms. In this chapter, we will develop numerical methods for the solution of the integral equations of electromagnetic radiation and scattering. We will also consider the advantages and disadvantages of integral equation methods relative to finite difference methods.

Integral operators and integral equations have wide-ranging applications and a vast literature. We will focus in this chapter on the integral equations of electromagnetic radiation and scattering and numerical methods for solving integral equations. Many books on the topic are available, including [1–4].

6.1 Integral Operators

The differential equations we have solved with the finite difference method have been defined using derivative operators. Integral operators and integral equations can also be used to model physical systems. The theory of integral operators is a broad field of mathematics, and there are many parallels between integral and differential operators. To emphasize the connections between derivatives and integral operators, in the mathematics literature integral operators are referred to as pseudodifferential operators. Derivative operators amplify changes in a function, so the derivative of a function becomes less smooth. Integral operators generally work in the opposite direction and increase the smoothness of a function, although when singular functions are involved, this simple intuition breaks down.

The general form for a one-dimensional (1-D) integral operator $\mathscr{L}$ is

Integral Operator

$$\mathscr{L}u(x) = \int_a^b \underbrace{K(x,y)}_{\text{Kernel}} u(y)\, dy \tag{6.1}$$

The function $K(x,y)$ is a fixed function of two variables that defines the integral operator $\mathscr{L}$ and is referred to as the kernel of the integral operator. Since the variable y is integrated out, the right-hand side is a function of x. The operator $\mathscr{L}$ transforms the "input" function $u(x)$ into a new "output" function of x that is given by the integral on the right-hand side of (6.1).

If the kernel is shift invariant, so that $K(x,y) = K(x-y)$, then $\mathscr{L}$ becomes a convolution operator, and the integral equation can be written using the convolution operator as $\mathscr{L}u = K \times u$. If the upper integration limit is x instead of a constant, we obtain a Volterra-type integral operator. Volterra integral equations are important for certain physical problems but do not usually arise in electromagnetic (EM) theory.

6.1.1 First and Second Kind Integral Equations

Given an integral operator $\mathscr{L}$, an integral equation arises if the function u is unknown and the function $\mathscr{L}u$ is given. Integral equations with smooth, square integrable kernels can be classified into two types:

Fredholm Classification

1. Fredholm integral equation of the first kind:
$$\mathscr{L}u = f \tag{6.2}$$
2. Fredholm integral equation of the second kind:
$$\mathscr{L}u + u = f \tag{6.3}$$
which can also be written as $(\mathscr{L} + \mathscr{I})u = f$, where $\mathscr{I}$ is the identity operator.

In these equations, u represents an unknown function and f is given. The second term in (6.3) is referred to as an identity term. This term makes the overall operator $\mathscr{L} + \mathscr{I}$ more like the identity operation. With a more identity-like behavior, second-kind integral equations are usually easier to solve numerically than first-kind equations.

In the integral equations associated with electromagnetic boundary value problems, u often represents an unknown current source and f is a given field. The relationship between the current and field in EM problems is given by a Green's function. Since Green's functions are generally singular, the Fredholm classification above breaks down and the relative ease of solving first and second-kind integral equations requires a deeper consideration.

6.1.2 Solving Integral Equations Numerically

An integral equation of the form

$$\mathscr{L}u = f \tag{6.4}$$

can be thought of as an infinite-dimensional generalization of the finite linear system

$$Ax = b \tag{6.5}$$

To solve an integral equation numerically, we transform the integral equation in (6.4) into a linear system of the form of (6.5) that approximates the original integral equation. The continuous operator $\mathscr{L}$ becomes a matrix A, the vector b represents samples of the given right-hand side f, and the unknown vector x represents samples or coefficients that give an approximation to the solution u.

The values in the vector x are sometimes referred to as degrees of freedom. Since it is generally not possible to represent a continuous function exactly with a finite number of degrees of freedom, the vector x leads to a numerical solution $\hat{u}$ that is not quite equal to the true solution u to the original integral equation. The magnitude of the error in the numerical solution and the factors that cause solution error will be discussed later in this chapter. Usually, the larger the size of the matrix A, the more accurate the numerical solution. A larger linear system requires more computation time to solve, so there is a direct correlation between computational cost and solution accuracy.

The process of transforming a continuous integral equation into a finite linear system is referred to as discretization. Discretizing an integral equation is similar to the process that we covered in Chapter 4 for discretizing a differential equation to form a finite number of difference equations. An integral equation is typically discretized using the method of weighted residuals or method of moments (MoM), which will be treated in Section 6.3.

6.1.3 Smooth Kernels and Operator Conditioning

To gain some intuition into the numerical properties of first- and second-kind integral equations, we will consider the behavior of the integral operator when applied to harmonic or sinusoidal functions. If u is rapidly oscillating, then the product of u and the kernel function is also rapidly oscillating. When the integral is performed, the positive and negative portions of the integrand cancel, and the value of the integral in (6.1) is small.

Because $\mathscr{L}u$ is small for rapidly oscillating functions, the operator $\mathscr{L}$ has eigenvalues that are small in magnitude and is numerically similar in behavior to a singular operator. If we think of u in terms of its Fourier series representation, the integral operator reduces the amplitudes of the rapidly varying components, which means that $\mathscr{L}u$ is a smoother function than u, and $\mathscr{L}$ might be referred to as a smoothing operator. Such an operator is referred to as "ill-conditioned" (see Section 7.5.3 for a quantitative measure for the degree of ill-conditioning). Ill-conditioning typically makes the integral equation more difficult to solve numerically than an integral equation with a well-conditioned operator.

Second-kind integral equations, on the other hand, are generally well conditioned. For rapidly oscillating functions and an operator $\mathscr{L}$ with a smooth kernel, $(I + \mathscr{L})u \simeq u$, so that the second-kind operator behaves like a perturbation of the

Table 6.1 Numerical solutions for formally first-kind integral equations

Kernel	Discretization	Linear system	Example
Smooth	Easy	Hard to solve	Blurring operator (Gaussian)
Singular	Hard	Easy to solve	Electric field integral equation (Green's function)

identity, and the eigenvalues of the operator are clustered around one. This makes a second-kind integral equation easier to solve numerically than a first-kind integral equation.

6.1.4 Singular Kernels and Conditioning for Non-Fredholm Operators

For the integral equations of electromagnetic radiation and scattering, this simple intuition breaks down, because the kernel functions that appear in the integral equations associated with radiation and scattering are singular and the Fredholm classification does not apply directly. In fact, if the kernel singularity is strong enough, an equation that appears to be first kind in its form actually behaves like a second-kind equation. If the kernel function $K(x,y)$ is the sum of a smooth function $K_0(x,y)$ and a delta function, so that

$$K(x,y) = K_0(x,y) + \delta(x-y) \tag{6.6}$$

then a formally first-kind equation with this kernel is really a second-kind integral equation.

For the integral equations of electromagnetics, the operators are in a sense "in between" the first- and second-kind classifications. For the electric field integral equation that will be introduced shortly, the kernel singularity is logarithmic for two-dimensional (2-D) field problems and of the form $1/r$ for three-dimensional (3-D) problems. A formally first-kind integral equation with a singular kernel behaves more like a second-kind integral equation. The resulting numerical method can be well conditioned. The price paid for the improved conditioning with singular kernels is that the integrals that must be evaluated to discretize the integral equation and solve it numerically are more challenging. These observations are summarized in Table 6.1.

6.2 Integral Equations in Electromagnetics

Maxwell's equations can be transformed into an integral equation using a Green's function. As we saw previously in Section 4.6.2.1, a Green's function is the solution to a boundary value problem with a point source. The basic procedure is to express the scattered or radiated fields in a problem as an integral of an unknown current source with the Green's function as the kernel, and then to match the scattered fields to an incident field using a boundary condition. There are two main types of integral equations in EM theory:

Types of Integral Equations in Electromagnetics

Surface integral equations typically arise when the scatterer or radiating body consists of a perfect electric conductor (PEC) or a homogeneous dielectric or conducting material. The unknown quantity in the integral equation is a surface current.

Volume integral equations are associated with inhomogeneous dielectric or conducting objects. The unknown quantity is a volume current.

Volume integral equations require samples of the unknown source on the surface and interior of the scatterer. The advantage of surface integral equations is that the unknown current source exists only on the surface of the scatterer, so the grid or mesh representation is one dimension smaller than the dimension of the problem. This leads to fewer unknowns than a volumetric method.

Finite difference methods lead to sparse matrices or update equations that depend only on nearby sample points, whereas integral equations lead to dense matrix representations, because a point source radiates fields that reach all other points on the scatterer. For a given number of unknowns, a dense matrix has a memory requirement on the order of N^2, whereas the memory requirement for a sparse matrix algorithm is of order N. The numerical solution of the large, dense matrices arising from integral equation methods can be computationally difficult. For large enough N, even generating and storing the elements of the dense matrix is challenging.

Most often, integral equations are used for time-harmonic or frequency-domain problems. Problems with pulsed or other broadband excitations can be analyzed using a frequency-domain algorithm by expanding the excitation as a sum of Fourier modes and analyzing each frequency component separately. Time-domain integral equations are difficult to solve numerically due to instability problems, but practical time-domain integral equation-based algorithms have been developed [5].

Integral equation-based numerical algorithms can be implemented for 2-D and 3-D radiation and scattering problems. For a 2-D problem, the fields, sources, and material structures are independent of one coordinate, so only the fields in a 2-D cross section must be solved for numerically (see Section 2.13.2). With electromagnetic integral equations, we must distinguish between the dimensionality of the physical problem and the dimensionality of the domain of the integral equation on which the unknown current is defined. If the integral equation is volumetric, then the domain on which the unknown current is defined has the same dimension as the radiation or scattering problem. For surface integral equations, the domain has dimension one less than the dimensionality of the radiation or scattering problem. If the unknown

Table 6.2 Boundary value problems, mesh dimension, and integral equation formulations

BVP	Mesh	Description	Kernel
2-D	1-D	2-D surface integral equation (2-D SIE)	EFIE, MFIE, CFIE (TM/TE)
2-D	2-D	2-D volume integral equation (2-D VIE)	EFIE (TM/TE)
3-D	1-D	Thin wire integral equation	Pocklington, Hallén
3-D	2-D	Surface integral equation (SIE)	EFIE, MFIE, CFIE
3-D	3-D	Volume integral equation (VIE)	EFIE

current flows on a thin wire, then the dimensionality of the domain of the integral equation can be further reduced to one by approximating the distribution of current azimuthally around the wire as constant, leading to a 1-D integral equation. 1-D integral equations are discussed in Section 6.2.5. Some of the variety of combinations of integral equation formulations, problem dimensions, and mesh dimensions are listed in Table 6.2. Many other variants of these integral equation formulations have been presented in the literature.

For a 3-D problem, sources and fields depend on all three spatial coordinates. Three-dimensional problems are of greatest interest in modern electromagnetics applications, but implementing a numerical algorithm for 3-D problems is typically much more difficult than for 2-D problems. For this reason, we will first consider 2-D integral equations, and then we will turn to 3-D integral equations in Section 6.8.

6.2.1 Electric Field Integral Equation, 2-D Transverse Magnetic Polarization (TM-EFIE)

We will first consider the analysis of scattering by a 2-D PEC object that has infinite extent in the z direction, with a TM^z polarized incident field. The unknown quantity in the integral equation is the current induced on the scatterer by an incident wave. Currents induced on a PEC object are confined to an infinitesimally thin surface layer on the surface of the scatterer. Because the unknown current on the scatterer is a surface current, the electric field integral equation for a PEC object is a surface integral equation.

The total field around the scatterer S is the sum of the incident and scattered field, so that the electric field intensity is $E_z = E_z^{\mathrm{i}} + E_z^{\mathrm{s}}$. If the scatterer is translationally invariant in the z direction, no depolarization occurs, and the incident and scattered fields are both oriented in the z direction, as illustrated in Figure 6.1. The boundary condition for the electric field intensity at the surface S of the object is

$$\hat{n} \times (\hat{z}E_z^{\mathrm{i}} + \hat{z}E_z^{\mathrm{s}})\big|_S = 0 \tag{6.7}$$

For a cylindrical object, the surface normal vector $\hat{n}$ is always orthogonal to $\hat{z}$, and the boundary condition reduces to

$$E_z^{\mathrm{i}}(\overline{\rho}) + E_z^{\mathrm{s}}(\overline{\rho}) = 0 \tag{6.8}$$

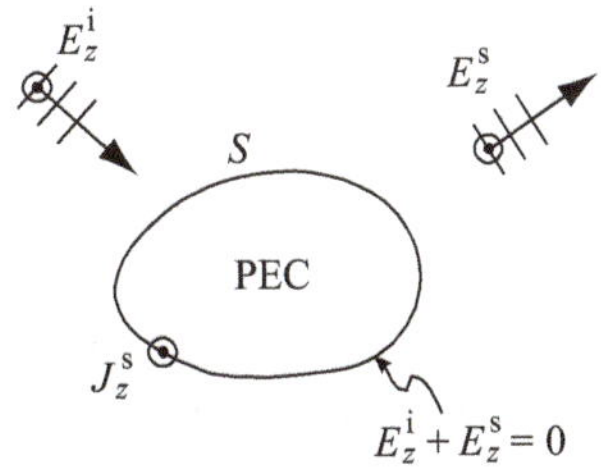

Figure 6.1 Incident field, scattered field, boundary condition, and surface current density associated with the TM-EFIE for a PEC scatterer

for all points $\overline{\rho}$ on the scatterer surface S, where $\overline{\rho}$ represents a point (x, y) in the $z = 0$ plane.

The incident field induces an electric current

$$\overline{J}(\overline{\rho}) = J_z(\overline{\rho})\hat{z} \tag{6.9}$$

on the surface of the scatterer. This surface current, if it were impressed in free space without the scatterer present, would radiate the scattered field E_z^s. Using the 2-D free-space radiation integral (4.137), the scattered field can be expressed in terms of the current induced on the scatterer as

$$E_z^s(\overline{\rho}) = -\frac{k\eta}{4} \int_S d\overline{\rho}' \, H_0^{(2)}(k|\overline{\rho} - \overline{\rho}'|) J_z(\overline{\rho}') \tag{6.10}$$

Substituting the radiation integral into the boundary condition leads to the first-kind integral equation

2-D TM-EFIE for a PEC Scatterer

$$\frac{k\eta}{4} \int_S d\overline{\rho}' \, \underbrace{H_0^{(2)}(k|\overline{\rho} - \overline{\rho}'|)}_{\text{Green's function}} \; \underbrace{J_z(\overline{\rho}')}_{\substack{\text{Unknown} \\ \text{source}}} = \underbrace{E_z^i(\overline{\rho})}_{\substack{\text{Known} \\ \text{incident field}}} \tag{6.11}$$

where the point $\overline{\rho}$ lies on the scatterer surface S. Since the fields and surface current do not vary in the z direction, the points $\overline{\rho}$ and $\overline{\rho}'$ are confined to the x–y plane, and the surface S is a path or contour along the cross section of the scatterer. This is the 2-D electric field integral equation for the transverse magnetic (TM) polarization, or TM-EFIE.

As mentioned earlier, the integral equations used in EM modeling are most often formulated in the frequency domain. The right-hand side of the TM-EFIE in (6.11) is a phasor representation of the incident wave. The surface current source J_z is a phasor as well. If the incident field is a plane wave, then it can be expressed in phasor form as in (4.105). Unlike the FDTD algorithm developed in Chapter 4, which deals with time-domain fields, both the right-hand side and the current source in the TM-EFIE are in general complex.

This integral equation can also be written in the operator form

$$\mathscr{L} J_z = E_z^{\mathrm{i}} \tag{6.12}$$

where the TM-EFIE integral operator is defined by

$$\mathscr{L} J_z = \frac{k\eta}{4} \int_S d\overline{\rho}' \, H_0^{(2)}(k|\overline{\rho} - \overline{\rho}'|) J_z(\overline{\rho}') \tag{6.13}$$

To solve the TM-EFIE numerically, the operator equation (6.12) is transformed into a linear system. The integral operator $\mathscr{L}$ becomes a matrix, the right-hand side E_z^{i} becomes a vector, and the linear system must be solved for an unknown vector that represents an approximate solution for the current J_z.

As discussed previously, the conditioning of the matrix in (6.5) is influenced by the kernel of the integral operator. For the EFIE, the kernel of the integral equation is the first-kind Hankel function of order zero, $H_0^{(2)}(x)$. Because the Hankel function is singular when its argument is zero, when we discretize the integral operator to solve the integral equation numerically as a linear system or matrix equation, the matrix discretization of the integral operator $\mathscr{L}$ is better conditioned and therefore easier to invert than if the kernel were a smooth function. When forming the matrix representation of the operator, however, we will need to compute integrals with singular integrands, which, as shown in Chapter 5, are generally more difficult than integrals of smooth functions.

6.2.2 Electric Field Integral Equation, 2-D Transverse Electric Polarization (TE-EFIE)

The electric field integral equation for a 2-D PEC object can also be formulated for a transverse electric (TE) polarized incident field. For the TE^z polarization, the current on the surface S of the PEC object is transverse to z and can be written in the form

$$\overline{J}(\overline{\rho}) = \hat{t} J_t(\overline{\rho}) \tag{6.14}$$

where $\hat{t}$ is a tangent vector to the scatterer surface S in the x–y plane. The TE-EFIE is

$$\mathscr{N} J_t = E_t^{\mathrm{i}} \tag{6.15}$$

where E_t^{i} is the tangential component of the incident electric field. The integral operator is given by [6]

$$\begin{aligned}\mathscr{N} J_t = {} & \frac{k\eta}{4} \hat{t} \cdot \int_S d\overline{\rho}' \, H_0^{(2)}(k|\overline{\rho} - \overline{\rho}'|) \hat{t}' J_t(\overline{\rho}') \\ & + \frac{1}{4k\eta} \hat{t} \cdot \nabla \int_S d\overline{\rho}' \, H_0^{(2)}(k|\overline{\rho} - \overline{\rho}'|) \nabla' \cdot [\hat{t}' J_t(\overline{\rho}')]\end{aligned} \tag{6.16}$$

where $\hat{t}$ is a unit vector tangential to S and lying in the x–y plane. The divergence operator ∇' inside the integrand acts on the primed coordinates $\overline{\rho}'$ whereas the divergence operator outside the integrand acts on the unprimed coordinates. Since the TE-EFIE operator includes derivative operators, the TE-EFIE is an integro-differential equation.

One difficulty with the TE-EFIE is that one of the derivatives in the second term on the right-hand side of (6.16) acts on the Hankel function and increases the strength of the singularity of the kernel. When the integral equation is discretized, the derivative can be moved using integration by parts so that it no longer acts on the kernel. This reduces the effective singularity of the integrand and makes the TE-EFIE easier to solve numerically.

6.2.3 Magnetic Field Integral Equation

Based on the magnetic field boundary condition, we can also derive a magnetic field integral equation (MFIE) for the TM and TE polarizations and for 3-D problems. For 2-D problems with a TM-polarized incident field, the TM-MFIE has the form

$$\tfrac{1}{2}J_z + \mathscr{M}_{\mathrm{TM}}J_z = H_t^{\mathrm{i}} \tag{6.17}$$

where H_t^{i} is the tangential component of the incident magnetic field. The integral operator is

$$\mathscr{M}_{\mathrm{TM}}J_z = \frac{jk}{4}\int_S d\overline{\rho}' \cos\psi H_1^{(2)}(k|\overline{\rho}-\overline{\rho}'|)J_z(\overline{\rho}') \tag{6.18}$$

where J_z represents the longitudinal component of the current density and ψ is the angle between the surface normal vector at $\overline{\rho}$ and the vector $\overline{\rho}-\overline{\rho}'$.

For the TE polarization, the TE-MFIE is

$$\tfrac{1}{2}J_t + \mathscr{M}_{\mathrm{TE}}J_t = H_z^{\mathrm{i}} \tag{6.19}$$

where the integral operator is

$$\mathscr{M}_{\mathrm{TE}}J_t = \frac{jk}{4}\int_S d\overline{\rho}' \cos\psi' H_1^{(2)}(k|\overline{\rho}-\overline{\rho}'|)J_t(\overline{\rho}') \tag{6.20}$$

In this expression, ψ' is the angle between the surface normal vector at $\overline{\rho}'$ and the vector $\overline{\rho}-\overline{\rho}'$. The MFIE for both the TE and TM polarizations is a second-kind integral equation and in some cases can yield more accurate numerical results than the EFIE [7]. The MFIE will arise later in this book as part of the finite element method–boundary element method (FEM–BEM) method developed in Section 8.5.

6.2.4 Combined Field Integral Equation

In some cases, it is useful to form a linear combination of the EFIE and MFIE to obtain a combined field integral equation (CFIE). For closed scatterers, the contour S encloses a two-dimensional region that is inside the scatterer cross section. Since a PEC object is impenetrable to time-varying fields, there is no interaction between the external illuminating field and the interior of S. At certain frequencies, however, the cavity inside S becomes resonant and could support a nonzero mode if there were a source inside the cavity and S were a hollow PEC shell. At these frequencies, the integral operator in the EFIE has a zero eigenvalue. The zero eigenvalue is a mathematical artifact of the shape of the surface S, so this phenomenon occurs whether the scatterer is hollow or solid. Even though no energy from the incident field can penetrate inside the scatterer, the zero eigenvalue causes the matrix discretization of the EFIE integral operator to become singular or ill-conditioned, and numerical results can be inaccurate. This is known as an internal resonance. The MFIE also suffers from internal resonances as well. Numerical difficulties associated with internal resonances can be overcome using the CFIE formulation. With the CFIE, the internal resonance frequencies are shifted off of the real axis, and numerical solution algorithms for closed scatterers become stable.

For 2-D scattering problems, the CFIE is

$$\left[\alpha\mathscr{L} + (1-\alpha)\eta(\tfrac{1}{2} + \mathscr{M}_{\mathrm{TM}})\right]J_z = \alpha E_z^{\mathrm{i}} + (1-\alpha)\eta H_t^{\mathrm{i}} \tag{6.21}$$

for the TM polarization. For the TE polarization,

$$\left[\alpha\mathscr{N} + (1-\alpha)\eta(\tfrac{1}{2} + \mathscr{M}_{\mathrm{TE}})\right]J_t = \alpha E_t^{\mathrm{i}} + (1-\alpha)\eta H_z^{\mathrm{i}} \tag{6.22}$$

The parameter α is known as the CFIE combination coefficient and lies between zero and one. A common choice for the combination coefficient is $\alpha = 0.2$. Reference [8] cites [9] as the source of this recommendation. It can be shown that $\alpha = 0.2$ minimizes the error in the computed 2-D scattering amplitude for a circular cylinder, independent of the cylinder radius and the number of unknowns used when discretizing the CFIE [10].

In this chapter, we will focus on numerical solution algorithms for the TM-EFIE and the 3-D EFIE, although the procedure for developing solution methods for the MFIE and CFIE are very similar to the approach developed here for the EFIE.

6.2.5 Thin-Wire Integral Equations

For structures made up of thin wires, such as a dipole antenna, another class of 1-D integral equations can be applied. Because solid conductors can be approximated by a wire grid, these thin-wire integral equations have a historical significance and have been used to develop quite powerful analysis tools, such as Numerical Electromagnetic Code (NEC) and its derivatives. Newer software tools rely on 3-D integral equations, but we will review the thin-wire integral equations here for completeness.

6.2.5.1 Pocklington's Integral Equation

Pocklington's integral equation relates the current distribution $I(z)$ as a function of position on a wire through an integral relationship to an electric field $\overline{E}^{\mathrm{i}}$ that is incident on the wire. To derive Pocklington's integral equation, we will begin with the magnetic vector potential radiated by a z-directed current source. The magnetic vector potential $\overline{A}$ is defined by the relationship [6]

$$\overline{B} = \nabla \times \overline{A} \tag{6.23}$$

It can be shown that the electric field is obtained from the magnetic vector potential by

$$\overline{E} = -j\omega\overline{A} - \frac{j}{\omega\mu\varepsilon}\nabla\nabla\cdot\overline{A} \tag{6.24}$$

Since the direction of the magnetic vector potential is the same as that of the current, $\overline{A}$ will have only a z component. In terms of the magnetic vector potential, the electric field intensity is then

$$E_z = -j\omega A_z - \frac{j}{\omega\mu\varepsilon}\frac{\partial^2 A_z}{\partial z} \tag{6.25}$$

The magnetic vector potential is given by the convolution of the free-space scalar Green's function with the current source, so that

$$\overline{A} = \mu \int d\overline{r}' \frac{e^{-jkR}}{4\pi R}\overline{J}(\overline{r}') \tag{6.26}$$

where $R = |\overline{r} - \overline{r}'|$. Combining this result with (6.25) leads to

$$\mu \int d\overline{r}' J_z(\overline{r}')\left[\frac{\partial^2}{\partial z^2} + k^2\right]\frac{e^{-jkR}}{4\pi R} = j\omega\varepsilon\mu E_z(\overline{r}) \tag{6.27}$$

On the surface of the conductor, the electric field boundary condition requires that

$$E_z + E_z^{\mathrm{i}} = 0 \tag{6.28}$$

where E_z^{i} is the incident electric field from the source at the feed gap that excites the antenna. This leads to the integral equation

$$\int d\overline{r}' J_z(\overline{r}')\left[\frac{\partial^2}{\partial z^2} + k^2\right]\frac{e^{-jkR}}{4\pi R} = -j\omega\varepsilon E_z^{\mathrm{i}}(\overline{r}) \tag{6.29}$$

We will now assume that the wire is thin, so that the current distribution does not vary significantly azimuthally around the wire. The current density can then be approximated as

$$J_z(\overline{r}) = \frac{I_z(z)}{2\pi a}\delta(\rho - a) \tag{6.30}$$

where a is the wire radius. Substituting this into (6.29) leads to

$$\begin{aligned} -j\omega\varepsilon E_z^{\mathrm{i}}(\overline{r}) &= \int_0^\infty \int_0^{2\pi} \int_{-l/2}^{l/2} \frac{I_z(z')}{2\pi a}\delta(\rho' - a)\left[\frac{\partial^2}{\partial z^2} + k^2\right]\frac{e^{-jkR}}{4\pi R}\rho'\,d\rho'\,d\phi'\,dz' \\ &\simeq \int_{-l/2}^{l/2} I_z(z')\left[\frac{\partial^2}{\partial z^2} + k^2\right]\frac{e^{-jkR}}{4\pi R}\,dz' \end{aligned}$$

where

$$\begin{aligned} R &= \sqrt{(x - x')^2 + (y - y')^2 + (z - z')^2} && (6.31) \\ &\simeq \sqrt{a^2 + (z - z')^2} && (6.32) \end{aligned}$$

Working the derivatives in this expression provides the final form of Pocklington's integral equation,

$$\int_{-l/2}^{l/2} I_z(z')\left[(1 + jkR)(2R^2 - 3a^2) + (kaR)^2\right]\frac{e^{-jkR}}{4\pi R^5}\,dz' = -j\omega\varepsilon E_z^{\mathrm{i}}(z) \tag{6.33}$$

One difficulty with this integral equation is the strong singularity of the kernel due to the R^{-5} term. To avoid dealing with this singularity, alternate integral formulations are available.

6.2.5.2 Hallén's Integral Equation

Another integral equation for thin-wire structures can be derived from the Helmholtz equation for the z component of the magnetic vector potential. On the surface of the conductor away from the driving source, A_z satisfies the PDE

$$\frac{\partial^2}{\partial z^2}A_z + k^2 A_z = 0 \tag{6.34}$$

In this equation, A_z represents the total field, including the driving source rather than only the radiated field as in the previous derivation. This differential equation has the solution

$$A_z(z) = -j\sqrt{\mu\varepsilon}[A\cos(kz) + B\sin(k|z|)] \tag{6.35}$$

where we have enforced the symmetry of the problem about $z = 0$. Using the Lorenz gauge relationship [6], the magnetic vector potential is related to the scalar electric potential by

$$\nabla \cdot \overline{A} = -j\omega\mu\varepsilon\phi \tag{6.36}$$

Since $\overline{A}$ depends only on z, the divergence evaluates to

$$\frac{\partial A_z}{\partial z} = \begin{cases} -j\sqrt{\mu\varepsilon}k[-A\sin(kz) + B\cos(k|z|)] & z > 0 \\ -j\sqrt{\mu\varepsilon}k[-A\sin(kz) - B\cos(k|z|)] & z < 0 \end{cases} \tag{6.37}$$

The electric potential therefore has the form

$$\phi = \frac{1}{-j\omega\mu\varepsilon}\begin{cases} -j\sqrt{\mu\varepsilon}k[-A\sin(kz) + B\cos(k|z|)] & z > 0 \\ -j\sqrt{\mu\varepsilon}k[-A\sin(kz) - B\cos(k|z|)] & z < 0 \end{cases} \tag{6.38}$$

To proceed, we must make an assumption about the source that excites the wire structure. We will model a dipole antenna with an infinitesimal feed gap at $z = 0$. The phasor voltage V_i is impressed across the feed gap. This is called a delta-gap source. Due to the source, ϕ must jump by V_i at $z = 0$, so that

$$V_i = \frac{1}{-j\omega\mu\varepsilon}\left[\left.\frac{\partial A_z}{\partial z}\right|_{z>0} - \left.\frac{\partial A_z}{\partial z}\right|_{z<0}\right] \tag{6.39}$$

Combining this with (6.38) leads to the result

$$-j\omega\mu\varepsilon V_i = -j\sqrt{\mu\varepsilon}k2B \tag{6.40}$$

from which we find that $B = V_i/2$. The radiation integral for $\overline{A}$ gives

$$A_z(z) \simeq \mu\int_{-l/2}^{l/2} I(z')\frac{e^{-jkR}}{4\pi R}\,dz' \tag{6.41}$$

Combining these expressions leads to the integral equation

$$\int_{-l/2}^{l/2} I(z')\frac{e^{-jkR}}{4\pi R}\,dz' = -\frac{j}{\eta}[A\cos(kz) + \frac{V_i}{2}\sin(k|z|)] \tag{6.42}$$

which has a less singular kernel than Pocklington's integral equation but includes an additional unknown coefficient A that must be solved for at the same time as the unknown current distribution $I(z)$.

6.3 Method of Weighted Residuals

The general approach for transforming a continuous integral equation into a finite linear system is the method of weighted residuals. As observed previously, an integral operator transforms an input function into an output function. If we can represent the input and output functions in terms of a finite number of coefficients or samples in a series expansion, then the operator can be approximated by a matrix that transforms the coefficients associated with the input function into the coefficients of the output function. The inverse of this matrix can be used to transform the sampled right-hand side into an approximation for the solution.

Suppose we want to solve the integral equation

$$\mathscr{L}u = f \tag{6.43}$$

We first choose N functions f_n, $n = 1, 2, \ldots, N$ with which to approximate the unknown function u. These are known as basis functions or expansion functions. We then write the solution u as a linear combination of the basis functions with unknown coefficients:

$$u(x) \simeq \sum_{n=1}^{N} a_n f_n(x) \tag{6.44}$$

where u is the exact solution to the integral equation. For convenience, we will introduce the notation

$$\hat{u}(x) = \sum_{n=1}^{N} a_n f_n(x) \tag{6.45}$$

for the approximate solution given by the linear combination of expansion functions.

The goal of a numerical solution method for the integral equation $\mathscr{L}u = f$ is to find a set of N numbers a_n such that the expansion (6.45) is as close to the exact solution u as possible. One way to do this is to minimize the residual error

$$r = \mathscr{L}\hat{u} - f \tag{6.46}$$

Because r is a continuous function, it must be sampled or discretized to obtain a numerical method. This can be done by forming inner products of the residual with M weighting or testing functions t_m, $m = 1, 2, \ldots, M$.

The relevant inner product here is the L_2 inner product on the function space associated with the right-hand side of the integral equation (6.43). For the surface integral equations considered in this chapter, the inner product is

$$\langle f, g \rangle = \int_S d\overline{\rho} f^*(\overline{\rho}) g(\overline{\rho}) \tag{6.47}$$

where the integral is over the surface of the scatterer. Since the fields and currents for frequency-domain integral equations are phasors, the integrand must include a complex conjugate on one of the functions. The testing functions are real-valued, so the complex conjugate can be omitted in this case.

If the residual error r were zero at every point in the range space of the integral operator, then $\hat{u}$ would be the exact solution to the integral equation. With a finite number of equations, we can only drive the residual to zero at a finite number of samples, which in this case are represented by M inner products of the residual with testing functions. Our goal with the method of weighted residuals is to choose the coefficients a_n in the approximate solution $\hat{u}$ to drive to zero the M inner products of the residual with the testing functions:

$$\begin{aligned} 0 &= \langle t_m, r \rangle \\ &= \left\langle t_m, \mathscr{L}\left(\sum_{n=1}^{N} a_n f_n\right) - f \right\rangle \end{aligned} \tag{6.48}$$

where the inner product is the L_2 inner product defined in (5.23) with weight function equal to one.

By the linearity of the operator $\mathscr{L}$, we obtain a system of M linear equations,

$$\sum_{n=1}^{N} \langle t_m, \mathscr{L} f_n \rangle a_n = \langle t_m, f \rangle, \quad m = 1, 2, \ldots, M \tag{6.49}$$

If we let A be the matrix with elements

$$A_{mn} = \langle t_m, \mathscr{L} f_n \rangle \tag{6.50}$$

and b be the vector with elements

$$b_m = \langle t_m, f \rangle \tag{6.51}$$

then (6.49) can be written in matrix form as

$$Ax = b \tag{6.52}$$

where x is a vector of the unknown coefficients a_n, and (6.50) is called the moment matrix. Solving the linear system produces a set of coefficients that can be used in (6.45) to obtain an approximate solution to the integral equation.

6.3.1 Basis Functions

Expansion and testing functions are referred to collectively as basis functions. One common choice for the expansion or testing functions is the pulse function

$$f(x) = \begin{cases} 1 & |x| \le h/2 \\ 0 & \text{otherwise} \end{cases} \tag{6.53}$$

This canonical pulse function of width h can be used to define a series of pulse functions along the scatterer surface. The scatterer is divided into many small mesh elements, and each pulse function is equal to one on one mesh element and zero on the others. For 3-D problems, mesh elements are represented by the coordinates of the vertices of triangles or other geometrical shapes, but for 2-D surface integral equations, the mesh is a 1-D contour, and it is convenient to represent the mesh using the midpoints of the elements and the element widths. If x represents a point on a contour along the surface of a cylindrical scatterer, the element centers are x_n, the element widths are h_n, and the nth element is the interval $x - h_n/2 \le x \le x + h_n/2$. Mesh elements can have different widths, but for a regular mesh, the element centers are evenly spaced and the element widths are equal, so that $h_1 = h_2 = \cdots = h_N = h$. The pulse functions on the mesh are obtained by shifting the pulse function (6.53) to each of the mesh elements according to

$$f_n(x) = \begin{cases} 1 & |x - x_n| \le h_n/2 \\ 0 & \text{otherwise} \end{cases} \tag{6.54}$$

If these pulse functions are used in expansion (6.45), the resulting approximate solution is piecewise constant. A pulse function is shown in Figure 6.2.

Another choice is the triangle function, which is zero for $x < x_{n-1}$, increases linearly to one at $x = x_n$, and is zero for $x > x_{n+1}$, as shown in Figure 6.2. For a regular discretization, the triangle functions can be written as $f_n(x) = t(x - x_n)$, where

$$t(x) = \begin{cases} 1 - |x|/h & |x| \le h \\ 0 & \text{otherwise} \end{cases} \tag{6.55}$$

This expansion function provides a piecewise linear approximation to $u(x)$.

Both the pulse function and the triangle function are interpolatory, which means that the value of the expansion function at the center of its region of support is unity. This implies that the value of the current expansion (6.45) at the mesh node points

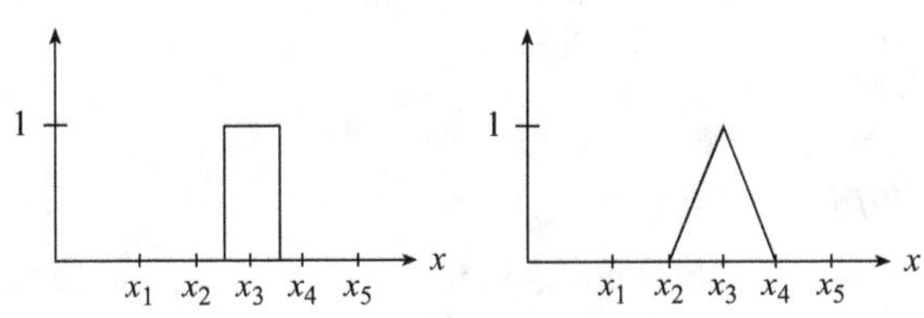

Figure 6.2 Pulse and triangle basis functions

is equal to the coefficient a_n, and the unknown coefficients represent samples of the current solution.

Higher order polynomials can also be used to increase the accuracy of the approximation in (6.45). For a smooth function, the best possible piecewise linear approximation that can be obtained using a triangle function expansion deviates less from the exact function than a pulse function or piecewise constant expansion. With quadratic basis functions, the deviation from the exact function is even smaller. Typically, choosing a set of basis functions that can more accurately represent the exact solution u to the integral equation translates to a more accurate solution from the method of weighted residuals. Higher order basis functions are an important area of research and development in the field of computational electromagnetics. Another type of basis functions mimic the oscillatory property of time-harmonic electromagnetic fields and currents. These basis functions typically have a much larger region of support than local basis functions like the pulse and triangle functions and are referred to as entire-domain basis functions.

By making various choices of the testing and expansion functions, we obtain different classes of numerical methods:

Classes of Weighted Residual Methods

Method of moments (MoM): General term for the numerical method obtained with the method of weighted residuals. The testing and expansion functions f_n and t_n may or may not be the same.

Galerkin's method: Testing and expansion functions are identical, so that $t_n = f_n$ and $M = N$.

Boundary element method (BEM): $\mathscr{L}$ is a surface integral operator. This is another name for the method of moments when it is applied to a surface integral equation.

Collocation or point matching: Testing functions are delta functions, so that $t_m(x) = \delta(x - x_m)$.

Least squares method: $t_m = \mathscr{L} f_m$.

In computational electromagnetics, the method of moments is the most common terminology for the method of weighted residuals, and the other variants can be viewed as special cases of the MoM algorithm.

6.3.2 MoM Implementation

Because of the various available options for the basis functions and other aspects of the MoM, an integral equation can be discretized in many different ways. A particular set of choices for the testing and expansion functions, the type of mesh on which the basis functions are defined, and the integration rules used to transform a continuous integral equation into a finite-dimensional linear system are referred to as a discretization scheme. A discretization scheme consists of the following components:

Components of an MoM Discretization

1. Mesh: A set of elements or patches that span the scatterer surface, typically specified in terms of a list of node points on the scatterer. For a regular discretization, the element width is often given the symbol h.
2. Expansion functions f_n on each element.
3. Testing functions t_n on each element.
4. Quadrature rules for computing the inner product integrals in (6.50) and (6.51).

As might be expected, the accuracy of the solution (6.45) obtained using the method of weighted residuals depends on all these aspects of the discretization method, including the fineness of the mesh, the smoothness of the testing and expansion functions, and the quadrature rule used to evaluate moment matrix elements. With a finer mesh, the solution is more accurate. Typically, if the basis functions are higher order polynomials, the solution also becomes more accurate. A poor quadrature rule can defeat the higher accuracy of smoother basis functions, so attention must be given to the integration rule or rules with which moment matrix elements are computed. If an approximate method for solving the linear system is used to improve the computational efficiency of the algorithm, linear system solution error can also be a factor.

6.3.3 Mesh Types

Common mesh types used with the method of weighted residuals, the method of moments, or other related numerical methods include the following:

Mesh Types

1-D surface mesh: A 1-D surface mesh consists of line segments or intervals that cover a path. The path may represent the surface of a 2-D object. The mesh elements can be described by the coordinate of the two endpoints of each element. The endpoints of each element are the mesh nodes. Alternatively, it is sometimes convenient to represent the mesh with the center point and the width of the element. A simple 1-D mesh is described in Section 6.4.1.

2-D volumetric mesh: For the volume method of moments, we mesh the interior of the scatterer, rather than just the surface. When the volume method of moments is implemented for 2-D problems (Section 6.6.1), the volumetric mesh is simple 2-D mesh of square elements.

2-D surface mesh: A 2-D surface mesh consists of two-dimensional shapes, typically triangles (sometimes quadrilaterals are used), that cover the surface of a 3-D object. A triangular mesh is represented with the coordinates of the corners of each triangle, together with a list that identifies the coordinates that make up each triangle. Triangular meshes are discussed in Sections 6.8.2 and 8.4.6.

3-D: A 3-D mesh can be constructed from tetrahedrons or hexahedrons.

The mesh elements and types used for the method of moments are similar to those used with the finite element method. Further discussion of mesh elements can be found in Section 8.2.1.

6.4 Method of Moments for the TM-EFIE

For the 2-D EFIE for TM-polarized fields, the elements of the moment matrix (6.50) are given by

$$A_{mn} = \frac{k\eta}{4} \int_S \int_S d\overline{\rho}\, d\overline{\rho}'\, H_0^{(2)}(k\,|\overline{\rho} - \overline{\rho}'|\,) t_m(\overline{\rho}) f_n(\overline{\rho}') \tag{6.56}$$

For a 2-D problem, the scatterer is an infinitely long cylinder and only the currents and fields in the x–y plane need to be modeled. Since the unknown current is contained to the surface of the scatterer, the domain of the integrals in (6.56) is a contour or path (labeled as S) that runs along the outer boundary of a cross section of the cylindrical scatterer.

6.4.1 1-D Mesh Generation for Simple Geometries

To define the expansion and testing functions in (6.56), a mesh or discrete representation of the contour S is needed. For a 2-D scattering problem, the mesh consists of arcs or segments along the contour S. A mesh is typically represented in terms of element endpoints or vertices. From (6.59), however, it is apparent that to compute the moment matrix elements, only the center point of each mesh element and the mesh element width must be known. For this reason, it is convenient to represent the mesh in terms of element centers $\overline{\rho}_n$, $n = 1, 2, \ldots, N$ and the width h_n of each element. For simple scatterers, the mesh is often regular, so $h_n = h$ is a constant. When creating a mesh, it is critical to ensure there are no repeated mesh elements, as a repeated element causes the moment matrix to become singular.

A mesh for a circular cylinder is shown in Figure 6.3. Mesh generation for more complex geometries is considered in Section 6.4.9.

6.4.2 Path Integrals

The integrals in (6.56) over the mesh elements for which f_n and t_m are nonzero can be evaluated in terms of parameterizations of the elements. A parameterization is a

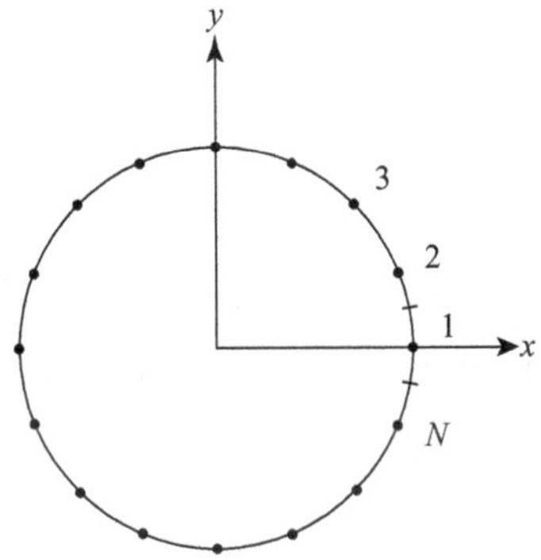

Figure 6.3 *Mesh representation of a circular cylinder. The cylinder is divided into equal length segments or mesh elements. For the surface MoM, it is convenient to represent the mesh using element center points. Mesh element 1 is indicated with tick marks.*

mapping between an interval of the real line and points in the plane. Such a mapping is commonly specified by two coordinate functions, such that $t \mapsto \overline{\rho}(t) = (g(t), h(t))$, where $a \le t \le b$. In terms of a parameterization, a path integral can be evaluated by transforming the variable of integration to the parameter of the path, so that

$$\int_S f(\overline{\rho})\, d\overline{\rho} = \int_a^b f(\overline{\rho}(t))\sqrt{\left(\frac{dg}{dt}\right)^2 + \left(\frac{dh}{dt}\right)^2}\, dt \tag{6.57}$$

There are uncountably many different parameterizations for the same path. If the path is parameterized by arc length, then the square root factor is unity, and the parameter is equal to arc length measured along the path.

A circle of radius a, for example, can be parameterized as $(a\cos t, a\sin t)$, $0 \le t < 2\pi$. If $f = 1$ in the integrand, the integral evaluates to $2\pi a$, the perimeter of the circle. Testing and expansion functions can be defined in terms of a parameterization, so that x in (6.53) and (6.55) becomes the parameter of the path.

Path integrals provide a rigorous method for evaluating moment matrix elements for basis functions defined on an arbitrary, conformal mesh. We will soon see, however, that by using a single-point integration rule for the integrations in (6.56), we can avoid having to explicitly specify a parameterization for the path S.

6.4.3 Testing Integration

The integrals in (6.56) can be evaluated numerically using quadrature rules. The $\overline{\rho}$ integration is commonly performed using a single integration point at the center of each mesh element. This is called point matching or the collocation method and is equivalent to choosing the testing functions to be delta functions located at the mesh element centers. This choice for the testing functions eliminates the integral over the observation point $\overline{\rho}'$ and reduces the moment matrix element expression to a single integration:

$$A_{mn} = \frac{k\eta}{4} \int_S d\overline{\rho}'\, H_0^{(2)}(k\,|\overline{\rho}_m - \overline{\rho}'|\,) f_n(\overline{\rho}') \tag{6.58}$$

In this expression, we have not included a weight for the quadrature rule, since the same rule will be applied to the right-hand side of the integral equation, and the weights will cancel. We will choose to use the same number of testing and expansion functions, with one testing function and one expansion function per mesh element.

6.4.4 Source Integration

If we choose the expansion functions to be pulse functions, then for the off-diagonal elements of the moment matrix, the expansion function f_n is zero at the testing point $\overline{\rho}_m$. This means that the singularity of the Hankel function in the integral operator kernel is driven to zero by the expansion function. Consequently, the integrand is nonsingular, and the $\overline{\rho}'$ integration can also be evaluated using a low-order quadrature rule. A single integration point suffices for reasonable accuracy. With the integration point for the source integration placed at the mesh element center point $\overline{\rho}_n$, the off-diagonal moment matrix elements evaluate to

Off-Diagonal Moment Matrix Elements for the 2-D TM-EFIE

$$A_{mn} = \frac{k\eta}{4} h H_0^{(2)}(k\,|\overline{\rho}_m - \overline{\rho}_n|), \quad m \neq n \tag{6.59}$$

The factor of h is the weight of the quadrature rule and is equal to the width of the mesh elements. We have assumed that the mesh is regular, so that all elements have the same width, but it is easy to modify (6.59) for an irregular mesh by replacing the mesh width h by a variable mesh width h_n indexed by the mesh element number.

For the diagonal elements A_{nn}, the testing point $\overline{\rho}_n$ is at the center of the mesh element on which the pulse function $f_n(\overline{\rho})$ is nonzero. This means that the argument of the Hankel function is zero where the expansion function is equal to one, and the integrand is singular. Since the integrand is singular a single-point integration rule cannot be used to evaluate the diagonal matrix elements. For the diagonal elements, (6.58) is

$$A_{nn} = \frac{k\eta}{4} \int_{S_n} d\overline{\rho}'\, H_0^{(2)}(k\,|\overline{\rho}_m - \overline{\rho}'|\,) \tag{6.60}$$

where the integral is over the nth mesh element S_n. If we transform the variable of integration to the arc-length parameter x for the nth mesh element, this becomes

$$A_{nn} = \frac{k\eta}{4} \int_{-h_n/2}^{h_n/2} dx'\, H_0^{(2)}(k\,|x'|\,) \tag{6.61}$$

The integral can be evaluated analytically using the first two terms of the small argument expansion of the Hankel function (5.54). This leads to the diagonal moment matrix element approximation

Diagonal Moment Matrix Elements for the 2-D TM-EFIE

$$A_{nn} = \frac{k\eta}{4} h_n \left[1 - j\frac{2}{\pi} \log \left(\frac{e^{\gamma - 1} k h_n}{4} \right) \right] \tag{6.62}$$

where $\gamma \simeq 0.577215665$ is Euler's constant or the Euler–Mascheroni constant. Formulas (6.59) and (6.62) allow the full moment matrix to be filled using an array of mesh element center points. It remains to fill the right-hand side of the linear system (6.52).

6.4.5 *Incident Field*

The incident field on the right-hand side of (6.11) must be discretized using the testing functions t_m according to (6.51) to form the M-element vector on the right-hand side of (6.52). This integration is evaluated with the same single-point quadrature rule as the testing integral for the moment matrix elements, so we simply evaluate the incident field at each mesh element center ρ_m on the scatterer contour, and obtain

Right Hand Side for the 2-D TM-EFIE

$$b_m = E_z^{\mathrm{i}}(\overline{\rho}_m) \tag{6.63}$$

for the elements of the right-hand side vector of the linear system (6.52).

6.4.6 *Physical Interpretation of the Method of Moments*

For the EFIE, the method of moments has a very simple physical interpretation. The continuous integral equation requires that the field radiated by the surface current J_z should be equal to the negative of the incident field on the scatterer surface. The discrete equation (6.52) has a similar meaning, except that instead of a continuous surface current, the scattered field is radiated by discrete sources.

The N expansion functions can be thought of as sources or transmitting antennas, each driven by a given input current. The N testing functions are receiving antennas. The matrix A relates the driving current strength at each source to the voltage induced at each receiver. The linear system requires that the current at each transmitter be such that the voltage at each receiver is the same as would be produced by the negative of the incident field. Therefore, the vector solution to the linear system represents the input currents into N transmitting antennas such that the radiated field induces a voltage at N receiving antennas that is the negative of the voltage induced by the incident field. For the EFIE, the matrix A has units of impedance, and is often called the impedance matrix and given the symbol Z.

Because of the discretization process used in deriving the method of moments, the current solution arising from the method of moments is not exactly equal to the current actually induced on a scatterer. This numerical error can also be given a physical interpretation in terms of the boundary condition at the surface of a scatterer. Error is contributed by several different effects. First, the current solution produced by the method of moments is constrained by the set of basis functions to be piecewise constant (with pulse expansion functions), piecewise linear (triangle expansion functions), or a combination of some other choice of a finite number of basis functions. Since the exact current solution does not lie in the approximation subspace spanned by these expansion functions, it is not possible for the numerical current solution to satisfy the required PEC boundary condition perfectly. Second, because moment matrix elements are evaluated with a finite-order quadrature rule, the moment matrix does not exactly represent the field radiated by each expansion function. Third, the radiated fields are measured only by a finite number of testing functions, so the boundary condition is not enforced exactly at every point along the scatterer surface. These and other factors influence the accuracy of the numerical solution obtained using the method of moments. The important topic of solution accuracy will be considered further in the following sections.

6.4.7 Mesh Element Density

Numerical error means that the continuous current distribution J_z is different from the approximate current $\hat{J}_z$ produced by the method of moments. In Section 6.5.2, we will consider the accuracy of the MoM in detail. The simplest consideration with regard to numerical error is that the surface current inherits oscillatory behavior from the incident field E_z^{i}, so that the current oscillates with a wave number on the order of k. To accurately sample the oscillatory current with narrow basis functions such as pulse or triangle functions, the width of the basis functions must be smaller than a half wavelength. This leads to the condition

$$h \ll \frac{\lambda}{2} \tag{6.64}$$

which is equivalent to the Nyquist criterion discussed in Section 4.2.9. The number of mesh elements per wavelength or the dimensionless mesh element density is

$$n_\lambda = \frac{\lambda}{h} \tag{6.65}$$

From (6.64), the mesh element density must satisfy the condition $n_\lambda \gg 2$. Because edges, corners, and curved regions of the scatterer cause the surface current to vary more rapidly with position than the incident field, to represent the current accurately using local basis functions, oversampling is required. A typical value of the mesh element density is

"Ten Points Per Wavelength" Discretization Rule

$$n_\lambda = 10 \tag{6.66}$$

which corresponds to ten mesh elements or current sample points per wavelength. This commonly used rule of thumb in computational electromagnetics is colloquially described as "ten points per wavelength."

Accuracy can be increased by using higher order polynomials or by choosing entire-domain basis functions with a longer region of support and with spatial behavior that matches the expected oscillatory character of the current solution. Entire-domain functions can provide better accuracy than local basis functions for a given number of basis functions. Because the size of the region of integration in (6.56) is larger for entire-domain basis functions than for local basis functions, more work is required to evaluate moment matrix integrals.

6.4.8 MoM Code Overview

A MoM code consists of several main functions: definitions, mesh generation, filling the moment matrix and solving for current unknowns, and postprocessing. A typical code outline is as follows:

MoM Code Outline

Physical constants: Constants needed in the algorithm are defined.

Problem definition: Define the scatterer geometry and composition. Set the frequency and direction of propagation of the incident field, or a list of directions for multiple incidence angles. Specify the parameters of the desired outputs to be computed in postprocessing, such as scattering angles.

Mesh density: For 2-D problems, it is common to specify the number of elements or unknowns per wavelength $n_\lambda = \lambda/h$, where λ is the wavelength of the incident field, and h is the width of a mesh element. This parameter is passed to the mesh generator and determines the fineness of the mesh.

Mesh generation: Generate a mesh representation of the scatterer, possibly using a high-level representation of the geometry such as a parameterization. For 2-D problems, this typically consists of a list of element midpoints and element widths, or a list of endpoints of elements. Mesh generation for complex objects can be done with a specialized software package.

Fill the moment matrix and right-hand side: The core of this code block consists of quadrature rules for the source and observation integrals, inside a loop over mesh elements. The incident field is also discretized to produce the right-hand side of the linear system (6.52). For multiple incidence angles, the right-hand side fill is typically done in the postprocessing portion of the code.

Solve for the current unknowns: This can be done using Gaussian elimination to factor the moment matrix into upper and lower triangular matrices, or using an iterative linear system solution algorithm. Iterative solvers can be faster for large matrix sizes, but Gaussian elimination is more efficient for multiple right-hand sides (incident fields).

Postprocessing: Compute antenna impedances, far fields, scattering amplitudes, radar cross section, or other quantities from the current solution.

The most time-consuming steps in the algorithm are typically the matrix fill and the linear system solution. In practice, however, mesh generation can also be difficult, especially for a complex object geometry. As with the finite difference time-domain (FDTD) method, postprocessing is important in generating quantities that are useful in a design or analysis problem. Mesh generation and postprocessing will be discussed further in the following sections.

6.4.9 1-D Mesh Generation for Complex Geometries

For 3-D problems, mesh generation can be very challenging. Even for 2-D problems, generating a mesh for a complex geometrical shape can be nontrivial. A general-purpose mesh generation software package creates a mesh from a high-level description of the geometry, such as a list of geometrical shapes and operations that add, subtract, or otherwise transform primitive shapes to produce a complex object. The high-level description may also be a computer-aided design (CAD) file that represents the mechanical design of an aircraft, antenna, or other object. Because the electromagnetic analysis requires a mesh that is fine enough to satisfy the condition (6.66), a high-level geometrical description must be processed to provide a mesh that is suitable for the method of moments.

Curved lines and surfaces in space can be described using parametric functions. A 1-D path in two dimensions can be described using $(x, y) = [f(t), g(t)]$, where f and g are functions and t is the parameter of the path. For each value of t, the parametric representation gives a point (x, y) in space.

A parametric representation is the most flexible way to describe the geometry of a scatterer. The parameterization must be integrated numerically or analytically to divide the surface into elements. This process is easy for simple curves but becomes challenging at inflection points where the parametric functions have vanishing derivatives or other mathematical irregularities.

A mesh can consist of flat facets or for a more faithful representation of the scatterer geometry the mesh elements can be curved and conformal to the scatterer surface. A conformal mesh is constructed using parametric polynomial functions. For low-order basis functions, a flat-facet mesh is adequate. With higher order basis functions, a flat-facet geometrical representation of a curved scatterer would destroy the accuracy of the higher order current representation [10]. To realize a high-accuracy solution, the polynomial orders of the basis functions, the integration rule used to evaluate moment matrix elements, and the polynomial order of the mesh representation must all be commensurate.

To deal with the challenges of meshing complex curves and surfaces, many ways for representing curved paths and shapes have been developed. These representations include polynomials defined on elements that are designed to have smoothness properties where elements meet. These are known as splines. A common type of spline used in 3-D modeling is the nonuniform rational B-spline or NURBS surface.

6.4.9.1 Polyarc Representation

As a compromise between the flexibility of a parametric representation and the simplicity of creating meshes for simple shapes such as straight lines, we give here a method for meshing 2-D scatterer contours that consist of straight segments and circular arcs, called a polyarc representation.

A polyarc representation consists of an array in which each row specifies one side of the scatterer, in the following form:

$$[x_{n1} \quad y_{n1} \quad x_{n2} \quad y_{n2} \quad \theta_n]$$

where $n = 1, 2, \ldots, K$, and K is the number of sides of the scatterer. The first coordinate $(x_{n1} y_{n1})$ gives one endpoint of the side, and the second coordinate (x_{n2}, y_{n2}) gives the other endpoint. For closed scatterers, by convention we choose a counterclockwise ordering of the endpoints. The angle θ_n allows the side to have curvature and specifies the angle subtended by a circular arc with the given endpoints. A negative angle causes the arc to be oriented inward with respect to a closed scatterer. $\theta_n = \pi$ corresponds to a semicircle, so that a circle of radius a has the polyarc description

$$\begin{bmatrix} a & 0 & -a & 0 & \pi \\ -a & 0 & a & 0 & \pi \end{bmatrix}$$

An array of this form creates a high-level representation of the scatterer geometry. One side of a geometry of this type is shown in Figure 6.4.

Curved sides ($\theta \neq 0$)

The next step is to create a mesh from the polyarc description that consists of elements of a specified width h. This is done using an algorithm that upsamples the polyarc description by dividing each arc into many small subarcs with length h. In implementing this algorithm, a few useful geometrical quantities are

$$d = \sqrt{(x_{n1} - x_{n2})^2 + (y_{n1} - y_{n2})^2} \tag{6.67a}$$

$$(x_m, y_m) = \left[\frac{(x_{n1} + x_{n2})}{2}, \frac{(y_{n1} + y_{n2})}{2}\right] \tag{6.67b}$$

$$\phi = \tan^{-1}\left(\frac{y_{n2} - y_{n1}}{x_{n2} - x_{n1}}\right) \tag{6.67c}$$

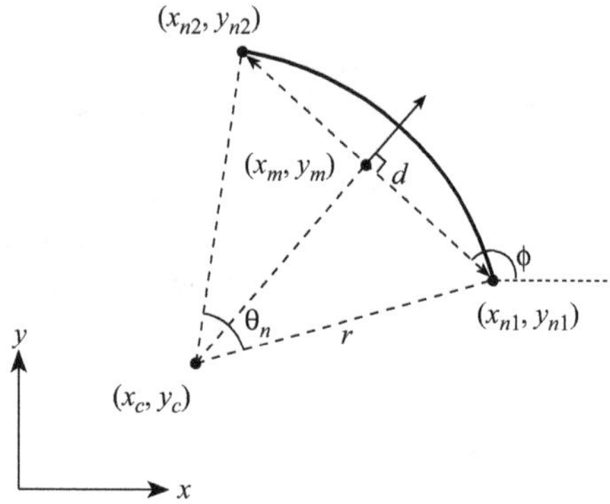

Figure 6.4 The nth side of a polyarc geometry representation

$$(n_x, n_y) = \left[\cos\left(\frac{\phi - \pi}{2}\right), \sin\left(\frac{\phi - \pi}{2}\right)\right] \tag{6.67d}$$

$$(x_c, y_c) = \left[x_m - rn_x \cos\left(\frac{\theta_n}{2}\right), y_m - rn_y \cos\left(\frac{\theta_n}{2}\right)\right] \tag{6.67e}$$

$$r = \frac{d}{2\sin(\theta_n/2)} \tag{6.67f}$$

$$l = r\theta = \frac{d}{\text{sinc}[\theta_n/(2\pi)]} \tag{6.67g}$$

where l is the length of the curved arc. These quantities are defined in Figure 6.4.

To create a mesh suitable for the method of moments, we want to divide the arc into $N_e = \lceil l/h \rceil$ subarcs, where $\lceil l/h \rceil$ is the smallest integer larger than l/h. This operation is implemented in the MATLAB® `ceil` function. In terms of these quantities, the smaller subarcs are defined by

$$\theta_e = \frac{\theta}{N_e} \tag{6.68a}$$

$$\theta_0 = \tan^{-1}\left(\frac{y_{n1} - y_c}{x_{n1} - x_c}\right) \tag{6.68b}$$

$$\theta_p = \theta_0 + p\theta_e, \quad p = 1, 2, \ldots, N_e \tag{6.68c}$$

$$x_{p1} = x_c + |r| \cos\theta_{p-1} \tag{6.68d}$$

$$y_{p1} = y_c + |r| \sin\theta_{p-1} \tag{6.68e}$$

$$x_{p2} = x_c + |r| \cos\theta_p \tag{6.68f}$$

$$y_{p2} = y_c + |r| \sin\theta_p \tag{6.68g}$$

The pth subarc has the representation

$$[x_{p1} \quad y_{p1} \quad x_{p2} \quad y_{p2} \quad \theta_e], \quad p = 1, 2, \ldots, N_e \tag{6.69}$$

The case for which θ is negative must also be handled by checking the sign of the angle and reversing the direction of the normal vector (n_x, n_y) in (6.67d).

Flat sides ($\theta = 0$)
Because of the singularity in the expression for r, flat edges need to be handled separately. This can be done using a linear combination of the endpoints of the arc according to

$$(x, y) = (1 - \lambda)(x_1, y_1) + \lambda(x_2, y_2) \tag{6.70}$$

where λ takes on N_e evenly spaced values in the range $0 \leq \lambda \leq 1$.

Alternatively, the formulas in (6.67a)–(6.67g) and (6.68a)–(6.67g) can be used to generate flat sides by using in the polyarc description a small subtended angle such as $\theta = 10^{-9}$. Care must be taken to ensure that rounding error does not lead to irregularities with the upsampled polyarc elements. This can be checked by plotting the upsampled mesh points.

Preparing the polyarc for MoM
When passing the upsampled polyarc to a method of moments code, a short preprocessing step can be added to compute the midpoint and length of each element of the upsampled polyarc, as these valued are what is required for the TM-EFIE implementation considered in this section. A common error is a mesh with a repeated midpoint. This must be avoided, as it causes the moment matrix to become singular.

6.4.10 Postprocessing

In postprocessing, the surface current solution produced by the MoM algorithm is used to compute other physical quantities that are the desired final results of the analysis. This may include scattering amplitudes, scattering widths, or the impedance at the input terminals of an antenna or other structure with a transmission line port. In the next section, we will discuss computation of scattering amplitudes with the method of moments.

6.4.11 Scattering Amplitudes

Computing far-field quantities such as a scattering amplitudes is simpler with MoM than FDTD, because the moment method yields a solution for currents rather than near fields, and the radiation integral can be used directly to find the far fields radiated by the currents. Using the radiation integral (4.136), the scattering amplitude of a conducting object with induced surface current J_z is

$$S(\phi^{\mathrm{s}}; \phi^{\mathrm{i}}) = \lim_{\rho \to \infty} \sqrt{\frac{\pi k \rho}{2j}} e^{jk\rho} \left(-\frac{k\eta}{4} \right) \int_S d\overline{\rho}' \, H_0^{(2)}(k \left| \overline{\rho} - \overline{\rho}' \right|) J_z(\overline{\rho}') \tag{6.71}$$

where ϕ^{i} is the angle of the incident plane wave that induced the surface current, and ϕ^{s} is the scattering direction where the far fields are measured. This expression for the scattering amplitude and those that follow in this section assume that the amplitude of the incident plane wave is unity. The angle of the point $\overline{\rho}$ is the scattering angle ϕ^{s}. By manipulating this expression, we can put it into a more convenient form for numerical computation.

For large ρ, we can use the large argument expansion of the Hankel function along with the far-field approximation $|\overline{\rho} - \overline{\rho}'| \simeq \rho - \hat{\rho} \cdot \overline{\rho}'$ to simplify the integrand of (6.71), as was done in deriving (4.141). This leads to the result

$$S = -\frac{k\eta}{4} \int_S d\overline{\rho}' \, e^{jk\hat{\rho}\cdot\overline{\rho}'} J_z(\overline{\rho}') \tag{6.72}$$

One way to understand this expression is to recognize that it is very much like a Fourier transform. Because of this, there is a Fourier transform-type relationship between near fields or current sources and the angular distribution of the far fields. Small, localized current distributions lead to broad, smooth far-field patterns, and a narrow far-field pattern implies a spatially large current distribution.

Another way to interpret (6.72) mathematically is to use the Hilbert space approach or geometric picture of functions as vectors in an infinite-dimensional vector space. The L_2 inner product for the space of functions defined on the scatterer surface is

$$\langle f, g \rangle = \int_S d\overline{\rho} f^*(\overline{\rho}) g(\overline{\rho}) \tag{6.73}$$

Using the inner product notation,

$$S = -\frac{k\eta}{4} \left\langle e^{-jk\hat{\rho}\cdot\overline{\rho}'}, J_z \right\rangle \tag{6.74}$$

The exponential term in this expression is a plane wave traveling in the $\hat{\rho}$ direction, which is the scattering direction, or the direction in which we are measuring the scattered field.

Often, the plane wave in the scattered direction appearing in (6.74) is called a scattered field but should not be confused with the scattered field E_z^{s} due to the presence of the object S, which is not a plane wave. To distinguish the plane wave in (6.74) and the scattered field E_z^{s}, we will use the notation

$$E_z^{\mathrm{s},\phi^{\mathrm{s}}}(\overline{\rho}') = e^{-jk\hat{\rho}\cdot\overline{\rho}'} \tag{6.75}$$

where ϕ^{s} in the superscript is a reminder that this is a plane wave in the direction ϕ^{s}, not the field scattered by the object. Using this notation, the scattering amplitude is

$$S = -\frac{k\eta}{4} \left\langle E_z^{\mathrm{s},\phi^{\mathrm{s}}}, J_z \right\rangle \tag{6.76}$$

So, to find the scattering amplitude at the angle ϕ^{s}, we compute the inner product of the current source with a plane wave propagating in the ϕ^{s} direction. In the Hilbert space point of view, this represents the amplitude of the component of the function J_z in the direction of the function $E_z^{\mathrm{s},\phi^{\mathrm{s}}}$.

If we substitute the approximate current solution (6.45) into this expression for the scattering amplitude, we obtain

$$S \simeq \hat{S} = -\frac{k\eta}{4}\left\langle E_z^{s,\phi^s}, \sum_{n=1}^{N} a_n f_n \right\rangle \tag{6.77}$$

Since the inner product is linear, this expression can be rearranged to

$$\hat{S} = -\frac{k\eta}{4}\sum_{n=1}^{N} a_n \left\langle E_z^{s,\phi^s}, f_n \right\rangle \tag{6.78}$$

The inner product term on the right-hand side represents the discretization of the plane wave E_z^{s,ϕ^s} by the expansion functions. This is analogous to the incident field vector in (6.63), which is discretized using the testing functions. If we define the vector c to have elements

$$c_n = \int_S d\overline{\rho}\, E_z^{s,\phi^s}(\overline{\rho}) f_n(\overline{\rho}) \tag{6.79}$$

then we can express the numerical scattering amplitude as

MoM Postprocessing—Scattering Amplitude

$$\hat{S} = -\frac{k\eta}{4} c^H x \tag{6.80}$$

where $x = A^{-1}b$. The superscript ^{H} denotes the conjugate transpose operation or Hermitian conjugate. This places the scattering amplitude solution into a simple, compact matrix formulation.

The numerically computed scattering amplitude can also be expressed as

$$\hat{S} = -\frac{k\eta}{4} c^H A^{-1} b \tag{6.81}$$

The continuous analogue of (6.81) is

$$S = -\frac{k\eta}{4}\left\langle E_z^{s,\phi^s}, \mathscr{L}^{-1} E_z^{i} \right\rangle \tag{6.82}$$

This expression highlights the connection between the inverse of the integral operator $\mathscr{L}$ and the inverse of the matrix discretization A in (6.81).

In practice, it is computationally expensive to form A^{-1}, so (6.80) is more efficient for implementation than (6.81). Because a plane wave is a smooth function, it is common to evaluate the integrals in (6.79) using a single-point integration rule, so that $c_n = hE_z^s(\overline{\rho}_n)$. To achieve best accuracy, the scattered plane wave should be discretized with the same basis functions and integration rule used for the expansion functions in (6.56).

If scattered fields for multiple incident fields are computed, the linear system solution step and the computation of scattering amplitudes can be combined to increase efficiency. The moment matrix can be factored into upper and lower triangular matrices using the MATLAB function

LU Factorization

```
[L,U,P] = lu(A);
```

For each incident field, the current solution can then be generated using

```
x = U\(L\(P*b));
```

Because of the triangular form of `L` and `U`, `L\y` and `U\y` where `y` is an arbitrary vector can be computed using backsubstitution in N floating point operations. This makes `x = U\(L\(P*b))` much faster than solving the original linear system using `A\b` for each incidence angle.

6.4.12 *MoM Implementation*

For the MoM algorithm, the core computational routine is the moment matrix filling step. A MATLAB matrix fill code fragment is illustrated as follows for the previously described MoM version with point matching for the testing functions and a single-point integration rule for the source integrations. In this code, the mesh is defined by the variables `sx`, `sy`, and `sdx`, which are arrays of length `N` of the mesh element center point x coordinate, mesh element center y coordinate, and the mesh element width, respectively. The constant `k0` is the wave number of the incident field, and `eta0` is the characteristic impedance of free space.

Filling Moment Matrix for TM-EFIE

```
% Fill moment matrix
Z = zeros(N,N);
fac = k0*eta0/4;
for m = 1:N,
    for n = 1:N,
        if m==n,
            % diagonal matrix element
            Z(m,m) = fac*sdx(m)*(1-j*(2/pi)* ...
                log(1.781*k0*sdx(m)/(4*exp(1))));
        else
            % off-diagonal matrix element
            rho = sqrt((sx(m)-sx(n))^2+(sy(m)-sy(n))^2);
            Z(m,n) = fac*(sdx(n))*besselh(0,2,k0*rho);
        end
    end % loop over n
end % loop over m
```

Following the matrix filling and the right-hand side filling routines, a vector of current unknowns can be found using

```
x = A\b;
```

Once the current unknowns have been computed, postprocessing is used to compute derived quantities such as scattering widths.

To compute the bistatic scattering width as a function of scattering angle, (6.80) can be used to compute the scattering amplitude, and the scattering amplitude is converted to the scattering width using the relationship (4.117). This procedure is implemented in MATLAB code as follows:

Postprocessing for Scattering Width

```
% Postprocessing
% Compute bistatic scattering width
NA = 361;
for m = 1:NA,
    % Scattering angle
    phi_scat = (m-1)*2*pi/NA;

    % Wave vector of scattered plane wave
    kscat = k0*[cos(phi_scat) sin(phi_scat)];

    % Plane wave in scattering direction
    Es = exp(j*(kscat(1)*sx+kscat(2)*sy));

    % Amplitude of scattered field
    Ez_far = sum(sdx.*Es.*x);

    % Scattering amplitude
    sa(m) = -fac*Ez_far;

    % Scattering width
    sw(m) = (4/k0)*abs(sa(m))^2;
    angles(m) = phi_scat;
end
```

6.4.12.1 Numerical Results

Figure 6.5 gives an illustration of numerical results for the method of moments algorithm applied to a PEC circular cylinder with radius $a = \lambda/2$. The scattering width of the cylinder is shown as a function of scattering angle. The incident field is a TM-polarized plane wave at the angle of incidence $\phi^{\mathrm{i}} = \pi$. The mesh density is $n_\lambda = 10$, which means that the mesh has 31 elements.

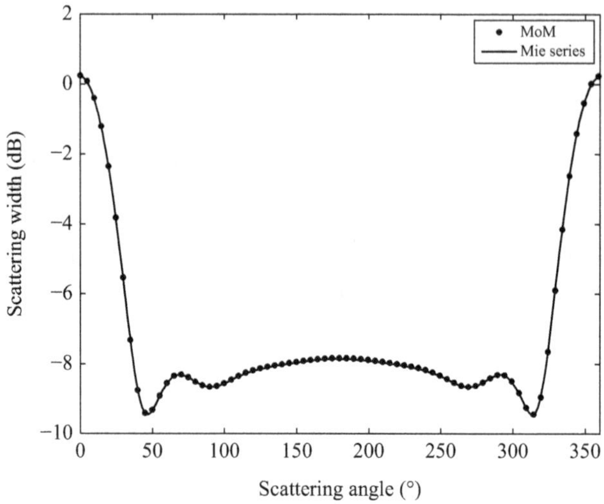

Figure 6.5 Scattering width of a PEC circular cylinder with half wavelength radius. The MoM results are based on the TM-EFIE with point matching and a single-point quadrature rule for the source integration.

Even though a low-order quadrature rule is used to evaluate moment matrix element integrations, the numerical results are quite accurate. The root mean square (RMS) error of the scattering width expressed in dB relative to the exact Mie series solution is 0.02 dB, and the maximum relative scattering width error is 0.7%. For other scatterers with edges, corners, or resonant cavity-like structures, the numerical solution error is significantly higher. The accuracy of the MoM algorithm and how it is influenced by the various aspects of the discretization scheme will be considered in greater detail in the next section.

6.5 Accuracy and Efficiency of the Method of Moments

For any algorithm, we need to have some assurance that the results are correct, and the algorithm needs to be fast enough to solve complex problems of practical interest. The former requires an accuracy analysis, and the latter is determined by the computational cost or computational efficiency of the algorithm. In this section, we will survey some of the basic principles of accuracy and efficiency for the MoM algorithm. The analysis will allow us to compare the performance of MoM for surface integral equations with that of the FDTD algorithm.

6.5.1 Computational Cost

Computational cost can be measured in floating point operations (flops), which translates fairly directly to computer run time. An algorithm typically has a problem "difficulty" parameter, which for the method of moments is the number of unknowns or degrees of freedom, N. For most CEM algorithms, the computational cost is well modeled by a power law of the form

$$C = c_1 N^{\alpha} \tag{6.83}$$

For the state-of-the-art electromagnetic analysis and design problems, the computation time can be many hours or days, so the development of improved algorithms is an important area of research. As more efficient numerical algorithms are developed for a given type of problem, either the constant c_1 or the exponent α in (6.83) becomes smaller. A reduction in the exponent is generally more valuable than a decreased constant factor.

For the method of moments, the number of unknowns N is dictated largely by the physical size of the region S on which the unknown current is defined in relation to the wavelength of the electromagnetic fields in the problem. N increases with both the physical size of the scatterer and the frequency of the incident field. Problems for which S is many wavelengths in size are referred to as electrically large. For electrically large problems, N can be in the hundreds of thousands or millions.

To analyze the computational cost of the MoM algorithm, we must determine the number of unknowns N for a given problem and then estimate the number of flops required for the MoM algorithm as a function of N. We will assume that the MoM algorithm is applied to a surface integral equation, since the computational cost analysis is different for a volume integral equation. Suppose we are modeling a square scatterer of side d. The perimeter is $4d$, and using the sampling density n_λ defined in (6.65), the number of unknowns required is

$$N \simeq 4d/h = \frac{2}{\pi} kd\, n_\lambda \quad \text{(2-D MoM)} \tag{6.84}$$

For any scatterer, we can say that N is on the order of kd, or $N = O(kd)$. Since $k = 2\pi/\lambda$, the quantity kd is proportional to the size of the scatterer measured in wavelengths (d/λ), so we refer to kd as the electrical size of the scatterer at a given frequency.

If the linear system arising from the moment method is solved using a direct algorithm such as Gaussian elimination (see Section 7.3), the computational cost is

$$C_{\text{solve}} = \frac{2}{3} N^3$$

The total computational cost is

$$\begin{aligned} C_{\text{MoM}} &= C_{\text{fill}} + C_{\text{solve}} \\ &= c_1 N^2 + \frac{2}{3} N^3 \\ &\simeq \frac{2}{3} N^3 \\ &\simeq 0.17 (kd)^3 n_\lambda^3 \end{aligned} \tag{6.85}$$

From this, we can see that the computation time required for the MoM algorithm with LU decomposition is $O(k^3 d^3)$. The memory storage requirement is $N^2 \simeq (kd)^2 n_\lambda^2$.

How does the computational cost of MoM compare with that of the FDTD algorithm? If we assume that the absorbing boundary condition is good enough that the extra space required around the scatterer is small, the number of grid points required is

$$N = \left(\frac{d}{h}\right)^2 = \frac{(kd)^2}{4\pi^2}n_\lambda^2 \quad \text{(2-D FDTD)} \tag{6.86}$$

where k is the maximum spatial frequency of the field variations for a given incident wave or source function. The number of time steps is roughly determined by the time required for a wave to cross the entire box, so that

$$N_t = \frac{d}{c\,dt} = \frac{d}{\alpha h} \simeq \frac{kd}{2\pi}n_\lambda \tag{6.87}$$

The total computational cost is

$$C_{\text{FDTD}} \simeq NN_t \simeq 0.004(kd)^3 n_\lambda^3 \tag{6.88}$$

and the storage requirement is $3N \simeq (kd)^2 n_\lambda^2/(4\pi)$. Equation (6.88) is an underestimate, because we have assumed only one flop per grid point per time step, whereas each grid point actually takes several steps. Since large radiation and scattering problems are of most interest in real-world applications, and the size of problems that are commonly solved continually increases, the order of the computational cost, or the exponent on the electrical size kd, is more important than the scale factor in the cost estimate.

By comparing (6.85) and (6.88), we can see that MoM and FDTD have computational costs of the same order in the electrical size of the scatterer. The constant in the FDTD cost is smaller, but a higher mesh density is often required to achieve the same accuracy with FDTD. For MoM, $n_\lambda = 10$ is typical, whereas for FDTD n_λ is often 15–20 or larger. In practice, for problems of modest electrical size, MoM is often faster due to practical considerations such as the use of efficient implementations of a linear system solution routine.

The method of moments can be accelerated by using an iterative solution method to reduce the N^3 term in the computational cost for the linear system solution or by using a "fast algorithm" such as the fast multipole method (FMM) or the fast Fourier transform (FFT)-based adaptive integral method (AIM) to reduce the matrix fill and solution time to an order that is roughly $N \log N$. Fast algorithms are of great importance in computational electromagnetics, since they reduce the computational cost and memory requirement of the MoM algorithm significantly for electrically large problems (see Section 7.6.1).

6.5.2 Error Analysis

The goal here is to analyze the error associated with a moment method solution and understand what factors improve solution accuracy. To study solution error, a quantitative measure of the accuracy of a solution is required. In analyzing dispersion error for the FDTD algorithm, the solution error in the field at a given point was adequate for the analysis. For the method of moments, a measure of the error in the solution for the current on a scatterer is required. An error measure must quantify the difference between the exact current and the approximate current solution obtained using MoM, both of which are functions of position on the scatterer surface. The magnitude of the difference between two functions can be measured using a norm.

6.5.2.1 Norms

A norm is a functional on an abstract vector space or linear space that maps an element in the space to a positive real number. For the purposes of this section, the vector space is the space of functions defined on the surface of a scatterer. If J denotes the surface current on a scatterer, the norm of the function J is denoted by $\|J\|$.

A norm satisfies several axioms:

1. $\|J\| \geq 0$ if $J \neq 0$, $\|J\| = 0$ if $J = 0$ (positive definiteness)
2. $\|\alpha J\| = |\alpha|\|J\|$ (homogeneity)
3. $\|J_1 + J_2\| \leq \|J_1\| + \|J_2\|$ (triangle inequality)

The zero function is the only function that has zero norm, and if a nonzero function is multiplied by a scalar the norm changes by the same scale factor. These properties collectively imply that the value of the norm reflects the "largeness" of the function.

Norms are not unique. For a given space, many norms can be defined. The most common norm on a space of functions is the L_2 norm

$$\|J\|_{L_2} = \int_S |J(\overline{\rho})|^2 \, d\overline{\rho} \tag{6.89}$$

Since the L_2 norm is the most common norm on a space of functions, often we will write it as $\|J\|$, without the subscript L_2. This norm is related to the L_2 inner product (6.73) by $\|J\|^2 = \langle J, J \rangle$.

6.5.2.2 Error Measures

Using the L_2 norm, the current solution error and the relative current solution error can be defined as

$$L_2 \text{ current error} = \|J - \hat{J}\| \tag{6.90a}$$

$$\text{Relative } L_2 \text{ error} = \frac{\|J - \hat{J}\|}{\|J\|} \tag{6.90b}$$

where J is the exact current solution and $\hat{J}$ is a numerical approximation obtained using the MoM algorithm. Relative error has the advantage that it is dimensionless and can be readily converted to a percentage error.

These error measures have two limitations. The first is a practical one: the numerical current solution is a linear combination of basis functions, which means that evaluating the L_2 norm requires integrating over expansion functions. This is certainly possible but is somewhat inconvenient. To avoid having to work integrals of the expansion functions, the discrete RMS current error

$$\text{Relative RMS current error} = \frac{\|J - \hat{J}\|_{\text{RMS}}}{\|J\|_{\text{RMS}}} = \frac{\left(\frac{1}{N}\sum_{n=1}^{N} |J_n - \hat{J}_n|^2\right)^{1/2}}{\left(\frac{1}{N}\sum_{n=1}^{N} |J_n|^2\right)^{1/2}} \tag{6.91}$$

is often used instead. The RMS value of a finite set of real numbers can be viewed as an L_2 norm on a finite-dimensional vector space. In the case of a 2-D MoM algorithm with pulse or triangle expansion functions or another interpolatory basis set, the basis function weights $\hat{J}_n$ are equal to samples of the current at mesh node points, so the RMS current error is simply the error in the unknown coefficients in the linear system (6.52) with respect to samples of the exact current solution. For noninterpolatory basis functions, the expansion (6.45) can be used to evaluate the numerical current solution at mesh node points to evaluate the RMS current error.

The second difficulty is harder to deal with. For scatterers with edges or sharp corners, the surface current can become infinite at the edge or corner. For a TM-polarized field incident on a PEC object with an edge, the current behaves like $x^{-1/2}$, where x is the distance to the edge. When the current is squared in the integrand of the L_2 norm, the resulting singularity is not integrable, and the L_2 norm is infinite. This means that the current solution is not in the space of L_2 functions, or the set of all functions with finite L_2 norm. One possible fix for this is to replace the L_2 norm with the L^1 norm. Another is to measure error in terms of scattered fields. There are a number of error measures for scattering amplitudes and radar cross section (RCS), including

$$\text{Relative RMS scattering amplitude error} = \frac{\left(\frac{1}{N}\sum_{n=1}^{N}\left|S(\phi_n^s) - \hat{S}(\phi_n^s)\right|^2\right)^{1/2}}{\left(\frac{1}{N}\sum_{n=1}^{N}\left|S(\phi_n^s)\right|^2\right)^{1/2}} \tag{6.92a}$$

$$\text{Maximum relative RCS error} = \max_{n=1,2,\ldots,N} \frac{\left|\sigma(\phi_n^s) - \hat{\sigma}(\phi_n^s)\right|}{\left|\sigma(\phi_n^s)\right|} \tag{6.92b}$$

$$\text{RMS dB RCS error} = \left(\frac{1}{N}\sum_{n=1}^{N}\left|10\log_{10}\left[\frac{\sigma(\phi_n^s)}{\hat{\sigma}(\phi_n^s)}\right]\right|^2\right)^{1/2} \tag{6.92c}$$

where $\phi_1, \phi_2, \ldots, \phi_N$ is a set of N scattering angles. These error measures are used for both bistatic and monostatic scattering.

These error measures require a knowledge of the exact current or scattered fields, which are available analytically only for cylinders, spheres, and other simple, canonical geometries. If the solution error is needed for a more complex scatterer, a reference solution can be generated using a very fine mesh with a solver code that is known to be accurate. Measured data can also be used as a reference solution.

6.5.3 Sources of Error

While the error measures listed in the previous section provide a simple way to quantify the accuracy of a computed MoM solution, predicting the solution error for a given problem is difficult due to the many factors that can influence the numerical behavior of the MoM algorithm. Solution accuracy is influenced by aspects of the physical problem, mesh representation, and MoM implementation.

6.5.3.1 Physical Problem

Scatterer geometry: Smooth scatterers typically lead to higher accuracy than scatterers with sharp edges, corners, or points. For scatterers with these types of geometrical singularities, the current solution is singular, which makes it harder to model with standard basis functions such as the pulse function or triangle function. It is possible to use specialized basis functions near the edge, corner, or point that match the form of the current singularity to improve accuracy, at the cost of an increase in the difficulty of implementation of the moment method.

Incidence angle: For a smooth, flat scatterer, a plane wave that is normally incident induces a current that is nearly constant over the scatterer. A constant or slowly varying current can be accurately represented by pulse and triangle functions. Near grazing incidence, the current on the scatterer oscillates with approximately the same period as the incident field. Oscillatory currents are less accurately samples by local basis functions, and error is larger.

Resonance: A concave, cavity type geometry at high frequencies leads to resonances, or peaks in the stored energy inside the cavity at certain frequencies. At resonance, the EFIE and MFIE integral operators are nearly singular, and error in numerical solutions is large. As observed in Section 6.2.4, closed scatterers can also have nonphysical internal resonances that reduce numerical accuracy. No energy from an external incident field can penetrate a closed PEC scatterer, but, due to discretization error incurred when the continuous integral equation is transformed into a finite matrix equation, energy "leaks" inside the scatterer and internal resonances have an impact on solution error. If solution error for a closed scatterer is plotted as a function of frequency, error increases near the internal resonance frequencies. This problem with internal resonances can be overcome using the CFIE, although real resonances due to open cavities are numerically challenging even with the CFIE.

6.5.3.2 Mesh Representation

Mesh element density: As discussed in Section 6.4.7, a large number of mesh elements per wavelength is required to model accurately the induced current on a scatterer. Typically, solution error decreases in proportion to a power of the mesh element density n_λ. For most implementations, the current solution error decays asymptotically as n_λ^{-1} or n_λ^{-2} as the number of elements increases.

Geometrical modeling error: In many implementations of MoM, the mesh elements are flat. For curved scatterers, this means that the mesh representation is not conformal to the scatterer geometry, and the resulting decrease in solution accuracy is referred to as flat facet error. Curved or conformal mesh elements can be used to reduce this source of error.

Irregular mesh: For a regular mesh, all elements have the same size. For complex geometries, it is often not possible to create a regular mesh. If the sizes of adjacent mesh elements are significantly different, solution error increases.

6.5.3.3 MoM Implementation

Expansion and testing functions. The choice of expansion and testing functions strongly influences solution accuracy. Generally, the smoother the basis functions or the higher the polynomial order, the more accurate the numerical solution, but there are certain "magic" cases for which low-order basis functions can produce a more accurate solution than higher order polynomial basis functions [7].

Quadrature rules. The choice of quadrature rules used to evaluate moment matrix elements and to discretize the incident field and scattered plane wave has a significant impact on solution accuracy. As might be expected, low-order quadrature rules generally lead to larger error, and accuracy improves with more integration points.

Linear system solution algorithm. For direct solution algorithms such as *LU* decomposition, there is essentially no linear system solution error. To speed up the linear system solution for large matrix sizes, iterative algorithms are often used, but the solution is not exact and an additional error contribution is introduced.

6.5.3.4 Current Solution Error

For a given physical problem, mesh representation, and algorithm implementation, how can we quantitatively estimate the actual error in a MoM solution? In general, this is a difficult problem, but if we make some simplifications useful, error estimates can be obtained.

A basic concept in error analysis of numerical methods is optimality. Once we choose a set of basis functions, the current solution can be no better in a norm sense than the linear combination of expansion functions that is closest to the exact solution:

$$\text{Error} \geq \min_{a_n} \left\| J - \sum_{n=1}^{N} a_n f_n \right\| \tag{6.93}$$

The solution corresponding to the set of weights a_n for which the error is minimized is called the optimal solution. If a particular numerical method produces a solution that is close to optimal, we can reduce the problem of error analysis to one of approximation theory: given the properties of the exact current solution, how well can we represent it using the expansion functions?

The assumption of optimality greatly simplifies the error analysis problem, because many of the details of the numerical algorithm can be ignored. All we need to determine is how well the basis functions can represent the current solution. Algorithms are almost always suboptimal to some degree, though, so error estimates based on optimality underestimate the actual solution error. We will see later that some aspects of the implementation of the method of moments, such as low-order, inaccurate quadrature rules, can cause a significant deviation from optimality, and the solution error is larger than predicted using the optimality principle.

Using the assumption of optimality, we will now analyze the current solution error for the case of a TM-polarized plane wave illuminating a flat PEC strip of width

d. The scatterer lies on the x-axis from $x = -d/2$ to $x = d/2$. The current induced on the strip in the physical optics (PO) approximation is [6]

$$J(x) \simeq 2\hat{n} \times \overline{H}^{\mathrm{i}} = \frac{2}{\eta} \sin(\phi^{\mathrm{i}}) e^{-jk_x x} \tag{6.94}$$

where $k_x = -k\cos(\phi^{\mathrm{i}})$. This approximation neglects the effect of edge scattering, which manifests itself as a current singularity of the form $x^{-1/2}$ near each edge, where x is the distance to one of the edges of the strip. If the strip is wide in relation to the electromagnetic wavelength, the singular edge currents are only a small part of the overall current distribution on the scatterer, so that (6.94) is a reasonable approximation for the current away from the edges of the strip.

Using this approximation for the current solution on a flat scatterer, we can use the optimality principle to obtain an estimate of the current error associated with the MoM algorithm. The best possible expansion of the current (6.94) in terms of the basis functions in (6.45) can be found by minimizing the L_2 error

$$\|u - \hat{u}\|^2 = \int \left(J(x) - \sum_{n=1}^{N} a_n f_n(x) \right)^2 dx \tag{6.95}$$

over the coefficients a_n. Setting the derivative of the L_2 error with respect to a_m to zero leads to

$$\int J(x) f_m(x)\, dx = \sum_{n=1}^{N} a_n \int f_n(x) f_m(x)\, dx \tag{6.96}$$

For pulse functions, the integral on the right-hand side evaluates to

$$\int f_n(x) f_m(x)\, dx = \begin{cases} h & m = n \\ 0 & m \neq n \end{cases} \tag{6.97}$$

since pulse functions on different mesh elements do not overlap and the integrand is zero unless $m = n$. This leads to the solution

$$a_n = \frac{1}{h} \int J(x) f_n(x)\, dx = \frac{1}{h} \langle f_n, J \rangle \tag{6.98}$$

These coefficients represent the best possible combination of pulse functions to represent $J(x)$ in the sense of minimum L_2 norm.

Inserting the current expression (6.94) into (6.98) and evaluating the integral leads to

$$\begin{aligned}
a_n &= \frac{1}{h}\int_{x_n-h/2}^{x_n+h/2} \frac{2}{\eta}\sin(\phi^{\mathrm{i}})e^{-jk_x^{\mathrm{i}}x} f_n(x)\,dx \\
&= \frac{2}{\eta}\sin(\phi^{\mathrm{i}})e^{-jk_x^{\mathrm{i}}x_n}\frac{1}{h}\int_{-h/2}^{h/2} e^{-jk_x x} f(x)\,dx \\
&= J(x_n)\underbrace{\left[\frac{1}{h}\int_{-h/2}^{h/2} e^{-jk_x x} f(x)\,d\right]}_{\substack{\text{Fourier transform of}\\ \text{basis function } F(k_x)}} \\
&= J(x_n)\underbrace{\frac{\sin(k_x h/2)}{k_x h/2}}_{\text{sinc}} \qquad (6.99)\\
&\simeq \hat{J}(x_n) \qquad (6.100)
\end{aligned}$$

The last approximation invokes the assumption that the MoM algorithm is optimal and produces a solution $\hat{J}$ that is close to the best L_2 expansion of $J(x)$. These values for a_n are different from the sample $J(x_n)$ of the current solution at the point x_n. Since a_n represents the solution that would be obtained from the MoM algorithm using pulse basis functions, this implies that the MoM solution for the surface current on the scatterer does not quite match the true current solution at the mesh node points. This deviation can be used to estimate the solution error for the MoM algorithm.

The relative solution error at the node point x_n is

$$\begin{aligned}
\frac{J(x_n)-\hat{J}(x_n)}{J(x_n)} &\simeq \frac{J(x_n)-J(x_n)F(k_x)}{J(x_n)} \\
&= 1-F(k_x) \\
&= 1-\frac{\sin(k_x h/2)}{k_x h/2} \\
&\simeq \frac{k_x^2 h^2}{24} \\
&= \frac{\pi^2\cos^2(\phi^{\mathrm{i}})}{6} n_\lambda^{-2} \quad \Rightarrow \quad \text{Second order} \qquad (6.101)
\end{aligned}$$

where $n_\lambda = \lambda/h$. The most important observation is that the error is second order. Another notable fact is that the error is largest for grazing incidence and decreases toward normal incidence.

This expression for the MoM solution error is similar to the result that we obtained for the 2-D FDTD method in (4.43), except that the error does not depend on the electrical size of the problem. This is important because it means that MoM does not suffer from dispersion error. The solution error is independent of the size of the scatterer.

6.5.3.5 Suboptimal MoM Implementations

The second-order current solution error estimate (6.101) is useful in gaining insight into the numerical behavior of the MoM algorithm, but it has several limitations. It neglects edge effects, which are not captured by the physical optics approximation and cause the current to be singular at the edges of the strip. The analysis also assumes that the numerical solution is optimal. The error estimate vanishes at normal incidence ($\phi^{\mathrm{i}} = \pi/2$) due to the cosine factor, yet the actual numerical error at grazing incidence is not zero due to suboptimality of the solution.

Suboptimality can be caused by many of the error factors listed in Section 6.5.3. These include resonance, error in evaluating moment matrix elements using a quadrature rule, and geometrical discretization error. For the TM-EFIE with analytically evaluated diagonal moment matrix elements and a single integration point for off-diagonal elements as implemented in this chapter, the solution is suboptimal because of quadrature error. In that implementation, we are integrating a function with a logarithmic singularity using a low-order quadrature rule and the solution is not as accurate as predicted by (6.101). It can be shown that the relative RMS current solution error for this discretization of the TM-EFIE is [7]

$$\frac{\|J - \hat{J}\|_{\mathrm{RMS}}}{\|J\|_{\mathrm{RMS}}} \simeq \frac{0.2}{n_\lambda} \tag{6.102}$$

The low-accuracy integration rule for the moment matrix integrals translates into a first-order error in the current solution, and the accuracy is significantly poorer than predicted by (6.101), which assumes ideal integration of the matrix elements.

A comparison of the error estimate (6.102) with observed numerical error for a PEC circular cylinder is shown in Figure 6.6. For small values of the mesh density, the mesh has only a few elements. At a mesh density of $n_\lambda = 1$, the mesh has three elements, and the moment matrix is three by three in size. The relative numerical error is near to unity in value, which means that the computed current solution is no better than simply guessing zero. As the number of mesh elements increases, the current solution error decreases at the first-order rate predicted by (6.102). At the commonly used mesh density of ten sample points or unknowns per wavelength ($n_\lambda = 10$), the relative RMS current solution error is about 3%. Improved implementations of the MoM algorithm with higher order moment matrix integration rules do not suffer from suboptimality and can achieve much lower error at $n_\lambda = 10$, at least for smooth scatterers. We will see shortly that suboptimality due to inaccurate integration or other effects is also important in analyzing error for scattering amplitudes and RCS.

6.5.3.6 Scattering Amplitude Error

The error analysis based on optimality that was used in Section 6.5.3.4 for the current solution cannot be applied to scattering amplitudes, scattering widths, or radar cross sections. Expression (6.81) for the scattering amplitude is referred to as variational, which means that a portion of the error in the current cancels when the scattering

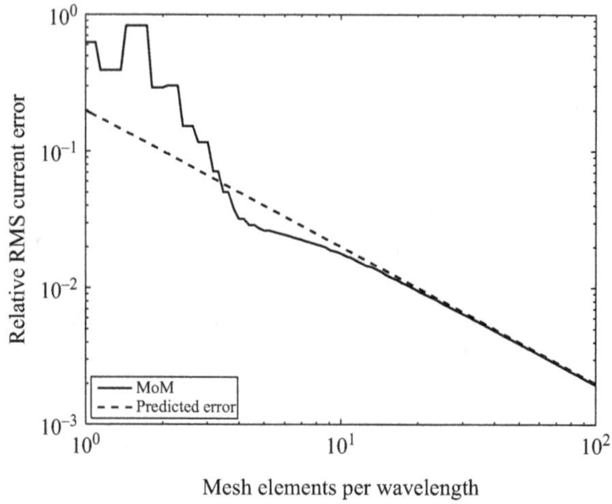

Figure 6.6 Relative RMS current solution error for MoM applied to the TM-EFIE for a PEC circular cylinder with half wavelength radius. The MoM implementation uses point matching and a single-point quadrature rule for the source integration. The predicted error is computed using (6.102). The slope of the error curve is −1, which means that the current solution computed with this MoM implementation is accurate to first order in the number of mesh elements.

amplitude is computed. Because of this, scattering amplitudes can be more accurate than the current solutions from which they are computed.

A simple interpretation of the variational property can be found by rewriting the expression (6.81) for the numerical scattering amplitude in the form

$$S = -\frac{k\eta}{4}[\langle f_m, E_z^{\mathrm{s}}\rangle]^H \, [\langle t_m, \mathscr{L} f_n\rangle]^{-1} \, [\langle t_n, E_z^{\mathrm{i}}\rangle] \tag{6.103}$$

where the angle brackets denote an inner product. If we think of the incident field as a sum of Fourier components, the inner product with t_n introduces a scale factor T on that Fourier component in the same way as in (6.99). The inner product of the scattered field likewise introduces a scale factor F on the corresponding Fourier component. But the kernel of the integral operator also has a Fourier decomposition, and the testing and expansion functions introduce scale factors FT on the eigenmodes of the integral operator as well (which for a circular cylinder are Fourier modes and for other smooth scatterers approximate Fourier modes). Since the operator is inverted, the overall scale factor for a given Fourier component is $F(TF)^{-1}T = 1$, and the effects of the expansion and testing functions cancel. This means that scattering amplitudes are often much more accurate than the current solution used to compute them.

If the MoM algorithm is implemented for the TM-EFIE with pulse expansion functions and point testing and if an accurate integration rule is used to evaluate moment matrix elements and discretize the scattered plane wave used to compute the scattering amplitude, the current solution error is second order with respect to the mesh element density. In this case, the current error is second order in the number of mesh elements as predicted using the optimality principle in Section 6.5.3.4, and the scattering amplitude error is third order. For a circular cylinder of radius a, the error estimate

$$\frac{|S - \hat{S}|}{|S|} \simeq 1.9(ka)^{-1} n_\lambda^{-3} \tag{6.104}$$

has been obtained for the backscattering amplitude [10]. The scattering amplitude accuracy is better by one order than the current solution.

Unfortunately, the variational property is destroyed by some sources of error, such as low-accuracy quadrature rules. Discretization error terms that are additive to the matrix term in (6.103) do not cancel and contribute to the scattering amplitude error. These additive terms make the computed current solution suboptimal and increase the error in the computed scattering amplitude. If the additive error terms are large enough in magnitude, the scattering amplitude accuracy is no better than that of the current.

For the single-point integration rule, the Hankel function expansion used in evaluating the diagonal moment matrix elements is accurate only to first order in the mesh element size, and for that discretization scheme both the current and scattering amplitude errors are only first order. This is similar to the increase of the current solution error from second- to first order that we observed in Section 6.5.3.5. For a PEC circular cylinder of radius a, with the single-point integration rule used in (6.4.4) to evaluate the moment matrix integrals, the backscattering amplitude error increases to

$$\frac{|S - \hat{S}|}{|S|} \simeq \frac{0.2}{n_\lambda} \tag{6.105}$$

A comparison of this theoretical estimate for the backscattering amplitude error to the actual computed error is shown in Figure 6.7. As with the current solution, for small values of the mesh element density, accuracy is poor, and the error decreases at a first-order rate as the mesh density becomes large. At ten points per wavelength ($n_\lambda = 10$), the relative error is about 3%.

This treatment of numerical accuracy has considered only some of the factors listed in Section 6.5.3. The circular cylinder is a smooth scatterer, which means that the current on the scatterer is well behaved. For scatterers with corners or edges, error can be significantly larger than for the circular cylinder. Electrically large, resonant scatterers with cavity-like inlets or ducts can also be extremely difficult to solve accurately. A good deal of research has been devoted to developing algorithms that are capable of solving challenging problems with complex geometries accurately and efficiently. For a more exhaustive treatment of numerical error for the MoM algorithm, see [7].

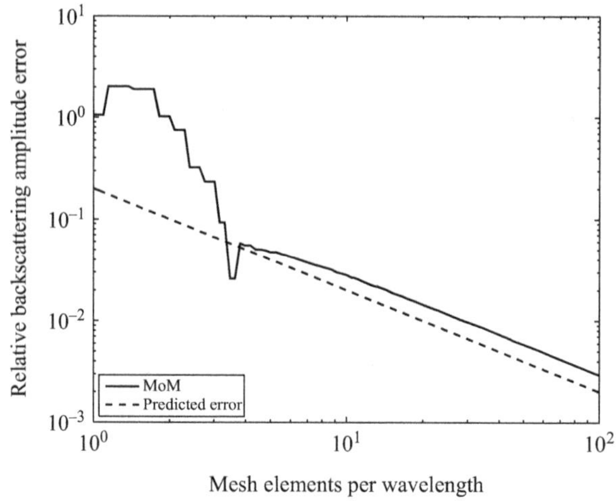

Figure 6.7 *Relative backscattering amplitude error for MoM applied to the TM-EFIE for a PEC circular cylinder with half wavelength radius. The MoM implementation uses point matching and a single-point quadrature rule for the source integration. The predicted error is computed using (6.105). The scattering amplitude computed with this MoM implementation converges at a first-order rate with respect to the number of mesh elements. The inaccurate single-point integration rule obscures the variational property of the method of moments. With a better integration rule for moment matrix elements, the accuracy of computed scattering amplitudes improves dramatically to third order*

6.6 Dielectric Structures

The method of moments can also be applied to dielectric and conducting scatterers as well as PEC objects. Unlike the PEC case, for which the unknown current in an integral equation is equal to the conduction current actually flowing on the conductor, the unknown currents in an integral equation for a dielectric object are fictitious. The equivalent current for a dielectric scatterer represents a current distribution that if impressed in free space would radiate the same fields as are scattered by the object for a given incident field.

For homogenous dielectric objects, either surface or volume integral equations can be used. The surface integral formulation for a homogeneous dielectric consists of two integral operators, one similar to the operator in the EFIE and the other to the MFIE operator, and both electric and magnetic equivalent surface currents are required. For inhomogeneous objects, volume integral equations must be used, and the unknown current is nonzero on the region of support of the scatterer. Since the surface integral operators for homogeneous dielectrics are fairly similar to the integral equations for PEC objects we treated in the previous section, we will focus here on the volume integral formulation.

6.6.1 2-D Volume Method of Moments

To derive the volume EFIE, we want to rearrange Maxwell's equations so that the permittivity and permeability are equal to their free-space values and the dielectric properties of the scatterer are lumped into an equivalent source. We will assume that the scatterer is nonmagnetic, so that $\mu = \mu_0$. By rearranging (4.114b) slightly, Faraday's and Ampère's laws can be written in the form

$$\nabla \times \overline{E}^{s} = -j\omega\mu_0\overline{H}^{s} \tag{6.106a}$$

$$\nabla \times \overline{H}^{s} = j\omega\varepsilon_0\overline{E}^{s} + \overline{J} \tag{6.106b}$$

where

$$\overline{J}(\overline{r}) = \{j\omega[\varepsilon(\overline{r}) - \varepsilon_0] + \sigma(\overline{r})\}\,\overline{E} \tag{6.107}$$

and $\overline{E}$ is the total field. With the governing equation (4.114b) used for the FDTD in the scattered field formulation, we had a known source radiating in the presence of the scatterer, whereas in (6.106a) and (6.106b) we have an unknown source radiating in the absence of the scatterer. This highlights a key difference between the FDTD algorithm, for which sources are known, and MoM, for which sources are unknown.

Because we have moved the effect of the scatterer into the equivalent current source, so that the source lies in free space, we can use the free-space radiation integral to express the scattered field in terms of the source. This leads to an integral equation for the unknown source. For a 2-D problem and TM^z-polarized incident field, the scattered field radiated by the source is

$$E_z^{s}(\overline{\rho}) = -\frac{k\eta}{4}\int_S d\overline{\rho}'\, H_0^{(2)}(k|\overline{\rho} - \overline{\rho}'|)J_z(\overline{\rho}') \tag{6.108}$$

Using the scattered field decomposition in (4.103), this can be rewritten to include the known incident field, so that

$$E^{i}(\overline{\rho}) = E_z(\overline{\rho}) + \frac{k\eta}{4}\int_S d\overline{\rho}'\, H_0^{(2)}(k|\overline{\rho} - \overline{\rho}'|)J_z(\overline{\rho}') \tag{6.109}$$

or, in terms of the unknown current,

$$E^{i}(\overline{\rho}) = \frac{J_z(\overline{\rho})}{j\omega\varepsilon_0[\varepsilon_r(\overline{\rho}) - 1] + \sigma(\overline{\rho})} + \frac{k\eta}{4}\int_S d\overline{\rho}'\, H_0^{(2)}(k|\overline{\rho} - \overline{\rho}'|)J_z(\overline{\rho}') \tag{6.110}$$

This integral equation has the form of a second-kind integral equation but is not a true second-kind equation in the Fredholm classification because the kernel is not square-integrable. Since $\varepsilon(\overline{r}) = \varepsilon_0$ and $\sigma(\overline{r}) = 0$ outside the scatterer, $\overline{J}$ is nonzero only inside the scatterer, so the domain of the integral equation is the volumetric region occupied by the scatterer.

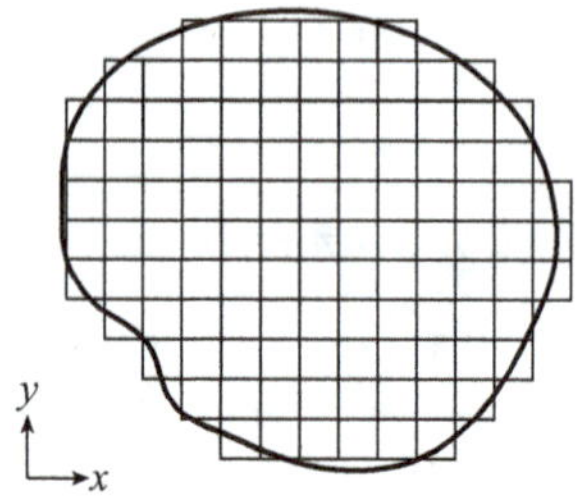

Figure 6.8 2-D inhomogeneous dielectric scatterer with grid cells for the 2-D volume method of moments

This integral equation can be solved by dividing the scatterer into N small cells as shown in Figure 6.8 and expanding J_z in terms of the 2-D pulse function

$$f_n(\overline{\rho}) = \begin{cases} 1 & \text{if } \overline{\rho} \text{ in cell } n \\ 0 & \text{otherwise} \end{cases} \tag{6.111}$$

We will test the $\overline{\rho}$ dependence with delta functions at the centers of each cell. The $\overline{\rho}'$ integrals can be evaluated using a single integration point for different testing and expansion cells, and the self-terms can be integrated analytically by approximating the cells as circular. If this is done, a linear system is obtained with matrix elements and the right-hand side given by

2-D Volume MoM for TM^z Polarized Fields

$$b_m = E_z^{\mathrm{i}}(\overline{\rho}_m) \tag{6.112}$$

$$A_{mn} = \begin{cases} \frac{\eta\pi s_n}{2} J_1(ks_n) H_0^{(2)}(k|\overline{\rho}_m - \overline{\rho}_n|) & m \neq n \\ \frac{\eta\pi s_n}{2} H_1^{(2)}(ks_n) - \frac{j\eta\varepsilon_{rn}}{k(\varepsilon_{rn}-1)} & m = n \end{cases} \tag{6.113}$$

where s_n is the radius of a circle with the same area as the nth mesh element. To account for the conductivity of a lossy scatterer, the relative permittivity in this expression is taken to be the complex permittivity

$$\varepsilon_{r,c} = \varepsilon_r + \frac{\sigma}{j\omega\varepsilon_0} \tag{6.114}$$

By using (6.107), the total field on the scatterer can be found directly from J_z. If the scattered far field is desired, that can be found in postprocessing using (6.108). Using the same approximations as in (6.113), the scattering width can be written as

$$\sigma(\phi^s) = \frac{k\eta^2}{4} \left| \sum_{n=1}^{N} a_n \frac{2\pi s_n}{k} J_1(ks_n) e^{j\overline{k}^s \cdot \overline{\rho}_n} \right|^2 \tag{6.115}$$

where a_n are the current unknowns resulting from the solution of the linear system with the right-hand side having elements (6.112) and matrix given by (6.113).

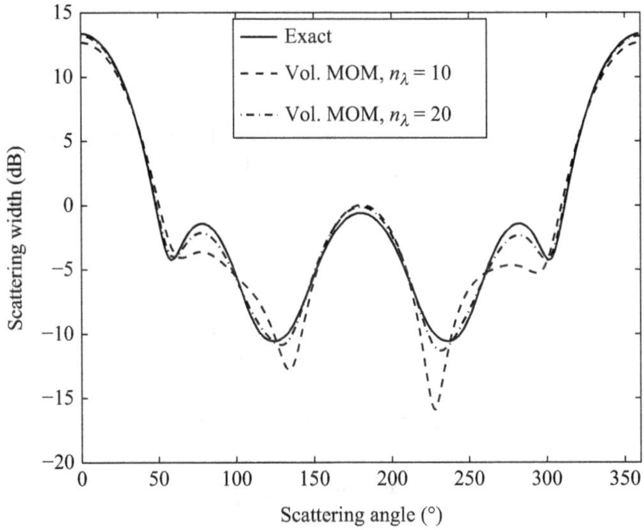

Figure 6.9 Bistatic scattering width of a dielectric cylinder with radius $a = 0.5\,\lambda$ and relative permittivity $\varepsilon_r = 2$. The analytical solution is compared with the numerical solution for two different mesh discretization densities

Using similar methods, we can derive the 2-D volume integral equation for the TE polarization and the 3-D volume electric field integral equation. Unlike the 2-D TM^z polarization that we have considered above, these integral equations involve derivative operators, which increases the singularity of the integrand, and specialized integration techniques are needed to implement the method of moments. For the 3-D volume integral equation, the mesh is three-dimensional and the mesh elements are generally cubes or tetrahedra.

6.6.1.1 Numerical Example

Computed and analytical bistatic scattering width values are shown in Figure 6.9 for a dielectric cylinder illuminated by a TM^z-polarized plane wave arriving from the incidence angle 180°. Numerical results are given for two values of the mesh discretization density. For $n_\lambda = 10$, the mesh has 73 elements, and for $n_\lambda = 20$, the mesh has 309 elements. Accuracy is not quite as good as the surface method of moments with the same mesh element density, because the volume MoM mesh is made up of square elements and the effect of geometrical modeling error is more significant than is the case for the surface MoM.

6.7 2.5-Dimensional Methods

All the numerical methods we have looked at so far are two-dimensional, because the fields and sources are constant in the z direction. Three-dimensional algorithms require vector basis functions, because the fields and currents cannot be treated as scalar quantities. Meshing a three-dimensional object can also be challenging. Before considering three-dimensional problems, it is worth observing that there is an intermediate class of problems for which the structures (conductors and dielectrics) are two-dimensional but the fields are three-dimensional. This means that the illuminating field strikes an infinitely long, cylindrical object obliquely with respect to the axis of symmetry of the object. Two-dimensional codes can be modified to accommodate the z variation of the incident field. These methods are sometimes referred to as 2.5-dimensional (2.5-D).

6.8 3-D Electric Field Integral Equation

By combining the 3-D radiation integral with a boundary condition at the surface of a PEC object, we can derive the three-dimensional electric field integral equation. The radiation integral for a 3-D time-harmonic surface current distribution $\overline{J}_s(\overline{r})$ in free space is

$$\overline{E}(\overline{r}) = -jk\eta\left[\int_S d\overline{r}'\, g(\overline{r},\overline{r}')\overline{J}_s(\overline{r}') + \frac{1}{k^2}\nabla\int_S d\overline{r}'\, g(\overline{r},\overline{r}')\nabla'\cdot\overline{J}_s(\overline{r}')\right] \tag{6.116}$$

where the 3-D scalar Green's function is

$$g(\overline{r},\overline{r}') = \frac{e^{-jk|\overline{r}-\overline{r}'|}}{4\pi|\overline{r}-\overline{r}'|} \tag{6.117}$$

For a PEC object illuminated by an incident field $\overline{E}^{\mathrm{i}}$, the electric field boundary condition at the surface of the object is

$$\hat{n}\times(\overline{E}^{\mathrm{i}}+\overline{E}^{\mathrm{s}}) = 0 \tag{6.118}$$

The scattered electric field can be expressed using the radiation integral in terms of the induced surface current on the scatterer. This leads to the electric field integral equation for 3-D fields,

$$\hat{n}\times\overline{E}^{\mathrm{i}}(\overline{r}) = \hat{n}\times jk\eta\left[\int_S d\overline{r}'\, g(\overline{r},\overline{r}')\overline{J}_s(\overline{r}') + \frac{1}{k^2}\nabla\int_S d\overline{r}'\, g(\overline{r},\overline{r}')\nabla'\cdot\overline{J}_s(\overline{r}')\right] \tag{6.119}$$

This expression is an integro-differential equation, because it includes both integral and differential operators.

Typically, the incident field is a plane wave polarized in the $\hat{\theta}$ or $\hat{\phi}$ direction, so that

$$\overline{E}^{\mathrm{i}}(\overline{r}) = \left\{\begin{matrix}\hat{\theta}\\ \hat{\phi}\end{matrix}\right\} e^{-j\overline{k}^{\mathrm{i}}\cdot\overline{r}} \tag{6.120}$$

where $\overline{k}^{\mathrm{i}}$ is the wave vector of the incident field. To characterize the scattering properties of an object, the MoM algorithm is run for each of the two possible incident wave polarizations.

Because the surface current is tangential to the surface, the unknown current has two independent components. To expand the current, two sets of scalar basis functions can be employed to expand each component separately, or a single set of nonseparable vector basis functions can be used. Two possible expansion functions are the rooftop and the Rao–Wilton–Glisson (RWG) functions.

6.8.1 Rooftop Functions

To simplify the analysis, we will develop rooftop functions for a flat PEC scatterer lying in the x–y plane. The surface current on the scatterer has the form

$$\overline{J}_s(x,y) = J_x(x,y)\hat{x} + J_y(x,y)\hat{y} \tag{6.121}$$

The divergence of the surface current in the integral on the right-hand side of the EFIE (6.119) evaluates to

$$\nabla \cdot \overline{J}_s = \frac{\partial J_x}{\partial x} + \frac{\partial J_y}{\partial y} \tag{6.122}$$

For the divergence to be finite, the basis functions used to expand J_x must be differentiable by x, and those for J_y must be differentiable by y. The simplest functions of this kind are the rooftop functions

$$t_x(x,y) = t(x)p(y) \tag{6.123a}$$

$$t_y(x,y) = p(x)t(y) \tag{6.123b}$$

where p and y are the triangle and pulse functions as defined in (6.53) and (6.55).

To generate N expansion functions, we shift t_x and t_y so that they are located at a set of rectangular grid points (x_n, y_n), so $t_{x,n}(x,y) = t_x(x - x_n, y - y_n)$ and $t_{y,n}(x,y) = t_y(x - x_n, y - y_n)$. The surface current can be expanded in terms of rooftop functions as

$$\overline{J}_s(x,y) \simeq \hat{x}\sum_{1}^{N} a_{x,n}t_{x,n}(x,y) + \hat{y}\sum_{1}^{N} a_{y,n}t_{y,n}(x,y) \tag{6.124}$$

where $a_{x,n}$ and $a_{y,n}$ are the unknown coefficients.

Rooftop functions are simple to define and are closely related to the 1-D pulse and triangle functions. Rooftop functions are rarely used in practice, however, because the support of the basis functions is rectangular, and it can be hard to mesh a complicated two-dimensional surface using rectangles. An expansion function defined on a triangular mesh is generally more convenient.

6.8.2 Rao–Wilton–Glisson (RWG) Basis

The Rao–Wilton–Glisson basis is a vector basis, meaning that instead of being constructed from scalar functions, it is defined directly as a vector field over a mesh element [11]. RWG functions are defined on pairs of triangular elements or patches. The RWG functions are edge basis functions, in the sense that each triangle edge in the interior of a mesh has an RWG function associated with it.

The mesh must be simplicial, meaning that each node is a vertex of each neighboring triangle. Such a mesh is commonly described in terms of three arrays:

$$\text{Nodes} = \begin{bmatrix} x_1 & y_1 & z_1 \\ x_2 & y_2 & z_2 \\ & \vdots & \\ x_{N_n} & y_{N_n} & z_{N_n} \end{bmatrix} \tag{6.125}$$

which gives the coordinates of each node in the mesh,

$$\text{Triangles} = \begin{bmatrix} n_{11} & n_{12} & n_{13} \\ n_{21} & n_{22} & n_{23} \\ & \vdots & \\ n_{N_p,1} & n_{N_p,2} & n_{N_p,3} \end{bmatrix} \tag{6.126}$$

which is an array of indices into the node array of the corners of each triangle in the mesh, and

$$\text{Edges} = \begin{bmatrix} n_{11} & n_{12} & p_{11} & p_{12} \\ n_{21} & n_{22} & p_{21} & p_{22} \\ & \vdots & & \\ n_{N_e,1} & n_{N_e,2} & p_{N_e,1} & p_{N_e,2} \end{bmatrix} \tag{6.127}$$

which is a list of the indices into the node array of the two endpoints of each edge and the indices into the triangle array of the two triangles adjacent to the edge. For a closed scatterer, the node indices in the triangle array are typically ordered so that they are counterclockwise relative to the outward surface normal (i.e., they are ordered according to the right-hand rule with respect to the outward direction). A mesh for a sphere is shown in Figure 6.10.

On a triangular mesh, the RWG functions are defined by

$$\overline{f}_n(\overline{r}) = \begin{cases} \frac{l_n}{2A^+}\overline{\rho}_n^+ & \text{if } \overline{r} \text{ is on the + triangle} \\ -\frac{l_n}{2A^-}\overline{\rho}_n^- & \text{if } \overline{r} \text{ is on the − triangle} \end{cases} \tag{6.128}$$

where the + and − triangles are the two triangles that share the nth edge, l_n is the length of the edge, and A^+ and A^- are the areas of the two triangles. The + triangle is usually the one indexed in the third column of the edge array, and the − triangle is the element in the fourth column. The vector $\overline{\rho}_n^+$ points from the vertex of the + triangle that is not on the edge to $\overline{r}$, so that

$$\overline{\rho}_n^+ = \overline{r} - \overline{r}_i^+ \tag{6.129}$$

where $\overline{r}_i^+$ is the position vector of the opposite vertex. Similarly, $\overline{\rho}_n^-$ is the vector from the vertex of the − triangle that is not on the edge to $\overline{r}$. An RWG basis function is illustrated in Figure 6.11.

To evaluate moment matrix elements for the EFIE, the divergence of the RWG functions must be computed. The divergence of the RWG function is

$$\nabla \cdot \overline{f}_n(\overline{r}) = \begin{cases} \frac{l_n}{A^+} & \text{if } \overline{r} \text{ is on the + triangle} \\ -\frac{l_n}{A^-} & \text{if } \overline{r} \text{ is on the } - \text{ triangle} \end{cases} \tag{6.130}$$

One important property of the RWG basis is that the vectors are tangentially continuous across patch edges, so that $\overline{f}_n(\overline{r}) = \overline{f}_m(\overline{r})$ if the two basis functions are associated with triangles that share an edge, and $\overline{r}$ lies on the shared edge. This property matches

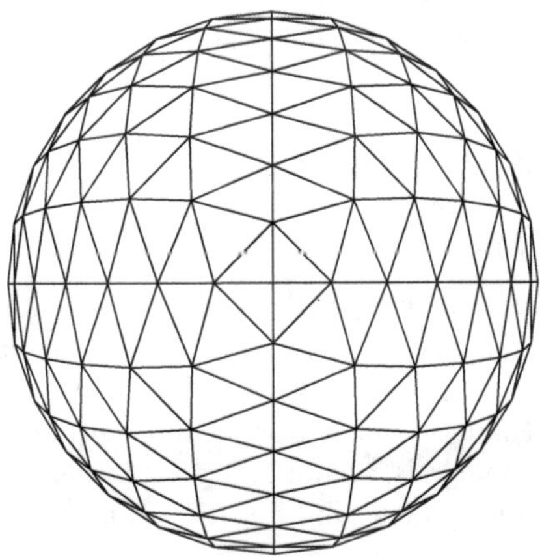

Figure 6.10 Triangular mesh for a spherical scatterer

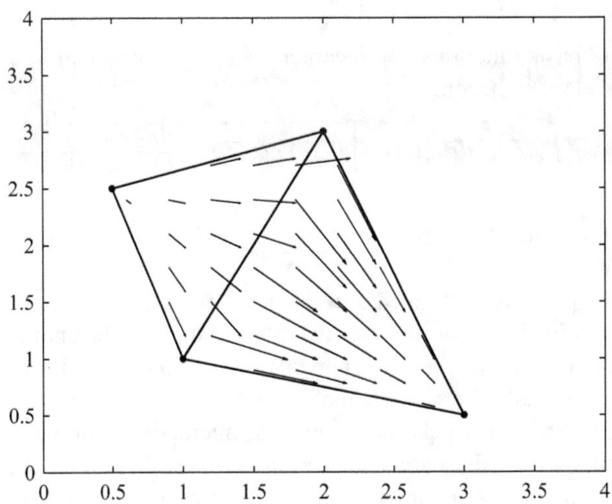

Figure 6.11 The vector field defined by an RWG vector basis function on a pair of triangular mesh elements. This basis function is associated with the edge shared by the two elements

the continuity of surface currents. We have defined the vector functions in terms of a mesh of flat patches, but the RWG basis can be modified for use on a curved mesh.

6.8.3 Method of Moments

In terms of a vector basis, the surface current can be expanded as

$$\bar{J}_s(\bar{r}) \simeq \sum_{n=1}^{N_e} a_n \bar{f}_n(\bar{r}) \tag{6.131}$$

where the a_n are unknown coefficients. Substituting this into the EFIE and using the same functions to test the observation point dependence leads to moment matrix elements of the form

$$A_{mn} = jk\eta \int d\bar{r} \int d\bar{r}' \left[\bar{f}_m(\bar{r}) \cdot \bar{f}_n(\bar{r}') g(\bar{r}, \bar{r}') + \frac{1}{k^2} \bar{f}_m(\bar{r}) \cdot \nabla g(\bar{r}, \bar{r}') \nabla' \cdot \bar{f}_n(\bar{r}') \right] \tag{6.132}$$

To arrive at this result, we have dropped the cross-product with the surface normal by assuming that the basis functions are tangential to the scatterer surface.

One problem with (6.132) is that, because the Green's function has a $1/r$ type singularity, its gradient is even more strongly singular ($1/r^2$). To avoid having to compute highly singular integrals, we must integrate by parts to move the ∇ operator from the Green's function to the testing function. This leads to the matrix element expression

$$A_{mn} = jk\eta \int d\bar{r} \int d\bar{r}' \, g(\bar{r}, \bar{r}') \left[\bar{f}_m(\bar{r}) \cdot \bar{f}_n(\bar{r}') - \frac{1}{k^2} \nabla \cdot \bar{f}_m(\bar{r}) \nabla' \cdot \bar{f}_n(\bar{r}') \right] \tag{6.133}$$

If we use RWG basis functions, the divergences can be evaluated in a simple form, and the matrix element becomes

$$\begin{aligned} A_{mn} &= jk\eta \int d\bar{r} \int d\bar{r}' \, g(\bar{r}, \bar{r}') \left[\bar{f}_m(\bar{r}) \cdot \bar{f}_n(\bar{r}') - \frac{1}{k^2} \frac{\pm l_m}{A_m^\pm} \frac{\pm l_n}{A_n^\pm} \right] \\ &= jk\eta \int d\bar{r} \int d\bar{r}' \, g(\bar{r}, \bar{r}') \frac{\pm l_m}{A_m^\pm} \frac{\pm l_n}{A_n^\pm} \left[\frac{\bar{\rho}_m^\pm \cdot \bar{\rho}_n^\pm}{4} - \frac{1}{k^2} \right] \end{aligned} \tag{6.134}$$

where the $\bar{r}$ integration is performed over the +, − triangles adjacent to the mth edge, and the $\bar{r}'$ integration is performed over the +, − triangles adjacent to the nth edge.

To compute the moment matrix elements, we use a quadrature rule defined on the triangles in the mesh. We separate the integrations into far interactions, for which the mth and nth edges are widely separated; near interactions, for which the triangles associated with the two edges are adjacent but do not overlap; and self-interactions, for which the two edges share an adjacent element or are the same edge and share both adjacent elements. The integrations are handled differently:

Far interactions: A low-order quadrature rule defined on the elements in the mesh is generally accurate enough.

Near interactions: Integrand is nearly singular, so more quadrature points or a special quadrature rule may be required.

Self-interactions: Integrand is singular, so singularity subtraction, nonclassical Gaussian quadrature, or a transformation such as Duffy's transform is used.

For far interactions, the midpoint rule can be used:

$$\begin{aligned} &\text{node:} \quad \bar{r} = \lambda_1 \bar{r}_1 + \lambda_2 \bar{r}_2 + \lambda_3 \bar{r}_3, \quad \lambda_1 = \lambda_2 = \lambda_3 = 1/3 \\ &\text{weight:} \quad w = \text{Triangle area} \end{aligned} \tag{6.135}$$

Another common rule is given by the nodes and weights

$$\begin{aligned} \text{nodes:} \quad & \lambda_{1i} = 1/3, 1/3, 1/3 \\ & \lambda_{2i} = 1/6, 1/6, 2/3 \\ & \lambda_{3i} = 1/6, 2/3, 1/6 \\ & \lambda_{4i} = 2/3, 1/6, 1/6 \end{aligned} \tag{6.136}$$

$$\text{weights:} \quad w_1 = w_2 = w_3 = w_4 = \text{Area}/4$$

6.8.3.1 Duffy's Transform

Many implementations of MoM for the EFIE have used singularity subtraction to evaluate self-interaction terms in the moment matrix, but more recently a better transformation-based approach has been developed. For near interactions, Duffy's transform can be used to remove the singularity of the integrand [12].

In the moment matrix element expression, the observation integral ($\bar{r}$) is evaluated by a standard quadrature rule. For each observation point, we break up the source triangle into subtriangles with the observation point as a vertex. For each subtriangle, we need to evaluate an integral with a singularity at one of the corners. On a canonical triangle in the plane with corners (0, 0), (1, 0), (1, 1), with the singularity at the origin as in Figure 6.12, we must evaluate an integral of the form

$$I = \int_0^1 \int_0^x \frac{f(x,y)}{r} \, dy \, dx \tag{6.137}$$

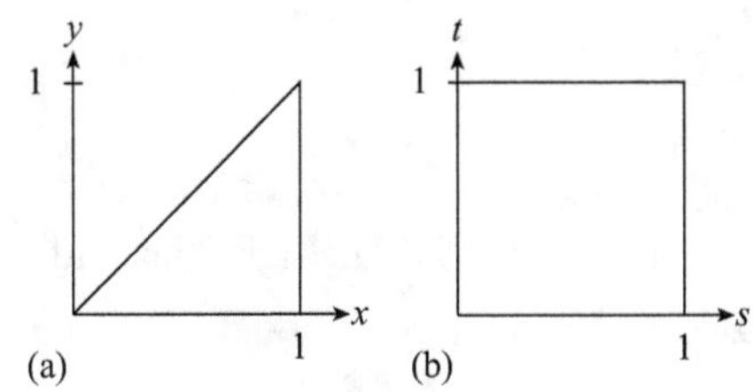

Figure 6.12 (a) Canonical triangle with singularity at origin. (b) Duffy plane. The origin maps to the unit interval on the t-axis, which spreads the singularity of the function $1/r$ *at the origin along the t-axis and makes the transformed integrand finite*

We then use the transformation $s = x$, $t = y/x$, so that $ds = dx$ and $dt = dy/x$, to obtain

$$I = \int_0^1 \int_0^1 \frac{f(s, st)}{\sqrt{s^2 + s^2t^2}} s \, dt \, ds \tag{6.138}$$

$$= \int_0^1 \int_0^1 \frac{f(s, st)}{\sqrt{1 + t^2}} dt \, ds \tag{6.139}$$

which is no longer singular. The integral is now over a square region shown in Figure 6.12(b), sometimes called the "Duffy plane."

In the Duffy plane, the nonsingular integrand can be evaluated using a simple product quadrature rule with nodes (s_n, t_n) on the unit square. These points transform to $(x, y) = (s_n, s_n t_n)$ on the canonical triangle. The points on the canonical triangle are related to points on the original mesh subtriangle by the barycentric coordinates

$$\begin{aligned} \lambda_1 &= 1 - \lambda_2 - \lambda_3 \\ \lambda_2 &= x - y \\ \lambda_3 &= y \end{aligned} \tag{6.140}$$

such that the source integration point in 3-D space is $\overline{r}'_n = \lambda_1 \overline{r}_1 + \lambda_2 \overline{r}_2 + \lambda_3 \overline{r}_3$ in terms of the vertices of the subtriangle. Barycentric coordinates are also used to define shape functions for the finite element method, as illustrated in Section 8.4.8.

The Jacobian required in the transformed integral is the product of Jacobians for the transformation from the subtriangle to the canonical triangle and from the canonical triangle to the Duffy plane:

$$\begin{aligned} |J| &= |J_1||J_2| \\ &= \frac{A_1}{A_2} \begin{vmatrix} \frac{\partial x}{\partial s} & \frac{\partial x}{\partial t} \\ \frac{\partial y}{\partial s} & \frac{\partial y}{\partial t} \end{vmatrix} \\ &= 2A_1 \begin{vmatrix} 1 & 0 \\ t & s \end{vmatrix} \\ &= 2A_1 s \end{aligned} \tag{6.141}$$

where A_1 is the area of the subtriangle and $A_2 = 1/2$ is the area of the canonical triangle.

6.8.3.2 Incident Field

The right-hand side of the EFIE can be discretized using the same integration rule as for far interactions in the moment matrix. The right-hand side elements are

$$\begin{aligned} b_m &= \int d\overline{r} \overline{f}_n \cdot \overline{E}^{\mathrm{i}}(\overline{r}) \\ &= \sum_{a=1}^{2} \int d\overline{r} \frac{l_m}{A_{m,a}} \overline{\rho}_{m,a}(\overline{r}) \cdot \overline{E}^{\mathrm{i}}(\overline{r}) \end{aligned} \tag{6.142}$$

where $a = 1, 2$ denotes the + and − triangles that share the mth edge, respectively. The integral can be evaluated with the triangle midpoint rule (6.135).

6.8.3.3 Postprocessing

After solving for the unknown coefficients, from the current solution (6.131), impedances, scattered fields, far fields, and RCS can be computed in postprocessing. It can be shown that the RCS is given by

$$\sigma_{3\text{-D}} = \frac{(k\eta)^2}{4\pi}\left|\int_S d\overline{r}\,\overline{E}^{s*}(\overline{r})\cdot\overline{J}_s(\overline{r})\right|^2 \tag{6.143}$$

where $\overline{E}^{s*}(\overline{r})$ represents a plane wave in the scattering direction. The scattered plane wave is polarized in the $\hat{\theta}$ or $\hat{\phi}$ direction, so that

$$\overline{E}^{s}(\overline{r}) = \begin{Bmatrix}\hat{\theta}\\ \hat{\phi}\end{Bmatrix} e^{-j\overline{k}^{s}\cdot\overline{r}} \tag{6.144}$$

where $\overline{k}^{s}$ is a vector in the scattering direction.

As with the 2-D MoM, the integral in (6.143) can be evaluated by substituting expansion (6.131) for the surface current density in terms of basis functions. This leads to

$$\sigma_{3\text{-D}} = \frac{(k\eta)^2}{4\pi}\left|\sum_{n=1}^{N} a_n \int_S d\overline{r}\,\overline{E}^{s*}(\overline{r})\cdot\overline{f}_n(\overline{r})\right|^2 \tag{6.145}$$

The integral over the basis function $\overline{f}_n$ should be evaluated using the same quadrature rule as used in Section 6.8.3 for the source point integration and far interactions. For a given pair of incident and scattering directions, there are four possible combinations of polarizations for the incident and scattered plane waves, so in general we must compute $\sigma_{\theta\theta}$, $\sigma_{\theta\phi}$, $\sigma_{\phi\theta}$, and $\sigma_{\phi\phi}$ to characterize the scattering properties of an object.

Problems

6.1 Apply the method of moments to the blurring operator

$$\int_{-1}^{1} e^{-(x-y)^2/a^2} f(y)\,dy = g(x)$$

In terms of the kernel and right-hand side of this integral equation, give simple, closed-form expressions for (a) the elements of a matrix A, and (b) the elements of a right-hand side vector b for a linear system $Ax = b$ that can be solved to find an approximation for f. (c) Give an expression for the resulting approximation for f in terms of the elements of the solution vector x to the linear system.

6.2 For the method of moments applied to the EFIE for the 2-D TM polarization, the diagonal elements of the moment matrix can be approximated by

$$A_{nn} = \frac{k\eta}{4}h\left[1 - j\frac{2}{\pi}\log\left(\frac{e^{\gamma-1}kh}{4}\right)\right]$$

where $\gamma \simeq 0.577215665$ is Euler's constant. Derive this result.

6.3 Implement the 2-D MoM to solve the TM-EFIE for the surface current on a PEC cylinder of a half wavelength radius due to a TM-polarized incident plane wave traveling in the $-x$ direction. Compare with the analytical solution

$$J_z(\phi) = \frac{2E_0}{\pi ka\eta} \sum_{n=-\infty}^{\infty} \frac{j^n}{H_n^{(2)}(ka)} e^{jn\phi}$$

where E_0 is the amplitude of the incident field. Does the MoM solution converge as the number of unknowns is increased? If so, how rapidly?

6.4 Add postprocessing to the TM-EFIE MoM code of the previous problem to compute the bistatic scattering amplitude and scattering width of the scatterer from the surface current. Check the results by comparing with the analytical solution for the circular cylinder.

6.5 Implement the polyarc method described in the notes for meshing shapes with flat or curved sides. Compute the bistatic scattering width for a square PEC cylinder of side length 1 λ, illuminated by a plane wave traveling in the $-x$ direction. Compare with results obtained using the FDTD method. Give a physical explanation for the behavior of the scattering width versus scattering angle.

6.6 Solve the MFIE for the TM and TE polarizations using the method of moments. Approximate the diagonal elements of the moment matrix as $1/2$ and evaluate off-diagonal matrix elements using a single-point integration rule. Test the codes for a circular cylinder.

6.7 (a) Run the TM-EFIE MoM algorithm for a PEC circular cylinder of a fixed radius in a loop over the frequency of the incident field. Plot the error in the backscattering width as a function of frequency, and identify one of the internal resonance frequencies of the cylinder. Hint: the first internal resonance of the circular cylinder occurs when $ka \simeq 2.405$, where a is the cylinder radius. (b) Combine the TM-EFIE and TM-MFIE MoM algorithms to solve the CFIE, and show that the solution error is smaller at the internal resonance frequency.

6.8 Implement the 2-D volume method of moments for the TM polarization. Compute the bistatic scattering width for a circular dielectric cylinder of radius $\lambda/2$ and relative permittivity $\varepsilon_r = 2$ at 300 MHz. Compare the bistatic scattering width in dB obtained from the volume MoM with results using FDTD. Give the values of n_λ used for each method.

6.9 Use the Duffy transformation to integrate e^{jkR}/R, where $R = |\overline{r} - \overline{r}'|$, over the triangle with vertices $(0, 0, 0)$, $(h, 0, 0)$, and $(0, h, 0)$, with $\overline{r} = \overline{r}' = (0, 0, 0)$. Validate the Duffy transform integration using a product midpoint rule. Hint: For the product midpoint rule, generate a dense rectangular grid of points and use `inpolygon` to select integration points lying inside the triangle.

6.10 Develop a meshing algorithm for a sphere of radius a. Divide the sphere into octants using the coordinate planes. Choose one of the octants and mesh the spherical triangle in that octant. The mesh density is controlled by the number n_t of triangle edges along one side of the spherical triangle. Generate node points corresponding to the corners of one mesh triangle at a corner of the octant, three in the next row, five in the third row, and so forth until the octant is fully meshed with n_t rows of triangles. At the same time, create edge and triangle arrays for the mesh. The remaining seven octants can be meshed by rotating the node points for the first octant to generate node points for the other octants and adding the nodes to the node, edge, and triangle arrays. A triangular mesh for a sphere is shown in Figure 6.10.

6.11 Implement the method of moments for the 3-D EFIE. Use Duffy's transform to evaluate diagonal moment matrix elements, and a low-order quadrature rule for off-diagonal moment matrix elements. The discretized incident and scattered plane waves can be computed using a single-point quadrature rule with integration point at the element center. Validate the code by computing the bistatic radar cross section of a PEC sphere and comparing with the exact solution shown in Figure 6.13.

6.12 Develop a code to create a mesh for a rectangular simulation domain. The inputs to the code are the mesh element length h and coordinates (x_1, y_1) and (x_2, y_2) of the lower left and upper right corners of the rectangular domain. (a) Generate a rectangular grid of points on the domain with spacing approximately equal to h, and use this to create the mesh node list or node array. (b) The rectangular grid defines a square tiling of the domain. For each square, determine the global node numbers (indices of the points in the node list) of each of the four corner points, and store the node numbers for the two triangles that span the grid square in the triangle array. (c) Loop through the mesh to create an edge array as defined by (6.127).

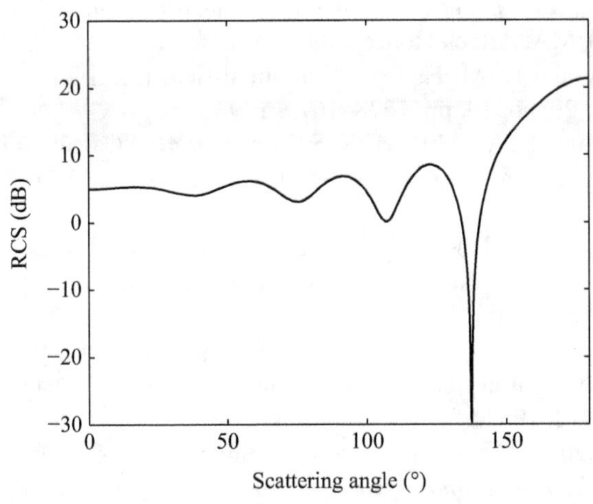

Figure 6.13 Exact solution for the RCS of a PEC sphere with radius 1 λ. The sphere is illuminated by a plane wave arriving from the θ = 0 direction, and the RCS is shown along a cut from θ = 0 to θ = 180°

6.13 Apply the MOM algorithm for the 3-D EFIE to find the bistatic RCS of a square plate. The algorithm of Problems 6.12 can be used to generate the required mesh (another approach to meshing a rectangle is given in Problem 8.11). Compare the results to the PO approximation for specular backscattering from a plate [6],

$$\sigma_{3\text{-D}} \simeq 4\pi\left(\frac{A}{\lambda}\right)^2 \tag{6.146}$$

where A is the plate area.

References

[1] R. C. Hanson, Ed., *Moment Methods in Antennas and Scattering*. Norwood, MA: Artech House, 1990.

[2] A. F. Peterson, R. Mittra, and S. L. Ray, *Computational Methods for Electromagnetics*. New York, NY: IEEE Press, 1998.

[3] K. Umashankar and A. Taflove, *Computational Electromagnetics*. Norwood, MA: Artech House, 1993.

[4] M. N. O. Sadiku, *Numerical Techniques in Electromagnetics with MATLAB*. Boca Raton, FL: CRC Press, 2009.

[5] D. Weile, G. Pisharody, N. Chen, B. Shanker, and E. Michielssen, "A novel scheme for the solution of the time-domain integral equations of electromagnetics," IEEE Trans. Antennas Propag., vol. 52, no. 1, pp. 283–295, 2004.

[6] C. A. Balanis, *Advanced Engineering Electromagnetics*. New York, NY: John Wiley & Sons, 1989.

[7] K. F. Warnick, *Numerical Analysis for Electromagnetic Integral Equations*. Norwood, MA: Artech House, 2008.

[8] A. Helaly and H. M. Fahmy, "Combined-field integral equation," Electron. Lett., vol. 29, no. 19, pp. 1678–1679, 1993.

[9] F. K. Oshiro, K. M. Mitzner, S. S. Locus, *et al.*, "Calculation of radar cross section," Air Force Avionics Laboratory, Tech. Rep. AFAL-Tr-7021, part II, 1970.

[10] C. P. Davis and K. F. Warnick, "Error analysis of 2D MoM for MFIE/EFIE/CFIE based on the circular cylinder," IEEE Trans. Antennas Propag., vol. 53, no. 1, pp. 321–331, 2005.

[11] S. M. Rao, D. R. Wilton, and A. W. Glisson, "Electromagnetic scattering by surfaces of arbitrary shape," IEEE Trans. Antennas Propag., vol. 30, no. 3, pp. 409–418, 1982.

[12] S. Wandzura, "Accuracy in computation of matrix elements of singular kernels," in *11th Annual Review of Progress in Applied Computational Electromagnetics*, Naval Postgraduate School, vol. II, Monterey, CA, 1995, pp. 1170–1176.

Chapter 7
Solving Linear Systems

Solving large linear systems is a key computational task that forms the core of many numerical methods. Algorithms from all the areas of scientific computations lead to large linear systems that must be solved for a vector of unknowns, or to the closely related problem of finding eigenvalues and eigenvectors of a large matrix. We have already seen that the finite difference method and the method of moments reduce a boundary value problem to a linear system of the form

Linear System

$$Ax = b \tag{7.1}$$

where A is a matrix, b is a given vector, and the vector x is unknown. A linear system is also produced by the finite element method to be discussed in Chapter 8. The field of study that deals with algorithms for solving linear systems and performing other types of matrix operations is referred to as numerical linear algebra.

Linear system solution methods can be divided into direct and indirect methods. Direct methods produce a matrix inverse or a factorization with which the solution to a linear system can be rapidly obtained. We will not spend much time discussing direct methods, since they are well understood and readily available in standard numerical code libraries. Indirect methods can be faster than direct solution algorithms but are more strongly dependent on the properties of the linear system to be solved. For this reason, users and code developers need a fairly deep understanding of matrix properties and how these properties interact with indirect solution algorithms to make educated decisions when implementing these algorithms.

7.1 Linear Spaces and Linear Operators

A square matrix is a linear operator on a finite-dimensional linear space. Because the theory of matrices and linear operators plays a central role in computational science, in this section we will develop some of the fundamental concepts associated with linear spaces and operators on linear spaces. In covering this material, the tone will be more abstract than that of earlier chapters. Despite the abstract nature of the topic, effort devoted to learning linear operator theory will be rewarded by a better understanding of not only the algorithms and techniques of numerical linear algebra but also mathematical methods used in many branches of engineering such as control theory, signal processing, and network analysis.

7.1.1 Linear Spaces

A linear space is a vector space combined with a scalar field, where the vectors are objects that can be added and multiplied by scalars. Linear space and vector space are often used to mean the same thing, so there is not really a distinction between the two terms. We will use the term linear space to emphasize that the same mathematical structure can represent many different physical quantities, including some that are quite unlike the geometrical picture of the arrows in space that we usually visualize when thinking of vector spaces. Matrices of a given dimension, for example, are members of a linear space.

The scalar field is often the real numbers but can also be complex numbers or another more general mathematical field. The most common type of vector consists of a direction and magnitude in three-dimensional (3-D) space. A three-dimensional vector of this type might represent the velocity of an object, an electric field intensity at a point, or a displacement from one point to another. A linear space can represent many different types of physical quantities but in mathematical terms is defined in terms of algebraic properties that are independent of any particular physical meaning of the vectors in the space.

Mathematically, a linear space is a collection of vectors that can be added, so that if x_1 and x_2 are vectors in the space, then the sum

$$x_3 = x_1 + x_2 \tag{7.2}$$

is also in the linear space. Addition of vectors must be associate and commutative, so that

$$x_1 + (x_2 + x_3) = (x_1 + x_2) + x_3 \quad \text{(associativity)} \tag{7.3}$$

$$x_1 + x_2 = x_2 + x_1 \quad \text{(commutativity)} \tag{7.4}$$

The linear space must include a zero element that when added to x yields x for all vectors in the space, and an additive inverse $-x$ that when added to x yields the zero vector. The linear space includes a field of scalars, represented by α, such that

$$x_4 = \alpha x_1 \tag{7.5}$$

is also in the linear space. Scalar multiplication must be associative and distributive, so that

$$\alpha(\beta x) = (\alpha\beta)x \quad \text{(associativity of scalar multiplication)} \tag{7.6}$$

$$(\alpha + \beta)x = \alpha x + \beta x \quad \text{(distributivity of scalar sums)} \tag{7.7}$$

$$\alpha(x_1 + x_2) = \alpha x_1 + \alpha x_2 \quad \text{(distributivity of vector sums)} \tag{7.8}$$

Finally, the scalar field must include an identity $\alpha = 1$ that when multiplied by x yields x for all vectors.

A particular notation for a given type of vector, such as the expansion of a three-dimensional vector in terms of unit vectors in (2.7), is referred to as a representation of the linear space. It is easy to verify that 3-D vectors of the form (2.7) with either the real numbers or complex numbers as the scalar field satisfies all of the required properties.

In the engineering literature, it is common to denote vectors with decorations such as overbars, arrows, or bold type. For mathematical treatments of linear algebra and operators, to avoid excessive typographical busyness, matrices and vectors are given simple roman italic symbols. We will follow the latter approach in this chapter, so that x represents a vector, rather than a coordinate in three-dimensional space. The difference between the coordinate and the vector should be clear from context.

7.1.2 Norms on Linear Spaces

A norm on a linear space is a mapping from vectors in the space to the nonnegative real numbers. A norm must have the properties given in Section 6.5.2.1, which we repeat here using a slightly different notation:

1. $\|x\| > 0$ if $x \neq 0$, $\|x\| = 0$ if $x = 0$ (positive definite)
2. $\|x + y\| \leq \|x\| + \|y\|$ (triangle inequality)
3. $\|\alpha x\| = |\alpha|\,\|x\|$ (scalar multiplication)

where x is a vector in the linear space and α is a scalar.

Norms and inner products are closely related. A norm can be written in terms of an inner product, so that

$$\|x\| = |\langle x, x\rangle|^{1/2} \tag{7.9}$$

It is also possible to derive an inner product from a norm. The inner product may be the L_2 inner product defined in (6.73), a weighted inner product as in (5.23), or any other inner product that satisfies the required properties.

For a finite-dimensional vector space, some examples of norms are

p-norm: $\|x\|_p = (\sum_{n=1}^{N} |x_n|^p)^{1/p}$
Weighted norm: $\|x\|_W = \|Wx\|$ where W is a Hermitian, positive definite linear operator

The p-norm with $p = 2$, or 2-norm, is the standard Euclidean distance, and in practice is the most common norm. The 2-norm of a vector is proportional to the root mean square (RMS) value of the elements of the vector.

Examples of linear spaces with norms, or normed linear spaces, include

N-dimensional Euclidean space R^N

Complex plane, $\|z\| = |z|$

N-dimensional complex vectors, C^N, with $\|x\| = |x^H x|^{1/2}$, where the superscript ^{H} denotes the conjugate transpose operation or Hermitian conjugate

L_2, the space of square integrable functions

Other function spaces such as continuous functions and continuously differentiable functions

An arbitrary function cannot be represented as a finite linear combination of basis functions, which implies that the last two linear spaces are examples of infinite-dimensional spaces. Since we are interested here in matrix operators, which operate on finite-dimensional spaces, the most important linear spaces for this chapter are R^N and C^N.

7.1.3 Linear Operators

An operator A maps a normed linear space into itself. This can be represented using the notation

Operator on a Linear Space

$$A : L \rightarrow L \tag{7.10}$$

where L is a linear space. This means that if x is a vector in the linear space, the result Ax obtained by applying the operator A to the vector x is also in the space.

The most important class of operators are the linear operators. An operator is linear if it satisfies

1. $A(\alpha x) = \alpha Ax$
2. $A(x_1 + x_2) = Ax_1 + Ax_2$

for all vectors x, x_1, and x_2. It is easy to show that linear operators are members of a linear space. If the linear space L has dimension N, A can be represented as an $N \times N$ matrix.

A transformation maps one linear space to another linear space and is more general than an operator, since the domain and range spaces can be different. For finite-dimensional spaces, a linear transformation can be represented as a nonsquare matrix.

7.1.4 Operator Norms

If the linear space on which the operator acts has a norm (which we will refer to as a vector norm), a norm on the space of operators can be constructed from the vector norm. This norm is referred to as the induced operator norm. A vector norm induces an operator norm by the relationship

$$\|A\| = \sup_x \frac{\|Ax\|}{\|x\|} \tag{7.11}$$

where sup means supremum, or the largest limiting value of the argument over all x.

Some operator norms are not induced by vector norms, such as the Frobenius norm

$$\|A\|_{\text{Fro}} = \left(\sum_{m=1}^{N}\sum_{n=1}^{N}|A_{mn}|^2\right)^{1/2} = \sqrt{\text{tr}(AA^H)} \tag{7.12}$$

The Frobenius norm is the RMS value of the elements of the matrix.

7.1.5 Range, Null Space, and Rank

For a finite-dimensional linear space, the range, null space, and rank are defined by

Range, Null Space, and Rank

$$\begin{aligned}\text{range}(A) &= \text{subspace spanned by vectors of the form } Ax\\ \text{null}(A) &= \text{all vectors satisfying } Ax = 0\\ \text{rank}(A) &= \text{dimension of range of } A\end{aligned}$$

where the span of a set of vectors is the linear space consisting of the set of all linear combinations of the vectors. The dimensions of the range and null spaces must satisfy

Fundamental Theorem of Linear Algebra

$$\text{rank}(A) + \dim(\text{null}(A)) = N \tag{7.13}$$

where N is the dimension of the linear space on which A operates.

If we represent A as a matrix,

$$\text{range}(A) = \text{span}\{\text{columns of } A\} = \text{span}\{\text{rows of } A\}$$

where $\text{span}\{x_1, x_2, \ldots\}$ is the space of vectors of the form $\sum_n a_n x_n$ for all possible sets of scalars a_n. If $Ax = y$, x can be viewed as the set of scalars that allows y to be expressed as a linear combination of the columns of A.

7.1.6 Operator Inverse and Adjoint

If $\text{null}(A) = \{0\}$, A maps each distinct vector x to a unique vector Ax, and the operator is said to be one to one. In this case, the operator can be inverted and A^{-1} exists. An invertible matrix is said to be nonsingular, and the inverse operator can be represented in terms of determinants of matrix cofactors, although this is typically used as a calculation method only for small matrices. For large matrices, as will be discussed in Section 7.2, the inverse operator is rarely computed explicitly.

The adjoint operator is often denoted as $A^{\dagger}$ and is defined as the operator that satisfies

$$\langle Av, w \rangle = \langle v, A^{\dagger} w \rangle \tag{7.14}$$

for all vectors v, w. The inner product can be any inner product associated with the linear space. The adjoint of an operator depends on the inner product. A given operator in general has a different adjoint with respect to each unique inner product.

By far, the most common choice for the inner product in applications of linear operators is the 2-norm discussed in Section 7.1.2. If the operator A is finite-dimensional and the norm on the underlying linear space L is the 2-norm, then it can be shown from this definition that the adjoint operator is the Hermitian or conjugate transpose of the matrix representation of A.

Other notations for the adjoint operator with respect to the 2-norm include A^*, A', and A^H. The notation A^H refers to the Hermitian conjugate and is commonly used for matrices. Since we are mainly interested in matrix operators, A^H will be the notation of choice in this book. The matrix transpose (without conjugation) is A^T. If the scalar field associated with the linear space is the real numbers (i.e., for real matrices), the matrix adjoint and the matrix transpose are the same.

7.1.7 Classes of Operators

Continuous linear operators can be grouped into several classes:

Classes of Linear Operators

1. Normal: $A^H A = AA^H$.
2. Self-adjoint or Hermitian: $H = H^H$.
3. Unitary: $U^H U = I$. A unitary matrix is norm-preserving: $\|Ux\| = \|x\|$.
4. Projection: $P^2 = P$. The projection is orthogonal if the range and null spaces are orthogonal.

Hermitian and unitary operators are found more commonly in applications than normal operators, but normal linear operators are also important. Normality is a necessary and sufficient condition for a matrix to be unitarily diagonalizable (see Section 7.1.10.1). A unitary matrix is often referred to as a rotation matrix, since it does not change the length of a vector in the norm for which the adjoint (7.14) is defined. A diagram that shows how these properties are related is shown in Figure 7.1.

Some physical problems, such as reciprocal electromagnetic systems, lead to symmetric operators with complex coefficients, or complex symmetric operators ($A = A^T$, $A \neq A^H$). This property has important physical meaning (electromagnetic reciprocity), but it is not a very restrictive or useful property from a linear operator

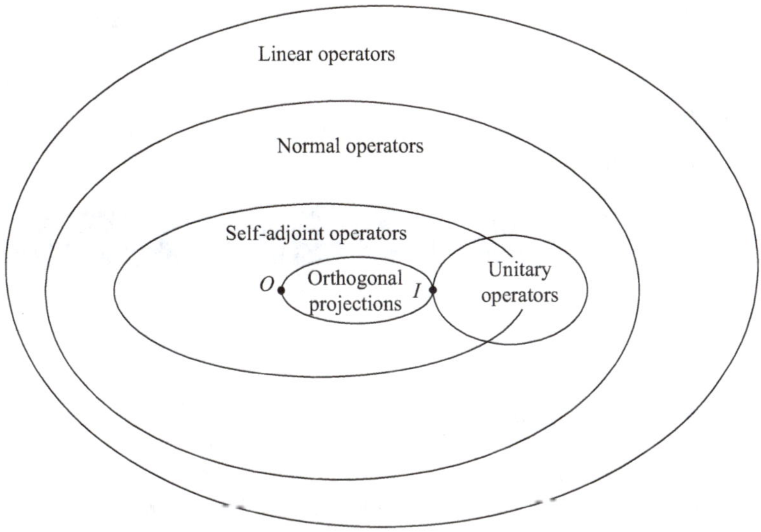

Figure 7.1 Venn diagram of operator classes

point of view. A real symmetric operator is self-adjoint, but a complex symmetric operator is not.

For matrix operators with real elements, the Hermitian and unitary properties reduce to

Classes of Real Linear Operators

1. Symmetric: $S = S^T$
2. Orthogonal: $O^T O = I$

If a two-dimensional (2-D) or 3-D vector is rotated by a given angle, the original and rotated vectors are related by an orthogonal matrix. If the coordinate system is rotated, the elements of the vector in the original and rotated coordinate systems are also related by an orthogonal matrix.

Other operator properties include the following:

Other Operator Properties

1. Positive: $\langle x, Ax \rangle \geq 0$ for all x
2. Positive definite: $\langle x, Ax \rangle > 0$ for all $x \neq 0$
3. Symmetric positive definite (SPD): $A = A^T$ and A is real and positive definite.

The last property will be used in Section 7.5.1 when the conjugate gradient iterative linear system solution method is presented.

7.1.8 Eigenvalues and Eigenvectors

If the nonzero vector x is a solution to

Eigenvalue and Eigenvector Equation

$$Ax = \lambda x \tag{7.15}$$

for some scalar λ, x is said to be an eigenvector of x, and λ is an eigenvalue. An operator on an N-dimensional space has N eigenvalues, but some may be repeated. For a nonsingular operator, all the eigenvalues are nonzero. If the scalar field is complex, the eigenvalues are in general complex. Hermitian operators have only real eigenvalues.

One way to find the eigenvalues of A is to find the zeros of its characteristic polynomial

$$p(\lambda) = \det [\lambda I - A] \tag{7.16}$$

This polynomial has order N and, therefore, has N zeros, which are the eigenvalues of the matrix A. Occasionally useful when working with matrices is that by the Cayley–Hamilton theorem the matrix A satisfies its own characteristic equation, so that $p(A) = 0$. Finding the eigenvalues and eigenvectors of a matrix is very important in matrix theory as well as in practical applications involving matrices and represents one of the fundamental tasks of matrix analysis. We will see later in this chapter that algorithms for finding eigenvalues and eigenvectors are closely related to algorithms that solve the linear system $Ax = b$.

7.1.9 Field of Values

The field of values (FOV) of an operator is the set of all complex numbers of the form $x^H Ax$, $\|x\| = 1$ or, alternatively, the set of numbers

$$f = \frac{x^H Ax}{x^H x} \tag{7.17}$$

for all vectors x. If A is normal, the FOV is the convex hull of the eigenvalues of A. If A is nonnormal, the FOV includes the convex hull of the eigenvalues as a subset.

7.1.10 Matrix Decompositions

Decompositions provide a way to represent a matrix in terms of a product of simpler matrices with well-understood properties. They are among the most important of all the tools provided by linear operator theory. Decompositions exist for both finite- and infinite-dimensional operators, but since the focus of this chapter is on matrices, we will present the decompositions in terms of finite-dimensional operators.

7.1.10.1 Unitary Diagonalization

The most important type of decomposition is unitary diagonalization. It arises from the eigenvalue–eigenvector expression (7.15), which we will modify to include an index for the eigenvalues and eigenvectors, so that

$$Av_n = \lambda_n v_n \tag{7.18}$$

where $n = 1, 2, \ldots, N$. If there exist N linearly independent solutions v_n to this equation,

Unitary Diagonalization

$$A = UDU^H \tag{7.19}$$

where

$$U = \left[\begin{array}{c|c|c|c} v_1 & v_2 & \cdots & v_N \end{array} \right] \tag{7.20}$$

$$D = \begin{bmatrix} \lambda_1 & & & \\ & \lambda_2 & & \\ & & \ddots & \\ & & & \lambda_N \end{bmatrix} \tag{7.21}$$

and the eigenvectors are normalized such that $\|v_n\| = 1$. It can be proved that the eigenvectors are orthogonal, which means that U is unitary.

Many theorems clarify which matrices are unitarily diagonalizable and which are not. A is unitarily diagonalizable if the eigenvalues are distinct or if the eigenvectors span R^N, where A is an $N \times N$ matrix. Most importantly, A is unitarily diagonalizable if and only if it is normal, so there is a one-to-one correspondence between the set of unitarily diagonalizable matrices and the normal matrices.

Physically, a unitarily diagonalizable operator corresponds to a system for which behavior is governed by a set of uncoupled modes. Since unitary diagonalization is the most common type of diagonalization, we sometimes say that a unitarily diagonalizable matrix is diagonalizable (i.e., without the modifier “unitarily”).

To find the unitary diagonalization of a matrix using MATLAB®, the command

```
[U,D] = eig(A);
```

can be used. The matrix `D` is diagonal, with the eigenvalues of the matrix `A` along the diagonal. If `A` is unitarily diagonal, the matrix `A` is given by

```
A = U*D*U';
```

If `A` is not unitarily diagonalizable, `U` is not a unitary matrix, and the relationship among `A`, `U`, and `D` becomes

```
A*U = U*D;
```

If only the eigenvalues are needed, the single-output form `e = eig(A);` can be used instead.

7.1.10.2 Other Matrix Decompositions

For nonnormal matrices, other types of decompositions can be applied:

Jordan canonical form: $A = UJU^H$, where U is unitary, and J is block diagonal with blocks that are either diagonal or are of the form

$$J_n = \begin{bmatrix} \lambda_n & 1 & & & \\ & \lambda_n & 1 & & \\ & & \ddots & \ddots & \\ & & & \lambda_n & 1 \end{bmatrix} \tag{7.22}$$

where λ_n is an eigenvalue of A. For a unitarily diagonalizable matrix, all the Jordan blocks have a size of one by one, so the matrix J is diagonal.

Schur: $A = UTU^H$, where T is triangular and has the eigenvalues of A along the diagonal. Algorithms for computing the Schur decomposition are more stable than those used to compute the Jordan canonical form, so the Schur decomposition is more commonly used in practice.

Similarity: $A = WDW^{-1}$, where W is not necessarily unitary.

Singular value decomposition (SVD): $A = USV^H$, U and V are unitary, and S is diagonal with the square roots of the eigenvalues of A^HA (singular values) on the diagonal. If A is normal, the singular values are the magnitudes of the eigenvalues.

The eigenvectors defined by (7.18) are right eigenvectors. Left eigenvectors can be defined by $A^Hw_n = \lambda_n w_n$. These can be used to create a decomposition similar to the SVD, so that $A = VDW$, where D has the eigenvalues of A along the diagonal rather than the singular values, and W may not be unitary.

7.1.10.3 Decomposition Using Projection Operators

Matrix decompositions can also be presented from a projection operator point of view. The rank one matrix

$$P = vv^H \tag{7.23}$$

where v is a vector normalized such that $\|v\| = 1$ is a projection operator that projects a vector onto the one-dimensional subspace spanned by the vector v. It is easy to show that $Px - x$ is orthogonal to x, so that P is also an orthogonal projection operator. A unitarily diagonalizable matrix can be expressed as a weighted sum of projections,

$$P_n = v_n v_n^H \tag{7.24}$$

$$A = \sum_{n=1}^{N} \lambda_n P_n \tag{7.25}$$

Any matrix can be expressed as

$$A = \sum_{n=1}^{N} s_n u_n v_n^H \tag{7.26}$$

using the SVD, where s_n are the singular values of the matrix, and u_n and v_n are sets of mutually orthonormal vectors.

7.1.11 Other Matrix Formulas

If A is nonsingular and v and w are vectors with commensurate dimensions,

$$(A + vw^H)^{-1} = A^{-1} - \frac{(A^{-1}v)(w^H A^{-1})}{1 + w^H A^{-1} v} \tag{7.27}$$

vw^H is a rank one matrix. This formula gives the inverse of a rank one perturbation of the matrix A.

Linear operator theory is a rich mathematical field. There are many other useful and interesting theorems, identities, and results that could be surveyed. This brief review of linear operator theory is adequate for the purposes of this chapter, but if a more in-depth treatment is desired, many excellent textbooks on the subject are available, including [1,2].

7.2 Direct and Iterative Solution Methods

Now that we have set the stage by reviewing basic concepts of linear operator theory, we are prepared to study the theory of linear system solution algorithms. The most obvious approach to solving the linear system

$$Ax = b \tag{7.28}$$

is to construct the matrix inverse A^{-1}. This can be done by forming ratios of determinants of submatrices of A. Computing the matrix inverse in this way is computationally very inefficient, since $O(N!)$ operations are needed to compute one $N \times N$ matrix determinant using expansion by minors. This would require an unacceptably long computation time for large matrices. For large matrices, therefore, direct computation of the inverse is rarely done in practice.

There is a class of relatively obscure algorithms that can give the inverse in less than $O(N^3)$ operations. These are based on Strassen's algorithm, which makes the use of repetitiveness in the computation of matrix multiplication to reduce computational cost. Algorithmic complexity orders as low as $O(N^{2.376})$ have been achieved [3]. Due to problems with instability, complexity of implementation, and a high constant factor in the computational cost, these methods are used in practice even less frequently than direct matrix inversion.

A better approach is to factor A into a product of upper and lower triangular matrices using Gaussian elimination. This results in the decomposition $A = LU$, where the matrix L is lower triangular, and U is upper triangular. This allows rapid solution of a linear system, since $A^{-1}b$ can be computed very rapidly if A is triangular using backsubstitution. The method of Gaussian elimination for LU decomposition requires on the order of N^3 operations for an $N \times N$ matrix. LU decomposition is referred to as a direct solution method, because it produces an accurate solution for a linear system in a fixed number of computational steps.

Although the computational cost of $O(N^3)$ for LU decomposition is lower than the cost of computing the matrix inverse using determinants, for very large linear systems even $O(N^3)$ operations can be prohibitively time-consuming. For many problems, iterative solution algorithms can be used to obtain an approximate solution with far fewer computational steps than LU decomposition. A major research direction in numerical linear algebra over the last couple of decades has been the search for indirect iterative solution algorithms with lower computational cost than LU decomposition. Because iterative algorithms are more sensitive to the particular properties of the matrix A than direct methods, many different iterative algorithms methods have been developed that have been found to work well for specific classes of matrices.

Examples of direct and iterative linear system solution methods include the following:

Linear System Solution Methods

1. *Direct*
    - LU decomposition by Gaussian elimination
    - Strassen's algorithm and modifications
2. *Iterative*
    - Stationary iterations
        - Jacobi
        - Gauss–Seidel
        - Successive overrelaxation (SSOR)
    - Nonstationary iterations
        - Krylov subspace and related methods
            - Conjugate gradient (CG)
            - Conjugate gradient on the normal equations (CGNE)
            - Biconjugate gradient (BCG)
            - Biconjugate gradient-stabilized (BCG-stab)

Generalized minimum residual (GMRES)
Quasi-minimum residual (QMR)
Conjugate gradient squared (CGS)
Other
Chebyshev iteration

7.3 *LU* Decomposition

The most important direct linear system solution method is *LU* decomposition using the Gaussian elimination algorithm. The main advantages of *LU* decomposition are its numerical stability and robustness, since Gaussian elimination is accurate for essentially all matrices that arise in scientific computation. Another advantage is that multiple right-hand sides (RHSs) can be solved in $O(N)$ operations per additional RHS.

The numerical stability analysis of Gaussian elimination is an interesting story. Some algorithms can be proved to be either accurate or numerically unstable with modest effort, but it took some years before the numerical behavior of Gaussian elimination was fully understood. A straightforward analysis led to the conclusion that rounding error would lead to unacceptably large error, but in practice, the algorithm worked wonderfully. Eventually Trefethen and Schrieber [4] found that Gaussian elimination fails only for a set of matrices so rare that one has never arisen in a real-world application of scientific computation (although simple examples are known).

We have already used *LU* decomposition in Section 6.4.11 to speed up the computation of bistatic scattering amplitudes. The MATLAB function

```
[L,U,P] = lu(A);
```

decomposes the matrix `A` into the product of a lower triangular matrix `L` and an upper triangular matrix `U`, with a sparse permutation matrix `P` associated with a pivoting step that improves the numerical stability of Gaussian elimination. The matrix `A` is factorized in the form

```
P*A = L*U
```

The linear system `A*x = b` can be solved by inverting the factorization to obtain

```
x = U\(L\(P*b));
```

Since `P` is sparse and `L` and `U` are triangular, this expression can be evaluated with a number of computational steps that is on the order of N, so that once the matrix is *LU* decomposed, it is easy to find the solution to the linear system (7.1).

7.4 Iterative Methods

While LU decomposition is widely used in practice and is very stable, if the matrix A is large enough, the required computation time becomes impractically long. The number of floating point operations required to construct the LU decomposition of an $N \times N$ matrix using Gaussian elimination is approximately

$$n_{LU} = \frac{2}{3}N^3 \tag{7.29}$$

The time required to LU decompose a matrix with $N = 100,000$ on a one Gigaflop computer is about

$$T_{LU} = \frac{2/3(N)^3 \text{ flops}}{10^9 \text{ flops/s}} \simeq 7 \times 10^5 \text{ s} \simeq 8 \text{ days} \tag{7.30}$$

If one were to increase the complexity of the problem being solved, so that N doubled, the time required increases by a factor of eight, to over 2 months. It is common to encounter matrices of this size in practice. A linear system with 100,000 unknowns for a 3-D surface method of moments code corresponds to a cubical scatterer that is only about 13 wavelengths on a side.

Suppose that a solution to the linear system could be constructed somehow from successive evaluations of the matrix–vector product Ay, where y is an arbitrary vector. This operation is matrix–vector multiplication, or matvec for short. A matvec requires N^2 operations, which for the previous example requires a time of only

$$T_{\text{matvec}} = \frac{N^2 \text{ flops}}{10^9 \text{ flops/s}} \simeq 10 \text{ s} \tag{7.31}$$

Doubling N increases the time required only by a factor of 4–40 s. If a combination of a relatively small number of matvecs could be used to solve a linear system, the computational cost required to find the solution x (or approximate it) would be dramatically reduced.

Suppose we begin with some vector x_1. If we then compute the matvec Ax_1, is there some way to combine x_1 and Ax_1 to obtain a better approximation x_2 to the exact solution of the linear system? It turns out that there are many prescriptions for doing this, which are referred to collectively as iterative linear system solution methods. The goal of an iterative method is to choose each approximation x_k as a combination of matvecs in such a way that the sequence $x_1, x_2, x_3, \ldots \rightarrow x$ is as rapidly convergent as possible. If the number of matvecs required to find a good approximation to the solution x is much smaller than the size N of the matrix, the computational cost of an iterative method is better than that of a direct solution algorithm.

7.4.1 Stationary Iterations

Historically, the first indirect linear system solution methods were the so-called stationary iterations. Stationary iterations are based on matrix splittings of the form

Matrix Splitting

$$A = M - R \tag{7.32}$$

If we have an approximation x_k to the exact solution x, the basic approach is to compute another approximation using

$$Mx_{k+1} = Rx_k + b \tag{7.33}$$

or

$$x_{k+1} = M^{-1}(Rx_k + b) \tag{7.34}$$

If $M = A$, this iteration converges in one step, but of course there is no efficiency gain. For the method to be efficient, M should be close in some way to A but such that $M^{-1}y$ is easy to find and (7.34) can be solved quickly. For this reason, M may be the main diagonal of A, a lower or upper triangular part, or some other simple matrix that is similar in some way to A but easier to solve in a linear system. There are a number of different types of stationary iterations, each with different choices for the matrix splitting.

Jacobi
$M = D$, where D is the main diagonal of A.

Gauss–Seidel
$M = D + R$, where R is the lower triangular part of A.

Successive overrelaxation (SSOR)
This method is a variant of Gauss–Seidel, for which one tries to speed up convergence by replacing the x_{k+1} with a linear combination of x_k and x_{k+1}:

$$x_{k+1} = wx_{k+1} + (1 - w)x_k \tag{7.35}$$

where w is a relaxation parameter. For certain types of matrices, there are ways to choose w so that convergence improves.

Methods of this type are called stationary because the iteration matrix $M^{-1}R$ that relates x_k and x_{k+1} is independent of the iteration count k. The iteration matrix governs the performance of the algorithm. The error vector at the kth iteration is

$$e_k = x_k - x \tag{7.36}$$

so that

$$\begin{aligned} Me_{k+1} &= Mx_{k+1} - Rx - b \\ &= Rx_k + b - Rx - b \\ &= R(x_k - x) \\ &= Re_k \end{aligned} \tag{7.37}$$

This means that the error from one iteration to the next satisfies the relationship

$$e_{k+1} = (M^{-1}R)e_k \tag{7.38}$$

If the initial error is e_0, the error at the kth step is

$$e_k = (M^{-1}R)^k e_0 \tag{7.39}$$

For this vector to be small for all possible initial error vectors, the matrix $(M^{-1}R)^k$ must be small, in the sense of having small induced matrix norm, so that the norm of e_k decreases with the iteration count k. If a singular value of $M^{-1}R$ is larger than one, the corresponding eigenvector component of e_0 will be amplified as the iteration proceeds and the error will grow.

If $M^{-1}R$ is a normal matrix, the singular values are the magnitudes of the eigenvalues, and we can reformulate the convergence condition for the iteration in terms of eigenvalues. For the iteration to converge, the spectral radius, or the magnitude of the largest eigenvalue of $M^{-1}R$, must be less than 1:

$$\rho(M^{-1}R) < 1 \tag{7.40}$$

This condition means that all the eigenvalues of the iteration matrix lie inside the unit circle. If this condition does not hold, the component of the initial error vector e_0 in the direction of an eigenvector with eigenvalue larger than one grows with each step and the iteration fails to converge.

7.4.1.1 Diagonal Dominance

One type of matrices for which stationary iterations can be rapidly convergent are diagonally dominant matrices. The term "diagonally dominant" is commonly used in a loose sense for matrices with large diagonal elements relative to the off-diagonal elements, but for a matrix to be diagonally dominant in the strict sense, the diagonal elements must be large enough that

$$|A_{mm}| > \sum_{n=1,\, n\neq m}^{N} |A_{mn}| \tag{7.41}$$

Physical problems for which local interactions are large and long-range interactions decay with distance can lead to this property, although many matrices arising from CEM algorithms like the method of moments have relatively large diagonal elements but are not strictly diagonally dominant.

For diagonally dominant matrices, splitting methods such as the Jacobi iteration converge rapidly, because the off-diagonal elements in the matrix N are small and the elements of M are large, making $M^{-1}N$ small. Gershgorin's circle theorem can be used to make this more rigorous: all the eigenvalues of a matrix lie in a combination of N circles with centers given by the diagonal elements of the matrix and radii equal to the sum of the magnitudes of the off-diagonal elements of the matrix. For the Jacobi iteration applied to a diagonally dominant matrix, the diagonal elements of $M^{-1}N$ are zero, the Gershgorin circles are centered at the origin, and the radii are

$$r_m = |A_{mm}|^{-1} \sum_{n=1, n \neq m}^{N} |A_{mn}| \tag{7.42}$$

which by comparison with (7.41) must be less than 1. This implies that spectral radius of $M^{-1}N$ is less than 1.

7.4.2 *Implementation of Iterative Algorithms*

Using MATLAB array variables and commands for working with arrays, the formulas that define the Jacobi and Gauss–Seidel iterations can be translated readily into code. For the Jacobi iteration, the `diag` command can be used twice in the line of code

```
M = diag(diag(A));
```

to create a matrix with main diagonal equal to that of `A` and off-diagonal elements equal to zero. The key instruction for all stationary iterations is

```
x2 = M\(N*x1 + b);
```

where `x1` is an approximation for the solution vector `x` from an initial guess or a previous iteration and `x2` is the updated approximate solution vector. The backslash operator is an efficient implementation of the solution to the linear system

```
M*x2 = N*x1 + b
```

If `M` is diagonal or triangular, the backslash operator computes the solution with an operation count equal to the number of rows or columns in the square matrix M, which is significantly faster than would be the case if M were an arbitrary matrix.

To complete the stationary algorithm, the update formula for `x2` in terms of the previous vector `x1` must be embedded in a loop, with a termination condition or stopping criterion for ending the iteration. A common stopping criterion is to continue the algorithm until the relative residual error is below a predefined tolerance (see Section 7.5.2). For stationary iterations, as we have seen, it is possible for the solution to diverge, so it is important to have a secondary check to make sure the algorithm terminates after a large, fixed number of iterations or if the residual error grows beyond a given upper limit.

7.4.2.1 Sparse Matrices

For some numerical methods, such as the finite difference algorithms developed in Chapter 4, the matrix in the linear system (7.28) is sparse. To avoid storing many zeros in memory, a sparse matrix data type can be used to store the matrix. This can significantly improve the efficiency of the matrix computations used in an iterative linear system solution algorithm. In MATLAB, a sparse matrix can be allocated using

```
A = spalloc(M,N,NNZ);
```

The resulting array variable is `M` by `N` in size and can store up to `NNZ` nonzero entries. Many MATLAB operators, scripts, and functions can be applied directly to sparse matrices without modification. Commands for generating specific types of sparse matrices include the following:

Sparse Matrix Types

```
B = spones(A);              % Replaces nonzeros with ones
C = speye(M,N);             % Identity matrix
D = sprand(M,N,density);  % Random matrix
E = spdiags(A,d);           % Extracts diagonals of A
```

For large matrices, the function `eigs` can be used to compute a few eigenvalues and eigenvectors in much less time than would be required for the `eig` command to compute all the eigenvalues and eigenvectors. Typically, this command is used to find the largest or smallest eigenvalues and the corresponding eigenvectors.

7.5 Krylov Subspace Iterations

Krylov subspace methods represent a major innovation in numerical linear algebra. These methods are nonstationary, meaning that the relationship between one iterate and the next is not determined by a fixed matrix. Krylov subspace methods were developed later than the stationary iterations and are generally more robust in the sense that they converge for a larger class of matrices.

7.5.1 Conjugate Gradient Method

The basic Krylov subspace iteration is the conjugate gradient method (CG), which can be used for linear systems with symmetric positive definite (SPD) matrices. At the kth step, the CG algorithm computes the vector x_k in the Krylov subspace

$$K_k = \text{span}\{b, Ab, A^2b, \dots, A^{k-1}b\} \tag{7.43}$$

that minimizes the functional

$$f(x) = \frac{1}{2}x^TAx - b^Tx \tag{7.44}$$

For an SPD matrix A, this functional is positive and can be viewed as a multidimensional paraboloid. If we take the vector derivative of $f(x)$ with respect to each element of x, we obtain $Ax - b$, which vanishes if x is the solution to $Ax = b$. This means that the functional reaches an extremal value (i.e., a local maximum or minimum) when the argument is $x = A^{-1}b$. If we take another derivative with respect to the vector x, we obtain the matrix A, which is positive definite, so that $f(x)$ can increase only if the vector x moves away from $A^{-1}b$. These observations imply that $A^{-1}b$ is the global minimum of the functional $f(x)$.

As the iteration count k increases, the Krylov subspace K_k increases in dimension. Consequently, the sequence of vectors x_k lie in increasingly large subspaces of the vector space R^N. Because the CG algorithm finds the vector x_k that makes the functional $f(x_k)$ as small as possible over these subspaces, the sequence of vectors tends to approach the solution $x = A^{-1}b$ as the iteration count k increases.

When the iteration count k is equal to N (the size of the linear system), as a consequence of the Cayley–Hamilton theorem, the Krylov subspace K_N is equal to the full vector space R^N. Theoretically, this implies that the vector x_N produced by the CG algorithm at the Nth iteration must be equal to $x = A^{-1}b$. (If b lies in a proper invariant subspace of the matrix A, exact convergence occurs sooner, and $x_k = x$ for $k < N$.) Thus, the CG algorithm can be viewed as a direct solution method when it is run for N iterations. When first implemented numerically, however, it was observed that accumulated rounding error caused the approximate solution x_k to fail to converge to x. For this reason, CG was more or less abandoned and received little use for years in scientific computation. Eventually, it was realized that CG was useful as an approximate iterative linear system solver when run for less than N iterations. Despite the rounding error, the solution can be quite accurate for many problems at relatively low iteration counts.

For an SPD matrix A and RHS vector b, the CG algorithm is as follows:

Conjugate Gradient Algorithm

Initialization ($k = 0$):

$$x_0 = \text{initial guess, almost always } x_0 = 0 \tag{7.45a}$$

$$r_0 = b - Ax_0 \quad \text{(initial residual vector)} \tag{7.45b}$$

$$d_0 = r_0 \quad \text{(initial search direction vector)} \tag{7.45c}$$

Loop over $k = 0, 1, 2, \ldots$

$$\gamma_k = \frac{r_k^T r_k}{r_k^T A d_k} \tag{7.46a}$$

$$x_{k+1} = x_k + \gamma_k d_k \quad \text{(update the approximation for } x\text{)} \tag{7.46b}$$

$$r_{k+1} = r_k - \gamma_k A d_k \quad \text{(next residual vector)} \tag{7.46c}$$

$$\eta_k = \frac{r_{k+1}^T r_{k+1}}{r_k^T r_k} \tag{7.46d}$$

$$d_{k+1} = r_{k+1} + \eta_k d_k \quad \text{(next search direction)} \tag{7.46e}$$

The vector d_k is the search direction, since x_{k+1} is obtained from the previous approximation x_k by adding a scalar multiple of d_k. The vector r_k is the residual error vector

Residual Error Vector

$$r_k = b - Ax_k \tag{7.47}$$

It can be shown rigorously that this algorithm minimizes the functional (7.44) over the Krylov subspace as previously described, but the proof is beyond the scope of this treatment.

7.5.2 Residual Error

Iterative algorithms need a stopping criterion that can be used to determine when the solution is good enough. Since we do not know the solution x, the error $\|x - x_k\|$ is not available during the iteration. The relative norm of the residual error,

$$\text{Relative residual error} = \frac{\|r_k\|}{\|r_0\|} \tag{7.48}$$

provides a convenient measure of error and can be used to determine whether the iteration has converged. When the relative residual error is below a predefined tolerance parameter, the iteration is terminated and the current vector x_k is returned as the approximate linear system solution.

The residual vector is not equal to the solution error $e_k = x - x_k$, but the L_2 norm of r_k is actually the weighted norm of the solution error. If we define the A-norm as

$$\|y\|_A = \|Ay\|, \tag{7.49}$$

the norm of the residual error is the A-norm of the solution error:

$$\|r_k\| = \|A(x - x_k)\| = \|e_k\|_A \tag{7.50}$$

We can also show that the residual error provides a bound on the norm of the solution error e_k. The L_2 norm of the residual error is related to the norm of the solution error by

$$\begin{aligned} \|r_k\| &= \|Ae_k\| \\ &\leq \|A\|\|e_k\| \\ &= |\lambda_{\max}|\|e_k\| \end{aligned} \tag{7.51}$$

where $\lambda_{\max}$ is the largest eigenvalue of A. The inequality follows if $\|A\|$ is the matrix norm induced by the vector norm, and the final equality assumes that A is a normal matrix and the vector norm is the L_2 or Euclidean norm. Similarly,

$$\begin{aligned} \|e_k\| &= \|A^{-1}r_k\| \\ &\leq \|A^{-1}\|\|r_k\| \\ &= \frac{1}{|\lambda_{\min}|}\|r_k\| \end{aligned} \tag{7.52}$$

where $\lambda_{\min}$ is the smallest eigenvalue of A. Combining the two inequalities leads to

$$\frac{1}{|\lambda_{\max}|}\|r_k\| \leq \|e_k\| \leq \frac{1}{|\lambda_{\min}|}\|r_k\| \tag{7.53}$$

This implies that the residual error norm closely follows the solution error if the ratio $|\lambda_{\max}/\lambda_{\min}|$ is close to one.

7.5.3 Condition Number

For a normal matrix, the ratio

$$\kappa(A) = \frac{|\lambda_{\max}|}{|\lambda_{\min}|} \tag{7.54}$$

is the condition number of the matrix in the L_2 norm. For a general operator and arbitrary operator norm, the condition number is defined to be

Operator Condition Number

$$\kappa(A) = \|A\|\|A^{-1}\| \tag{7.55}$$

This quantity is a measure of how near A is to a singular operator. If A is the identity, the condition number is one. If A is singular, the condition number is infinite. An operator with a large condition number is said to be ill-conditioned.

If the condition number of A (with respect to the L_2 norm) is close to 1, from (7.53) the residual error is always close to the solution error. If the condition number is very large, the residual error may not be a good measure of solution error. In some cases, the condition number can be large, but residual error is still a good indicator of solution error, because the residual and error vectors may not be close to the eigenvectors with very small or very large eigenvalues and the extreme limits in the inequality (7.53) are not reached.

As we will see later, the condition number has a strong influence on the convergence rate of the CG algorithm, so the most serious impact of an ill-conditioned operator is that it typically means slow convergence of the CG algorithm.

7.5.4 CGNE Algorithm

For the functional (7.44) to have a minimum at the solution x, the matrix must be SPD. If A is not SPD, we can instead solve the normal equation

$$A^H Ax = A^H b \tag{7.56}$$

The resulting algorithm is conjugate gradient on the normal equation (CGNE).

Another possibility is to apply CG to the equation

$$AA^H y = b \tag{7.57}$$

and then compute $x = A^H y$, with y the final iterate returned by the conjugate gradient algorithm. This is known as conjugate gradient on the normal residuals (CGNR).

For both CGNE and CGNR, since matrix multiplication is computationally expensive, instead of using the CG algorithm with AA^H or $A^H A$ we modify the algorithm so that only matvecs with A and A^H separately are used in the iteration.

The CGNE algorithm is as follows:

Conjugate Gradient on the Normal Equation (CGNE)

Initialization ($k = 0$):

$$r_0 = A^H b - A^H Ax_0 \tag{7.58a}$$

$$d_0 = r_0 \tag{7.58b}$$

Loop over $k = 0, 1, 2, \dots$

$$y = Ad_k \quad \text{(matvec by } A\text{)} \tag{7.59a}$$

$$\gamma_k = \frac{r_k^H r_k}{y^H y} \tag{7.59b}$$

$$x_{k+1} = x_k + \gamma_k d_k \tag{7.59c}$$

$$y = A^H y \quad \text{(matvec by } A^H\text{)} \tag{7.59d}$$

$$r_{k+1} = r_k - \gamma_k y \tag{7.59e}$$

$$\eta_k = \frac{r_{k+1}^H r_{k+1}}{r_k^H r_k} \tag{7.59f}$$

$$d_{k+1} = r_{k+1} + \eta_k d_k \tag{7.59g}$$

7.5.5 Other Krylov Subspace Methods

With minor variations, Krylov subspace iterations can also be used to find eigenvalues as well as solve linear systems. The CG and Lanczos algorithms are essentially the same algorithm applied in different ways. The Lanczos algorithm generates an approximate tridiagonalization of A, which can be used to estimate the eigenvalues of A. If the tridiagonalization is used to approximate A in a linear system, the CG solution

Table 7.1 Krylov subspace and related iterative algorithms

	$Ax = b$	$Ax = \lambda x$
$A = \text{SPD}$	CG	Lanczos
$A \neq \text{SPD}$	CGNE, CGNR BCG BCG-Stabilized GMRES QMR, TFQMR CGS	Bi-Lanczos Arnoldi

is obtained. The Lanczos algorithm was applied in Section 5.4.1 to the problem of generating nonclassical Gaussian quadrature rules.

One problem with CGNE is that it may converge more slowly than CG, because $A^H A$ is generally more poorly conditioned than A. The condition numbers are approximately related by

$$\kappa(A^H A) \simeq \kappa(A)^2 \tag{7.60}$$

To overcome this difficulty, other Krylov-subspace type iterations have been developed that work directly with non-SPD matrices. Several of these algorithms are listed in Table 7.1.

BCG and Bi-Lanczos are very similar to CG and Lanczos but are modified to handle arbitrary (non-SPD) matrices. For non-SPD matrices, the convergence of BCG is sometimes erratic. Modifications of BCG such as BCG-stabilized seek to reduce the erratic convergence behavior. Conjugate gradient squared (CGS) is an alternative algorithm that generalizes CG to non-SPD matrices. Transpose-free quasi-minimum residual (TFQMR) requires only matvecs with A (and not A^H), which is advantageous for numerical methods for which the full matrix A is not actually filled and a matvec with A^H is hard to implement.

The generalized minimum residual method (GMRES) is typically the most robust of the Krylov subspace iterations. Its disadvantage is that at the kth iteration one must store k vectors, as opposed to two or three for CG, so the memory requirement is larger. Restarted GMRES, or GMRES(n), throws away all but n of these stored vectors, reducing the memory requirement but also slowing convergence.

7.5.6 Convergence of Krylov Subspace Iterations

Analyzing the convergence properties of Krylov subspace iterations is somewhat more difficult than for stationary iterations. The treatment given here follows that of [5]. The fundamental observation is that because the residual vector r_k lies in a Krylov subspace

$$\text{span}\{r_0, Ar_0, A^2 r_0, \ldots, A^k r_0\} \tag{7.61}$$

it can be expressed as

$$r_k = p(A) r_0 \tag{7.62}$$

where p is a polynomial of degree k. For fast convergence, we would like the polynomial p to be small in some sense for low iteration counts k.

Let S be the spectrum of A, or the set of its eigenvalues in the complex plane. If N is very large, S consists of many eigenvalues, and we can approximate it as a continuum in the plane. The spectrum may also include a few outlier eigenvalues that are separate from the main group. The residual vector $p(A)r_0$ will be small if $p(\lambda_m)$ is small, where $\lambda_m \in S$. Thus, the key question is "For a given order k, how small can $|p(z)|$ be on the spectrum S?" Instead of trying to determine exactly what the polynomial p is, we will obtain a bound on how small an arbitrary polynomial can be on S in terms of its order.

If the zeros of p are denoted as z_n, we can write

$$|p(z)| = \prod_{n=1}^{k} |z - z_n| \tag{7.63}$$

Taking the log of this expression gives

$$\log|p(z)| = \sum_{n=1}^{k} \log|z - z_n|. \tag{7.64}$$

Because the potential due to a negative line charge in the plane (a 2-D point charge) is equal to the log function, this can be interpreted physically as the potential due to k negative charges located at each of the zeros z_n.

In the limit as k becomes large, which polynomial p is smallest over the extent of the spectrum S? It is not hard to see that the minimum potential arises when all of the charges are distributed around the boundary of S. If one of the negative charges were inside S, we could decrease the average potential on S by moving the charge to the boundary.

This property leads to a physical interpretation of the convergence rate of Krylov subspace iterations. Moving to the boundary is precisely what charges on a conducting body do—the charges distribute themselves evenly over the surface of the body so that the equilibrium positions minimize the total energy of the system, which is determined by the potential. Thus, the polynomial p for which $|p(z)|$ is smallest over S has zeros at the locations that k equal 2-D point charges would take on a PEC body with the same shape as S.

The charge and capacitance picture can be used to make a quantitative prediction of the iterative solution convergence rate. The smallest polynomial on S minimizes the ratio

$$\left|\frac{p(z)}{p(0)}\right|_{z\in S} \tag{7.65}$$

This is minimized when the potential

$$\log|p(z)| - \log|p(0)| \tag{7.66}$$

is smallest for z lying on the surface of a PEC body with the shape of S. In the limit as k goes to infinity, the charges will arrange themselves so that the potential is constant on the surface of the PEC. Thus, (7.66) is approximately the potential difference between S and the origin. Since the polynomial p has order k, there are k 2-D point charges on S, so the total charge is k Coulombs. The capacitance (per unit length) of S with respect to the origin is

$$C = \frac{Q}{V} = \frac{k}{-\log|p(z)/p(0)|} \tag{7.67}$$

where the capacitance is normalized such that it is dimensionless. Solving for the ratio $|p(z)/p(0)|$ gives

$$\left|\frac{p(z)}{p(0)}\right| \leq e^{-k/C} = \left(e^{-1/C}\right)^k = \rho^k \tag{7.68}$$

The constant $\rho = e^{-1/C}$ is known as the asymptotic convergence factor and lies in the interval $0 \leq \rho \leq 1$.

The meaning of this result can be summarized as follows. The convergence of a Krylov subspace iteration can be estimated by considering the capacitance of a conductor with the shape of the spectrum of A with respect to a point conductor at the origin. If the capacitance C is small, the asymptotic convergence factor is close to zero, and the residual error rapidly becomes small. If the capacitance is large, $\rho \simeq 1$, and convergence of the iteration is slow.

7.5.6.1 Examples

Well-conditioned matrix ($\kappa \simeq 1$)

If the eigenvalues of A are confined to a small region that is far away from the origin, the capacitance of a conductor with the shape of S relative to the origin is small due to the wide separation between S and the origin. In this case, $\rho = e^{-1/C} \ll 1$ and the convergence of a Krylov iteration is rapid.

Ill-conditioned matrix ($\kappa \gg 1$)

If some of the eigenvalues are very close to the origin, the capacitance is large, so $\rho \to 1$ and convergence is slow.

SPD case

If A is SPD, the eigenvalues lie in some interval $[\lambda_{\min}, \lambda_{\max}]$ on the positive real axis. The dimensionless capacitance per unit length of an infinite strip on this interval relative to an infinite wire is

$$C = -\left(\log\frac{\sqrt{\lambda_{\max}/\lambda_{\min}} - 1}{\sqrt{\lambda_{\max}/\lambda_{\min}} + 1}\right)^{-1} \tag{7.69}$$

The convergence factor is

$$\rho = \frac{\sqrt{\lambda_{\max}/\lambda_{\min}} - 1}{\sqrt{\lambda_{\max}/\lambda_{\min}} + 1} \tag{7.70}$$

In terms of the condition number κ, this becomes

$$\rho = \frac{\sqrt{\kappa} - 1}{\sqrt{\kappa} + 1} \tag{7.71}$$

If κ is large,

$$\rho \simeq \frac{\sqrt{\kappa} - 1}{\sqrt{\kappa}} \left(1 - \frac{1}{\sqrt{\kappa}}\right) \simeq 1 - \frac{2}{\sqrt{\kappa}} \tag{7.72}$$

If we want to iterate until the relative residual error norm is ε,

$$\varepsilon = \frac{\|r_k\|}{\|r_0\|} \simeq \rho^n \tag{7.73}$$

so that

CG Convergence Estimate

$$k = \frac{\log \varepsilon}{\log \rho} \simeq \sqrt{\kappa}\frac{|\log \varepsilon|}{2} \quad \text{(CG)} \tag{7.74}$$

This provides a useful an estimate of the number of iterations required to achieve a residual error of ε.

For CGNE or CGNR, κ is the condition number of $A^H A$, which is approximately the square of the condition number of A. This leads to the estimate

CGNE/CGNR Convergence Estimate

$$k \simeq \kappa(A)\frac{|\log \varepsilon|}{2} \quad \text{(CGNE/CGNR)} \tag{7.75}$$

Spectrum confined to a disk (e.g., BCG)

For non-SPD matrices, the spectrum is not confined to the real axis. If the spectrum is approximated as a disk in the complex plane, the asymptotic convergence factor becomes

$$\rho = \frac{\kappa - 1}{\kappa + 1} \tag{7.76}$$

so that the iteration count estimate becomes

$$k \simeq \kappa(A)\frac{|\log \varepsilon|}{2} \quad \text{(Disk)} \tag{7.77}$$

which is the same as the estimate for CGNE/CGNR. This result implies that the common argument based on (7.60) that CGNR or CGNE converge more slowly than CG may not hold in practice.

7.5.6.2 Other Factors Affecting Convergence Rates

The basic conclusion from the previous examples is that the matrix condition number strongly influences the convergence rate of a Krylov subspace iteration. If the condition number is small, convergence is rapid, and the iterative algorithm is a very efficient way to solve a linear system. If the condition number is large, convergence is slow.

While the matrix condition number is important, other factors can influence convergence rates. If the spectrum includes a few, isolated outlier eigenvalues, the polynomial $p(z)$ has to place only one zero at each outlier to converge to the same residual error as if the outliers were not there. These zeros can be placed in roughly one iteration per outlier.

The convergence estimates in Section 7.5.6.1 were developed for normal matrices, but when expressed in terms of the condition number κ, they actually take into account to a degree the effects of nonnormality. The condition number (7.55) in the L_2 induced operator norm is the ratio of the largest and smallest singular values, which implies that if the matrix is strongly nonnormal, the condition number is much larger than the ratio $|\lambda_{max}/\lambda_{min}|$ and the predicted convergence rates become slower than would be expected by examining only the matrix spectrum. In fact, for a given spectrum almost any convergence curve is possible, depending on the degree of nonnormality of the matrix [6]. Fortunately, CEM algorithms usually lead to matrices that are either normal or close enough to normal that convergence is reasonable.

One way to diagnose the degree of nonnormality is to compare the singular values and eigenvalues of a matrix. For a normal matrix, the singular values are equal to the magnitudes of the eigenvalues. If the extremal singular values are close to the magnitudes of the extremal eigenvalues, the matrix is close to normal.

Other contributing factors include the right-hand side, numerical rounding error, and the details of different Krylov subspace iterations. Further details can be found in the literature on numerical linear algebra.

7.5.7 Preconditioners

As the numerical linear algebra community has searched for new iterative algorithms that are robust and rapidly convergent, many variants of the basic Krylov subspace iterations have been developed. The goal is to find an algorithm that converges quickly for many different linear systems. While there has been progress toward this goal, negative theoretical results seem to preclude a single "black box" algorithm that works well for all linear systems [6]. Because it appears that no single algorithm can converge rapidly for all matrices, attention in the research community has largely shifted in recent years to finding ways of speeding the convergence of existing iterative algorithms.

One approach for improving convergence is to transform the linear system so that the matrix effectively has a lower condition number. This can be done using preconditioning matrices, which reduce the effective condition number of a matrix without substantially increasing the workload of the iterative algorithm.

If we apply a left preconditioner M to the linear system $Ax = b$, the linear system is transformed to

$$M^{-1}Ax = M^{-1}b \tag{7.78}$$

If CG or some other algorithm is applied to the preconditioned linear system, the condition number that governs convergence is that of $M^{-1}A$. The goal is to choose M so that

$$\kappa(M^{-1}A) \ll \kappa(A) \tag{7.79}$$

which implies that the convergence of the preconditioned iterative algorithm is faster.

In implementing a preconditioned iteration, we do not actually form M^{-1} or its product with A. An iteration can be modified to include a preconditioner by solving

$$My = c \tag{7.80}$$

for y at each iteration. If M were equal to A, this would take as much work as solving the original linear system, and the iteration would require only one step. We want M to be similar to A, so that the preconditioned matrix has a condition number as close to one as possible but is structured in such a way that the linear system (7.80) is easy to solve. This is quite similar to the conditions for choosing the matrix splitting (7.32) for a stationary iteration.

The preconditioner either can be constructed blindly from the matrix A, or the physics of the problem that led to A can be used to suggest a form for M. In CEM, a common preconditioner is a matrix of near-neighbor interactions, or moment matrix elements involving nearby testing and expansion functions. For 2-D scattering, this leads to a banded matrix. In 3-D, the matrix is not banded but is typically sparse. The physical radius of the near-neighbor region determines the effectiveness of the preconditioner, but the larger the radius, the more nonzero elements M has and the harder it is to solve (7.80) at each step in the iteration. Solving a linear system with a banded or sparse matrix requires more computation time than if the matrix were diagonal or triangular. To improve efficiency, approximate inverse methods have been proposed for sparse and banded preconditioners, so that (7.80) is solved approximately at each step in the iteration with lower computational cost than a direct solution method.

7.6 Multiscale Problems

For a given structure, the difficulty of solving an electromagnetic boundary value problem is strongly dependent on the frequency of the excitation. At the low-frequency end of the spectrum, electromagnetic radiation and scattering can be analyzed with so-called exact or error-controllable numerical methods such as FD, FDTD, MoM, and FEM. These methods all have a tuning parameter (grid spacing or mesh element density) that can be adjusted to produce a solution of arbitrary accuracy. As the frequency increases, the Nyquist criterion dictates that the number of samples or degrees of freedom used to represent unknown currents or fields must grow. For 2-D surface integral equation methods, the number of unknowns is proportional to the size of the computational domain in wavelengths, so that $N \sim kd$. For 3-D surface integral equations, $N \sim (kd)^2$. The computational cost of the linear system solution is proportional to N^3 for a direct method or N^2 for an iterative solution in the case of a well-conditioned linear system. It follows that exact numerical methods are efficiency-limited and require more computation time as the frequency increases.

At high frequencies, the number of mesh elements required for the method of moments or grid points for finite difference algorithms is large, and computational analysis can be impractical or even impossible. Since diffraction effects become less important at high frequencies, wave propagation can be approximated by ray-like behavior, and consequently large problems can be solved rapidly using high-frequency asymptotic approximations such as the geometrical optics (GO) and ray tracing that avoid sampling fields or currents on a subwavelength scale. These methods typically work best for structures that are large and smooth on the scale of the wavelength. GO can be modified or extended to add corrections for diffraction and improve accuracy, yielding such formulations as geometrical theory of diffraction (GTD), physical theory of diffraction (PTD), uniform theory of diffraction (UTD), and incremental length diffraction coefficients (ILDC). These methods are computationally efficient, but as frequency becomes smaller, they are increasingly inaccurate. Moreover, asymptotic approximations become increasingly complex and difficult to apply as additional correction terms are added, so these methods are inherently limited in accuracy and are not error-controllable in the same sense as FD, MoM, and other mesh- or grid-based algorithms.

These considerations show that there is a gap between the problems for which exact algorithms and asymptotic approximations can be used. For small and moderately sized problems, the error-controllable algorithms are adequate. For large, smooth, and relatively simple structures, the high-frequency approximations are applicable. There is an intermediate regime of problems that are electrically large but also include fine features that limit the accuracy of high-frequency approximations. These problems cannot be solved accurately or quickly by either class of methods and are referred to as multiscale problems.

To solve multiscale, electrically large problems with exact methods like MoM, how can we increase the computational efficiency of "low-frequency" algorithms like the method of moments? There are two basic approaches:

1. Reduce the $O(N^2)$ cost of filling the moment matrix and evaluating matvecs.
2. Reduce N by using better basis functions.

We will discuss each of these in the following sections.

7.6.1 Fast Algorithms

Modifications of the MoM for integral equations that reduce the $O(N^2)$ computational cost of filling the moment matrix and computing matvecs are referred to in CEM as fast algorithms. These algorithms replace the matrix with a "black box" algorithm that takes a vector x as an input and returns Ax, where A is a moment matrix, without actually filling the full matrix. Since these algorithms can solve a linear system without filling the matrix explicitly, they are also referred to as matrix-free.

One type of fast method is based on the observation that convolution with a shift-invariant Green's function can be implemented using the FFT algorithm. For volume integral equations, this is relatively straightforward. For surface integral equations, current basis functions must be shifted so that they lie on a rectangular grid and the FFT algorithm can be applied. This shift is accomplished by transforming a basis function to a new source that lies on a rectangular grid but radiates the same fields away from the source. One implementation of this idea is the adaptive integral method (AIM) [7]. Precorrected FFT [8] is another method of this type.

Another approach is the fast multipole method (FMM). This algorithm is one of the most important new developments in scientific computation in the last few decades. It was developed by Vladimir Rokhlin of Yale University and collaborators and is used for fast computation of interactions in problems from molecular dynamics to astrophysics. The extension of FMM to electromagnetic fields was accomplished at the University of Illinois at Urbana-Champaign by Weng Cho Chew and Cai-Cheng Lu [9].

The fundamental idea of FMM is to group all expansion and testing functions into a tree hierarchy of groups or boxes and to represent the fields radiated by all sources in a box with a single multipole source. This leads to a quad-tree structure in 2-D and an oct-tree in 3-D. When computing the field received at a distant testing function, all sources in a group are lumped together into an equivalent source, and the field at the testing function can be computed for all the sources in the group at the same time.

A single-level FMM algorithm with only one level of groups can be viewed as an approximate sparse matrix factorization of the moment matrix, of the form

$$\begin{aligned} A &\simeq A_n + \underbrace{\begin{bmatrix} V_1 & & & \\ & V_2 & & \\ & & \ddots & \\ & & & V_N \end{bmatrix}}_{\text{block diagonal}} \underbrace{\begin{bmatrix} 0 & T_{12} & \cdots & T_{1N} \\ T_{21} & 0 & \ddots & T_{2N} \\ \vdots & \ddots & \ddots & \vdots \\ T_{N1} & T_{N2} & \cdots & 0 \end{bmatrix}}_{\text{diagonal blocks}} \begin{bmatrix} V_1 & & & \\ & V_2 & & \\ & & \ddots & \\ & & & V_N \end{bmatrix}^H \\ &= A_n + VTV^H \end{aligned} \tag{7.81}$$

where A_n consists of all moment matrix elements for expansion and testing functions that are closer than some threshold distance. The computational savings arises because a dense matrix of far interactions is replaced by a product of sparse matrices. The matrix V_n represents the far-field radiation patterns, or essentially the scattering amplitudes, of the sources in the nth group. The diagonal matrices T_{mn} are called translation operators, and map outgoing plane waves radiated away from the nth group to incoming plane waves received by the mth group. The translation operators are computed from a series of terms that are related to the fields radiated by multipole sources (dipole, quadrupole, and higher order multipoles), which motivates the name of the algorithm.

When FMM is implemented, the matrix product VTV^H is not actually computed, because this would eliminate the computational savings of the approach. Instead, an approximate matvec is computed using

$$Ax \simeq A_n x + V(T(V^H x)) \tag{7.82}$$

which reduces the computational cost to less than the N^2 cost of a matvec with a dense matrix. Since FMM computes matvecs, it must be used in conjunction with an iterative linear system solution method. This is one of the reasons that iterative algorithms are so important in CEM.

From (6.72), it can be seen that if a source is moved, the scattering amplitude changes by the phase term $e^{-jk\hat{r}\cdot\overline{d}}$, where $\overline{d}$ is the displacement of the source. This fact can be used to efficiently combine the far-field patterns V_n for nearby groups (usually four in 2-D or eight in 3-D) into the radiation pattern for a larger group using a computation that is very much like an FFT. This generalizes the single-level FMM to the multilevel fast multipole algorithm (MLFMA). The key insight that makes this method efficient is that far-field patterns for small groups at lower levels require fewer sample points than groups at higher levels, because the physical size of the group determines how rapidly the far-field pattern oscillates in angle. The computational cost of MLFMA for a single matvec is roughly

$$C_{\text{MLFMA}} = cN \log N \tag{7.83}$$

where the constant is such that MLFMA becomes faster than a direct solver for unknown counts in the tens of thousands. MLFMA has allowed EM problems with many tens of millions of unknowns to be simulated numerically.

7.6.2 Reduced-Order Representations

Although MLFMA improves the efficiency of a matvec, it does not reduce the size of the moment matrix or the number of unknowns. Another approach to improving the efficiency of MoM for large problems at high frequencies is to choose a basis that can accurately represent the current on a scatterer with fewer total expansion functions.

As we have seen in earlier chapters, since surface currents are oscillatory, Nyquist sampling considerations imply that with localized basis functions, several basis functions are required per wavelength. If the current is expanded using entire-domain or large-support basis functions that have the same kind of oscillatory spatial variation as the current, the current can be accurately represented using fewer unknowns than would be required with local basis functions.

To leading order, the phase of the current on a large, smooth scatterer is the same as that of the incident field. This can be used to construct basis functions that are oscillatory and more like the current than pulse, triangle, and other low-order basis functions. From a Fourier point of view, the current is approximately bandlimited, so it can be accurately represented using fewer degrees of freedom or expansion coefficients than required by the Nyquist limit. Basis functions that represent a current solution with fewer unknowns than localized basis functions are an example of a class of numerical techniques known as reduced-order representations.

The goal of a reduced-order current representation for the MoM is to develop a numerical method for which the number of unknowns required for a given solution accuracy tends to a constant as frequency increases. If this can be done, the numerical method can solve a problem with bounded computation time as frequency becomes infinite. This is sometimes referred to as an $O(1)$ algorithm, since the computational cost grows as $(kd)^0 = 1$, where kd is the electrical size of the scatterer along one dimension.

There are many types of reduced-order representations, including the following:

Asymptotic phase: A known leading-order phase is factored out of the unknown current solution using $J = \tilde{J}e^{ik(x)x}$, and the smoother function $\tilde{J}$ is expanded with fewer unknowns [10].

Wavelets: Some types of wavelets can be used as basis functions that offer a compromise between being oscillatory, bandlimited, and having a localized region of support.

Plane wave decomposition: The current is represented as a sum of plane wave type complex exponentials with a set of discrete directions of propagation. In the finite element community, this is referred to as microlocal analysis.

There are two major difficulties with these methods. First, reduced-order representations can be nonrobust. Multiple scattering, edge diffraction, or sharp bends in the scatterer surface cause the current to deviate from the expected oscillatory behavior. If the spatial variation of the current is not like that of the basis function expansion assumed by the reduced-order representation, the solution can lose accuracy.

The second difficulty is that computing moment matrix elements requires many quadrature points, since the basis functions are not localized. Computing just one moment matrix element (for a naive implementation) can be as costly as filling an entire moment matrix.

Many methods of the reduced-order class have been proposed over the years, but due to these drawbacks, performance has been less than satisfactory. High-frequency asymptotic approximations like geometrical optics and the geometrical theory of diffraction have been far more popular in application areas. Hybridizations of high-frequency approximations with the method of moments have also received significant attention. Still, progress on numerical methods based on reduced-order representations has continued at a slow but relatively steady pace. Promising results on asymptotic integration of moment matrix elements to improve the efficiency of matrix element computation have been obtained recently by researchers at the University of Illinois, Caltech, the University of Reading, and elsewhere [11–13].

Problems

7.1 Verify that vectors of the form of (2.7) satisfy the axioms of a linear space given in Section 7.1.1.

7.2 (a) Find the eigenvalues of the matrix

$$A = \begin{bmatrix} 2 & 1 \\ 1 & 2 \end{bmatrix} \tag{7.84}$$

using the zeros of the characteristic polynomial (7.16). (b) Find the eigenvectors of A by identifying the null space of the matrix $\lambda I - A$ for each eigenvalue λ. (c) Use the eigenvectors and eigenvalues to construct the unitary diagonalization of A. (d) Check your result using the MATLAB `eig` command.

7.3 For the volume MoM as implemented in Problem 6.8 with a cylinder of radius $\lambda/2$ and $\varepsilon_r = 2$, (a) solve the linear system by replacing the MATLAB slash operator with the Jacobi iteration. Does it converge? Next, try the Gauss–Seidel iteration. Plot on a semilogy scale the relative residual error norm

$$\text{relerr}_k = \frac{\|Ax_k - b\|}{\|b\|}$$

versus the iteration count k for both iterations. (b) Raise the relative permittivity of the cylinder to 10 and repeat the Jacobi and Gauss–Seidel iterative solutions.

7.4 Implement the conjugate gradient method for the normal equation (CGNE). Use the algorithm to solve a linear system produced by the volume moment method as implemented in Problem 6.8.

7.5 Compare the Jacobi and Gauss–Seidel iterations with CGNE and some of the built-in MATLAB iterative solvers (e.g., BCG, QMR, and GMRES) for a linear system produced by a MoM code. Plot the normalized residual error as a function of iteration count on a semilogy scale for all of the algorithms. Use different line types or annotations and a legend so the curves can be identified. Which algorithm converges the fastest?

7.6 Compare LU decomposition and CG for dense linear systems.

(a) Using a method of moments code, increase the frequency or mesh density so that the size of the linear system increases. For each value of the linear system size N, time the MATLAB backslash command and CG algorithm (terminated at a given relative residual error). Use the MATLAB `tic` and `toc` commands to measure the execution times. Let the linear system size N range from a few tens of unknowns to at least 5,000–10,000 or as large as can be run in a minute or two of execution time.

(b) Plot the run times as a function of the number of unknowns on a log-log scale and comment on the results.

(c) Can you find a run-time crossover point between the two algorithms?

(d) How do the results compare with theoretical convergence estimates? Add lines to the plot for the computation time required for one matvec and the time required for Gaussian elimination based on the number of flops (7.29). The computation time for one flop can be estimated by dividing the time for one matvec by N^2.

(e) How do the matrix condition number and iteration count required for convergence of the CG algorithm vary with N?

7.7 Repeat Problem 7.6 for sparse linear systems. Use a finite difference code for a shape such as the rectangular waveguide from a previous homework to generate a sparse matrix, with a simple RHS such as $[1\ 0\ 0\ \cdots\ 0]^T$.

References

[1] G. Strang, *Introduction to Linear Algebra*. Wellesley-Cambridge Press, 1993.

[2] A. W. Naylor and G. R. Sell, *Linear Operator Theory in Engineering and Science*. New York, NY: Springer-Verlag, 1982.

[3] D. Coppersmith and S. Winograd, "Matrix multiplication via arithmetic progressions," J. Symbolic Comput., vol. 9, no. 3, pp. 251–280, 1990.

[4] L. N. Trefethen and R. S. Schrieber, "Average-case stability of Gaussian elimination," SIAM J. Matrix Anal. Appl., vol. 11, pp. 335–360, 1990.

[5] T. A. Driscoll, K.-C. Toh, and L. N. Trefethen, "From potential theory to matrix iterations in six steps," SIAM Rev., vol. 40, pp. 547–578, 1998.

[6] A. Greenbaum, V. Ptak, and Z. Strakos, "Any nonincreasing convergence curve is possible for GMRES," SIAM J. Matrix Anal. Appl., vol. 17, pp. 465–469, 1996.

[7] E. Bleszynski, M. Bleszynski, and T. Jaroszewicz, "AIM: Adaptive integral method for solving large-scale electromagnetic scattering and radiation problems," Radio Sci., vol. 31, no. 5, pp. 1225–1251, 1996.

[8] X. Nie, L. Li, and N. Yuan, "Precorrected-FFT algorithm for solving combined field integral equations in electromagnetic scattering," J. Electromagn. Waves Appl., vol. 16, no. 8, pp. 1171–1187, 2002.

[9] C. C. Lu and W. C. Chew, "A multilevel algorithm for solving a boundary integral equation of wave scattering," Microwave Opt. Technol. Lett., vol. 7, no. 10, pp. 466–470, 1994.

[10] D.-H. Kwon, R. J. Burkholder, and P. H. Pathak, "Efficient method of moments formulation for large PEC scatterers using asymptotic phasefront extraction (APE)," IEEE Trans. Antennas Propag., vol. 49, no. 4, pp. 583–591, 2001.

[11] O. P. Bruno, C. A. Geuzaine, J. A. M. Jr., and F. Reitich, "Prescribed error tolerances within fixed computational times for scattering problems of arbitrarily high frequency: The convex case," Philos. Trans. R. Soc. London, A, vol. 362, pp. 629–645, 2004.

[12] S. N. Chandler-Wilde, S. Langdon, and L. Ritter, "A high wavenumber boundary element method for an acoustic scattering problem," Philos. Trans. R. Soc. London, A, vol. 362, pp. 647–671, 2004.

[13] S. Langdon and S. N. Chandler-Wilde, "A wavenumber independent boundary element method for an acoustic scattering problem," Isaac Newton Institute for Mathematical Sciences Preprint, Tech. Rep. NI03049CPD, 2003.

Chapter 8
Finite Element Method

The finite element method (FEM) was developed by researchers solving problems in fluid dynamics and structural analysis. FEM has become an important method for computational electromagnetics applications and is implemented in several prominent commercial software packages. FEM is essentially a method for defining basis functions on mesh elements and using those basis functions to discretize a partial differential equation (PDE) in functional form using the Rayleigh–Ritz approach. Because of the breadth of its applications, the literature on the finite element method is vast. Tutorials that focus on EMs include [1,2]. Other useful references include [3,4].

FEM has similarities to both the finite difference method and the method of moments. With FEM, the unknown solution is expanded as a linear combination of basis functions, as is the case with the method of moments. FEM is usually applied to differential equations, but it can also be used with integral equations. When the integral equation is a surface integral equation, this is referred to as the boundary element method (BEM). When applied to integral equations, FEM and the method of moments are essentially identical, since they result in the same linear system for the same choice of basis functions.

When applied to a differential equation, the matrix resulting from the Rayleigh–Ritz procedure is very similar to the matrix resulting from the finite difference method, and in some simple cases, these two algorithms are also identical. We will show in Section 8.3.9 that for Laplace's equation with a simple mesh and triangle basis functions, the linear system obtained with the finite element method is equivalent to the finite difference equations for the same PDE. Using piecewise linear or triangular FEM basis functions to discretize a PDE with the Rayleigh–Ritz method is closely related to discretizing the PDE using the central difference approximation. The main advantage of FEM over FD is that because it is element-based rather than grid based, the algorithm is more geometrically flexible and can be readily applied to nonrectangular, conformal meshes.

In this chapter, we will first develop the machinery associated with variational principles and the Rayleigh–Ritz procedure. This will be used to derive the finite element method for a simple one-dimensional Laplace problem. FEM algorithms will then be developed for two-dimensional (2-D) Helmholtz problems, including modal eigenvalue problems and radiation and scattering. FEM will then be combined with a surface integral equation to develop the FEM–BEM algorithm.

8.1 Variational Principles in Mathematical Physics

In previous chapters, the finite difference method and the method of moments were applied as recipes for generating a numerical algorithm from a continuous integral or differential equation without much rigor or attention to assumptions about the continuous problem. In this chapter, we will see that the discretization process can be placed in a more general mathematical framework using the theory of variational methods.

Some of the methods for discretizing boundary value problems are as follows:

Discretization Methods

General problems
- Method of weighted residuals
  - Method of moments
  - Boundary element method
  - Galerkin's method
  - Least squares method
- Finite difference (FD) method

Variational problems
- Rayleigh–Ritz method
  - Finite element method (FEM)

The discretization methods that require the fewest assumptions about a problem are the method of weighted residuals and the finite difference method. Although empirical methods or numerical analysis can be used to characterize solution accuracy for specific boundary value problems, weighted residuals and finite difference methods generally come with no particular assurance that the numerical solution will be a good approximation to the exact solution.

If the continuous problem can be placed into a variational form, it can be discretized instead with the Rayleigh–Ritz method. In this case, there are many theorems and results in the numerical analysis literature that provide bounds on solution accuracy and insight into the numerical behavior of the method. As we will see in this chapter and in the following, for a given boundary value problem, differently named discretization methods can actually lead to the same numerical method, so there is significant overlap between the various approaches listed previously and the distinctions can be more philosophical than practical. For the purposes of this book, the value of variational methods and the Rayleigh–Ritz procedure is both practical, since the finite element method naturally fits into the framework of variational problems, and pedagogical, because variational principles provide a fruitful way to understand algorithms for solving differential and integral equations.

8.1.1 Operators and Functionals

Understanding variational methods requires some background on function spaces, operators, and functionals. The review of linear operator theory in Section 7.1 was oriented toward finite-dimensional linear spaces and matrix operators, but here we are interested in infinite-dimensional linear spaces for which the elements of the space are functions.

An operator on such a linear space has the space of functions as its domain and range:

$$\mathscr{L}u = f \tag{8.1}$$

Both integral and differential operators fit into this framework. Examples include the differential operator $\mathscr{L} = \nabla^2$ and the convolution operator $\mathscr{L}u = K \times u$ studied in Chapter 6. Some "operators" are actually transformations, because the domain and range spaces are different. If the kernel function that defines a convolutional integral operator is smooth, for example, the range space consists of smoother functions than the domain.

Many of the concepts of finite-dimensional matrix operator theory generalize directly to operators on function spaces. The adjoint $\mathscr{L}^\dagger$ of the operator $\mathscr{L}$ is defined by (7.14). As with a matrix operator, $\mathscr{L}$ is self-adjoint if $\mathscr{L} = \mathscr{L}^\dagger$. If this equality holds but $\mathscr{L}$ is a transformation and its domain and range spaces are not the same, then $\mathscr{L}$ is said to be formally self-adjoint. The transpose of an operator satisfies a relationship of the form of (7.14), with the inner product replaced by an integral similar to that in (6.73) but without a complex conjugate:

$$\int_S (\mathscr{L}^T u) v \, ds = \int_S u(\mathscr{L}v) \, ds \tag{8.2}$$

The operator is symmetric if $\mathscr{L} = \mathscr{L}^T$. The adjoint is much more common in operator theory than the transpose, but operator symmetry is related to the electromagnetic reciprocity principle, so symmetric operators arise often in computational electromagnetics. A positive definite operator satisfies

$$\langle \mathscr{L}u, u \rangle > 0 \text{ if } u \neq 0 \tag{8.3}$$

A functional is a map from a function space S into the real line,

$$I(f) : S \to R \tag{8.4}$$

The symbol I is used because many functionals involve the integral of the function in the argument. Some functionals are maps to complex numbers, but real-valued functionals are most common. A simple example of a functional is path length. The argument of the path length functional is a function that defines a parameterization of a path, and the result is the path length (see Problem 8.1). Another common type of functional is the energy stored in a field, so that if ϕ represents the electric potential, $I(\phi)$ is the energy stored in the electric field.

8.1.2 Variational Principles

For many functionals, there is one "point" in the range space (i.e., a function) for which the value of the functional takes on a maximal or minimal value. This is analogous to a maximum or minimum of a function. Such an extremal point f may be a maximum

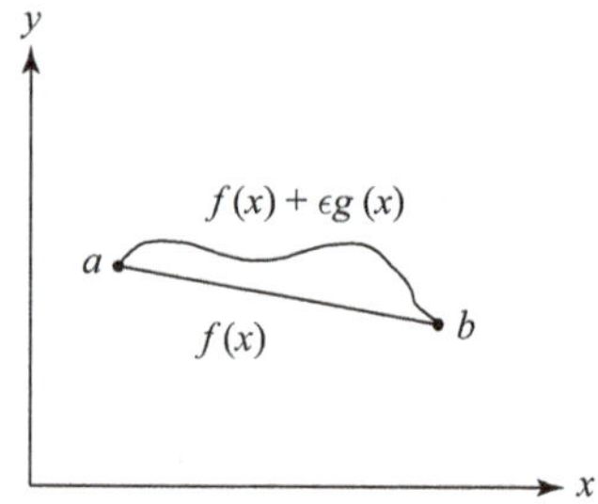

Figure 8.1 Two paths from point a to b. The straight line minimizes the path length, whereas the perturbed path has a longer length

or minimum of $I(f)$, but minimization is most common, so we typically think of minimizing the functional.

For the path length functional, the extremal point in the function space of path parameterizations is a straight line. If $x = t, y = f(t), a \leq t \leq b$ is a parameterization of the straight line from $(a, f(a))$ to $(b, f(b))$, then

$$I(f + \varepsilon g) = I(f) + O(\varepsilon^2) \tag{8.5}$$

where $g(t)$ satisfies $g(a) = g(b) = 0$ and ε is a small parameter. The function f is a minimum of the functional in the sense that $I(f) < I(f + \varepsilon g)$ if $\varepsilon \neq 0$. This condition is illustrated in Figure 8.1. For any path other than the straight line, the path length functional has a larger value.

Equation (8.5) can be better understood by analogy with the expansion of a function (not a functional) at the point x:

$$h(x + \varepsilon) \simeq h(x) + \frac{\partial h(x)}{\partial x}\varepsilon + \frac{1}{2}\frac{\partial^2 h(x)}{\partial x^2}\varepsilon^2 \tag{8.6}$$

If the linear term in the expansion vanishes, so that

$$\frac{\partial h(x)}{\partial x} = 0 \tag{8.7}$$

then x is a maximum or minimum of the function $h(x)$. Similarly, (8.5) has a zeroth-order term, $I(f)$, and a second-order term represented by $O(\varepsilon^2)$, but the linear term vanishes. Since the value of the functional varies only quadratically with the parameter ε, if ε is small, the value of $I(f + \varepsilon g)$ is close to that of $I(f)$, and the functional is said to be stationary at its extremal point. The stationarity condition can be expressed in a more compact notation using the variational symbol δ:

Variational Principle

$$\delta I(f) = 0 \tag{8.8}$$

If this condition holds, f is a stationary point of the functional $I(f)$.

Equation (8.8) is called a variational principle. In mathematical physics, the idea of a variational principle is a unifying concept for a diverse range of problems. From the variational point of view, the solution to a physical problem governed by continuous integral or differential equations, whether the path of a projectile, an electromagnetic field, or some other measurable quantity, is a stationary point of a governing functional. There is a direct correspondence between functionals and boundary value problems, since the solution to a PDE is generally also the stationary function of a functional:

Variational Principles and Boundary Value Problems

Stationary function of functional	$\Longleftrightarrow$	Solution to a boundary value problem (PDE or integral equation and boundary conditions)

This means that variational principles and differential equations are two different ways of modeling the same system. In physics, a variational principle is called the principle of least action. Physical systems tend to evolve in such a way that the total energy or action required to make the change is minimized. Many different types of systems can be described using variational principles. Examples are found in dynamics, electromagnetics, general relativity, and other fields.

8.1.3 Variational Calculus

A variational principle is solved by finding the function that makes the functional stationary. How can we find the stationary points of a functional? We can use a technique that is a generalization of the procedure for finding the extremal points of a function. For a function, if the condition (8.7) holds, x represents an extremum of h. To implement this approach for a functional $I(f)$, we have to take the derivative not by the variable x, but rather with respect to the argument f of the functional $I(f)$, which is a function.

The basic approach for computing a functional derivative was suggested in the previous section. We define a parametric family of functions and take the derivative of the functional with respect to the parameter using ordinary calculus. The parametric family of functions is

$$f_\varepsilon(x) = f(x) + \varepsilon g(x) \tag{8.9}$$

where $g(a) = g(b) = 0$. This allows us to transform a derivative with respect to a function to a partial derivative with respect to the parameter ε, since $I(f_\varepsilon)$ is no longer a functional, but instead is just a function of ε.

In terms of the parametric function, the condition for a stationary point of the functional I becomes

$$\left.\frac{\partial I(f_\varepsilon(x))}{\partial \varepsilon}\right|_{\varepsilon=0} = 0 \text{ for all perturbations } g(x)$$
$$\Rightarrow f(x) \text{ is a stationary point of } I(f) \tag{8.10}$$

The quantity on the left is the functional derivative of I evaluated at the function f. Equation (8.8) implies that (8.10) holds for all perturbation functions g.

The extremal function f makes the functional stationary, in the sense that the functional does not change very much for any function that is a small perturbation of f, much like the minimum of a function has a vanishing first derivative and is nearly flat at the minimum. For a different function h that is not a stationary point, the value of the functional changes linearly with respect to ε when applied to the perturbation $h + \varepsilon\eta$ of g.

Using the definition of the variational symbol, a system of algebraic rules for the functional derivative can be derived, which is referred to as the variational calculus. For our purposes, however, these rules are not of major importance, since we are most interested in transforming PDEs into variational principles and in using variational principles as a tool for developing numerical methods.

8.1.4 Euler–Lagrange Equation

As observed previously, there is a correspondence between variational principles and partial differential equations. The PDE associated with a variational principle can be derived using the rules of the variational calculus and is referred to as the Euler–Lagrange equation for the functional I:

$$\delta I(f) = 0 \quad \Longrightarrow \quad \text{Euler–Lagrange equation (PDE)} \tag{8.11}$$

If f is the stationary solution to $\delta(f) = 0$, f is a solution to the PDE. The Euler–Lagrange equation, therefore, provides a way to transform a variational principle into a PDE. The process can also be reversed to find a variational principle from a given PDE.

We can work out the details of this process by treating functionals of a particular form. Consider a functional defined on functions $f(x, y)$ of two variables given by an integral of the form

$$I(f) = \int\int F(f, f_x, f_y, x, y)\, dx\, dy \tag{8.12}$$

where f_x is the partial derivative of f by x. By setting the variation δI to zero and using the definition of the variational symbol, it can be shown that the Euler–Lagrange equation is

Euler–Lagrange Equation

$$\frac{\partial F}{\partial f} - \frac{\partial}{\partial x}\frac{\partial F}{\partial f_x} - \frac{\partial}{\partial y}\frac{\partial F}{\partial f_y} = 0 \tag{8.13}$$

The solution $f(x, y)$ to this PDE is an extremum of the functional (8.12).

Example: Laplace's Equation

The functional corresponding to Laplace's equation is

$$\begin{aligned} I(\phi) &= \frac{1}{2}\int_V \|\nabla\phi\|^2\, dV \\ &= \frac{1}{2}\int_V \left(\phi_x^2 + \phi_y^2\right)\, dx\, dy \end{aligned} \tag{8.14}$$

in the 2-D case. The factor of $1/2$ is included so that the functional can be identified as the energy stored by the field ϕ (for a particular physical problem there may also be an additional constant to make the units work out).

The Euler–Lagrange equation is

$$\frac{1}{2}\frac{\partial(\phi_x^2 + \phi_y^2)}{\partial\phi} - \frac{1}{2}\frac{\partial}{\partial x}\frac{\partial(\phi_x^2 + \phi_y^2)}{\partial\phi_x} - \frac{1}{2}\frac{\partial}{\partial y}\frac{\partial(\phi_x^2 + \phi_y^2)}{\partial\phi_y} = 0 \tag{8.15}$$

Working the partial derivatives gives

$$0 - \frac{1}{2}\frac{\partial}{\partial x}2\phi_x - \frac{1}{2}\frac{\partial}{\partial y}2\phi_y = 0 \tag{8.16}$$

Notice that the partial derivatives are evaluated symbolically as if ϕ, ϕ_x, and ϕ_y were all independent variables. This can be understood intuitively by thinking of forming a parameterized variation of each of these functions with respect to a different small parameter and then using (8.10) separately for each function. Since the subscripts on ϕ_x and ϕ_y represent partial derivatives, (8.16) becomes

$$\frac{\partial^2\phi}{\partial x^2} + \frac{\partial^2\phi}{\partial y^2} = 0 \tag{8.17}$$

which is Laplace's equation. The functional can easily be modified to include a source term, so that Poisson's equation is obtained.

The variational principle provides a beautiful physical interpretation for Laplace's equation. The solution to the Euler–Lagrange equation is the one for which the functional (8.14) is smallest. Since the value of the functional is the energy stored in the field ϕ, the solution to Laplace's equation (8.17) is the particular function satisfying an appropriate set of boundary conditions for which the energy stored in the field is smallest.

8.1.5 Variational Principles for PDEs

Functionals can be constructed directly from a boundary value problem. The basic procedure is to multiply the PDE by a variation $\delta\phi$ and integrate over the region of interest:

$$\int_V dV \underbrace{(\mathscr{L}\phi - f)}_{\text{PDE}} \delta\phi = 0 \tag{8.18}$$

Integration by parts is then used to transfer the derivatives in the operator $\mathscr{L}$ to $\delta\phi$, and the δ symbol is brought outside the integral. Generally, there is a boundary term that arises from the integration by parts, which vanishes by the boundary conditions of the boundary value problem.

8.1.6 Variational Principles for Self-Adjoint, Positive Definite Operators

Variational principles can be formulated for integral equations as well as PDEs. If $\mathscr{L}$ is a self-adjoint, positive definite, and real operator, whether differential or integral, then it can be shown that the functional corresponding to $\mathscr{L}\phi = f$ is

$$I(\phi) = \langle \mathscr{L}\phi, \phi \rangle - 2\langle \phi, f \rangle \tag{8.19}$$

8.1.7 Functionals in Mathematical Physics

The idea of a functional and finding extremal functions is a unifying concept in physics. It expresses a sort of "economy in nature" in which solutions to physical problems often correspond to minima in an energy or action quantity.

Examples of common functionals include the following:

$$I(\phi) = \frac{1}{2}\int_V \|\nabla\phi\|^2 \, dV \longleftrightarrow \nabla^2\phi = 0 \quad \text{(Laplace's equation)} \tag{8.20}$$

$$I(\phi) = \frac{1}{2}\int_V \left[\|\nabla\phi\|^2 - k^2|\phi|^2 + 2f\phi^*\right] dV \longleftrightarrow (\nabla^2 + k^2)\phi = f \quad \text{(Helmholtz equation)} \tag{8.21}$$

$$I(\phi) = \frac{1}{2}\int_0^t \int_V \left[\|\nabla\phi\|^2 - \frac{1}{c^2}\frac{\partial^2\phi}{\partial\tau^2}\right] dV\, d\tau \longleftrightarrow \left(\nabla^2 - \frac{1}{c^2}\frac{\partial^2}{\partial t^2}\right)\phi = 0 \quad \text{(Wave equation)} \tag{8.22}$$

More complex functionals used in mathematical physics include the functional for Maxwell's equations, which is expressed in tensor form as

$$\begin{aligned}
x &= (\overline{r}, ict) \quad \text{(four-dimensional spacetime point)} \\
A &= (\overline{A}, ic\phi) \quad \text{(scalar and vector potentials)} \\
j &= (\overline{J}, ic\rho) \quad \text{(charge and current densities)} \\
F_{\mu\nu} &= \frac{\partial A_\nu}{\partial x_\mu} - \frac{\partial A_\mu}{\partial x_\nu} \\
I(A) &= \int\limits_{\Omega} \left(\frac{1}{16\pi} F_{\lambda\rho} F_{\lambda\rho} + \frac{1}{c} j_\lambda A_\lambda \right) d\Omega
\end{aligned} \tag{8.23}$$

where A is a four-dimensional combination of the magnetic vector potential and the electric potential. The integrand of the functional $I(A)$ consists of products of derivatives of the components of the potentials with respect to space and time, which are proportional to the components of the electric field intensity $\overline{E}$ and the magnetic flux density $\overline{B}$. For the source specified by the charge and current density four-vector j, if the electromagnetic field $\overline{E}$, $\overline{B}$ that solves Maxwell's equations is represented by the potential A, then A is a stationary point of the functional $I(A)$.

The functional for Newton's laws of motion is

$$\begin{aligned}
T &= \text{Kinetic energy} \\
V &= \text{Potential energy} \\
I &= \int\limits_{t_1}^{t_2} (T - V)\, dt
\end{aligned} \tag{8.24}$$

Perhaps the most complex functional corresponds to Einstein's equation of gravitation:

$$R = \text{tensor that describes the four-dimensional curvature of spacetime}$$

$$I = \int R\, dV\, dt \tag{8.25}$$

Even quantum mechanics uses the concept of a functional. The complex probability amplitude of a quantum particle following a particular path from point a to point b is

$$\psi(f) = e^{jI(f)/\hbar} \tag{8.26}$$

where f describes the path and $\hbar$ is Planck's constant. The functional $I(f)$ is the action associated with the classical (nonquantum) equations of motion for the particle. One interpretation of quantum mechanics is that a particle actually follows all possible paths from a to b, and the particle's statistical behavior is governed by a sum or integral of the amplitude over all possible paths. Since $I(f)$ changes rapidly for paths away from the stationary or classical path of the particle (for which $\delta I(f) = 0$), when integrated over all paths, the probabilities cancel for paths well away from the classical path, and the particle is much more likely to follow a patch that is near the classical path. Consequently, classical mechanics can be viewed as the stationary phase point of quantum mechanics.

8.1.8 Rayleigh–Ritz Method

The Rayleigh–Ritz method transforms a variational principle into a numerical method. Instead of minimizing a functional over an infinite-dimensional function space, we instead find an approximate minimum in a finite-dimensional approximation subspace, such as a finite linear combination of basis functions. When restricted to the finite-dimensional subspace, the stationarity condition becomes a linear system that can be solved numerically. Since an arbitrary function cannot be exactly represented in the approximation subspace, the functional cannot, in general, be exactly minimized. When designing a numerical method based on this approach, the goal is to choose the approximation subspace so that the minimum of the functional when restricted to the subspace is as close as possible to the minimum over the full function space.

We will begin by approximating the function u using the linear combination of expansion functions introduced in (6.140), so that

$$u \simeq \hat{u} = \sum_{n=1}^{N} a_n f_n \tag{8.27}$$

The approximation subspace is the span of the basis functions f_n, or simply the set of all functions $\hat{u}$ that are linear combinations of the basis functions. We wish to find the stationary point of the functional $I(u)$ when restricted to the subspace spanned by the basis functions, which means finding the set of coefficients $a_1, a_2, \ldots, a_N$ such that the functional restricted to the approximation subspace is stationary. This can be expressed using the variational symbol as

$$\delta I\left(\sum_{n=1}^{N} a_n f_n\right) = 0 \tag{8.28}$$

The stationary solution for this restricted functional will not, in general, be the exact minimum of $I(u)$ over the full function space but should approximate it, so that

$$I(\hat{u}) \geq I(u) \quad \text{and} \quad I(\hat{u}) \simeq I(u) \tag{8.29}$$

The condition (8.28) minimizes the functional over the subspace spanned by the basis functions f_n, so $\hat{u}$ is the best approximation to the stationary point that can be found in the subspace. If we change $\hat{u}$ by adding a function that is orthogonal to the subspace, we could make the value of the functional slightly smaller, but if the basis functions are well chosen, this change will be small.

We can transform (8.28) into a matrix-based numerical method using the definition of the variational symbol δ. When restricted to the finite-dimensional subspace, since the basis functions are themselves fixed, the only degrees of freedom are the coefficients $a_1, a_2, \ldots, a_N$. This means that the restricted functional is a function of N variables:

$$I(\hat{u}) \Longrightarrow I(a_1, a_2, \ldots, a_N) \tag{8.30}$$

With the functional, we found the stationary point by taking the variation, or a derivative with respect to a function. For a function of N variables, we can find the stationary point by taking N partial derivatives, so that

$$\frac{\partial}{\partial a_m} I(a_1, a_2, \ldots, a_N) = 0, \quad m = 1, 2, \ldots, N \tag{8.31}$$

This provides a linear system of N equations, which can be solved numerically for the N unknown coefficients a_n. This is the basis of the Rayleigh–Ritz method.

By considering a specific type of functional, we can derive expressions for the system of equations resulting from the Rayleigh–Ritz method. For an integral or differential equation of the form $\mathscr{L}u = f$, such that the corresponding functional is of the form of (8.19), then

$$\begin{aligned} I(\hat{u}) &= \left\langle \mathscr{L} \sum_{\ell=1}^{N} a_\ell f_\ell, \sum_{n=1}^{N} a_n f_n \right\rangle - 2 \left\langle \sum_{\ell=1}^{N} a_\ell f_\ell, f \right\rangle \\ &= \sum_{\ell=1}^{N} \sum_{n=1}^{N} \langle \mathscr{L} f_\ell, f_n \rangle \, a_\ell a_n - 2 \sum_{\ell=1}^{N} \langle f_\ell, f \rangle \, a_\ell \end{aligned} \tag{8.32}$$

Performing the partial derivative with respect to a_m leads to

$$\underbrace{\sum_{n=1}^{N} \langle \mathscr{L} f_m, f_n \rangle \, a_n}_{\ell = m} + \underbrace{\sum_{\ell=1}^{N} \langle \mathscr{L} f_\ell, f_m \rangle \, a_\ell}_{n = m} - 2 \langle f_m, f \rangle = 0 \tag{8.33}$$

The first term arises from the double summation in (8.32) when the first summation index, ℓ, is equal to m and the second term arises when the second summation index, n, is equal to m.

Since $\mathscr{L}$ must be self-adjoint for the variational principle to exist, the first two terms are equal, and we obtain the system of linear equations

$$\sum_{n=1} \langle \mathscr{L} f_m, f_n \rangle \, a_n = \langle f_m, f \rangle \tag{8.34}$$

This can be put into matrix form $Ax = b$ using

$$A_{mn} = \langle \mathscr{L} f_m, f_n \rangle \tag{8.35}$$

$$x_m = a_m \tag{8.36}$$

$$b_m = \langle f_m, f \rangle \tag{8.37}$$

The matrix A is referred to as the Rayleigh–Ritz matrix.

The resulting linear system is the same as the linear system that would be obtained from the method of weighted residuals with testing functions identical to the expansion functions, or Galerkin's method. In this case, the Rayleigh–Ritz method reduces to the same numerical method as the method of weighted residuals. But the fact that the method comes from a variational principle provides some additional desirable properties of the method, such as a guarantee that the solution converges as N is increased given reasonable choices for the basis functions. Another advantage of the Rayleigh–Ritz method is that there are physical problems for which the governing integral or differential equations are very complicated, but the corresponding functional has a simpler form, so applying the Rayleigh–Ritz procedure to the functional is easier than constructing a numerical method directly from the governing equations.

The fact that Galerkin's method arises from the Rayleigh–Ritz procedure has been interpreted to mean that Galerkin's method is fundamentally more accurate than the more general method of moments (with different expansion and testing functions). It has been shown that this is not the case [5]. The method of moments can be derived from a generalized variational formulation (see Problem 8.4) and when implemented properly is just as accurate as Galerkin's method.

8.2 Overview of the Finite Element Method

The finite element method (FEM) arises when the Rayleigh–Ritz method is applied to a functional with basis functions defined on finite-sized elements that span a simulation domain. Together the elements make up a mesh representation for the simulation domain. FEM can be applied to many different PDEs, as well as integral equations and combinations of PDEs and integral operators. 1-D, 2-D, 3-D, and even higher dimensional problems can be solved.

The basic outline of the FEM approach is as follows:

Outline of the Finite Element Method

1. Create a mesh by dividing the simulation domain into elements. For problems defined on a 2-D surface, the elements are typically triangular.
2. Define basis functions on the mesh.
3. Use the functional for the PDE to be solved to compute a set of matrix elements for the basis functions on one canonical mesh element.
4. Assemble the element matrix entries into a global Rayleigh–Ritz matrix (also referred to as a stiffness matrix) for the entire mesh.
5. Create a right-hand side vector from sources or boundary conditions.
6. Solve the linear system for the basis function weights corresponding to the approximate solution to the boundary value problem (BVP).
7. Apply postprocessing to the solution to compute derived physical quantities.

8.2.1 Mesh Types and Mesh Generation

Meshes for FEM are similar to the meshes used with MoM algorithms (see Section 6.3.3). For 1-D problems, the mesh elements are simply intervals. For 2-D problems, triangular elements are most common, although quadrilaterals (quads) are occasionally used. For 3-D problems, the natural generalization of a triangle is a tetrahedron (tets). Hexahedrons (hexes) are also used, as well as other types of volumetric elements.

Problems often include small geometrical features as well as large, smooth structures. To achieve ideal solution accuracy with the smallest number of mesh elements, the mesh should typically be finer near small features of the structure to be modeled, and elements can be larger elsewhere. There are two approaches to generating meshes with variable degrees of refinement. The mesh generator can generate a mesh in a single pass with element size tied to the size and scale of problem features, or the FEM algorithm can be run in stages beginning with an initial coarse mesh, followed by several adaptive mesh refinement steps that subdivide elements in regions where the solution exhibits rapid spatial variation. Adaptive mesh refinement is an important part of major FEM software packages.

As noted in Section 1.2.3, mesh generation for complex structures is a major area of research and development, and many meshing techniques, tools, commercially available software packages, and highly capable freeware mesh generators [6] are available. The problems at the end of the chapter use MATLAB® mesh generation functions as well as other basic mesh generation techniques. Many technical articles on mesh generation are available, including [7,8].

8.2.2 Basis Functions

Once the elements are selected, basis functions must be defined on the elements. The most common type of basis function is a multidimensional generalization of the 1-D triangle function shown in Section 6.3.1. Other types of basis functions have been developed for a variety of applications, including higher order polynomial basis functions and vector basis functions [4,9]. There are two general classes of basis functions: (1) nodal, for which each mesh node point has an associated basis function and (2) edge, for which basis functions are associated with element edges.

8.2.3 Variational Principle and Rayleigh–Ritz Procedure

Once elements and basis functions are defined, a variational principle for a given PDE is typically used to derive a numerical solution algorithm for a boundary value problem. The unknown solution is approximated as a linear combination of basis functions, as with the method of weighted residuals or the method of moments, and the Rayleigh–Ritz procedure is applied to transform the variational principle into a linear system. A distinguishing feature of FEM is that the Rayleigh–Ritz procedure is applied to the basis functions on one mesh element to form a small, single-element Rayleigh–Ritz or stiffness matrix. The element matrices are then assembled to produce a matrix for the full simulation domain.

8.2.4 Linear System Solution

For the algorithms to be developed in this chapter, MATLAB's internal linear system solution routines are adequate, but for large linear systems, specialized solvers are more efficient. When applied to PDEs, the matrix produced by the finite element method is typically sparse. This means that iterative linear system solution algorithms are particularly attractive. Because of the importance of solving large, sparse linear systems in scientific computation, highly efficient software libraries have been developed for solving large-scale, sparse matrices, including LAPACK and SPARSKIT, several of which are freely available online.

8.3 Laplace's Equation: 1-D FEM

The simplest system to which FEM can be applied is a 1-D electrostatic problem, for which the unknown potential is governed by Laplace's equation. By taking the divergence of Ampere's law and using the fact that the divergence of a curl is zero, we obtain the continuity equation

$$\nabla \cdot \overline{J} = -\frac{\partial \rho}{\partial t} \tag{8.38}$$

This relationship means that the local flow of charge away from a point must be balanced by a decrease in the charge density at that point. If moving negative charges in a conducting medium are balanced by fixed positive charges, the right-hand side vanishes, and the current must flow in such a way that net charge does not accumulate ($\nabla \cdot \overline{J} = 0$).

The relationship between electric field intensity and electric potential for static fields was given in (4.65). In a conductor, the current and electric field are related by the point form of Ohm's law, which is

$$\overline{J} = \sigma \overline{E} \tag{8.39}$$

Combining these equations leads to a modified form of Laplace's equation,

$$\nabla \cdot \sigma \nabla \phi = 0 \quad \text{(conducting medium, static fields)} \tag{8.40}$$

If the conductivity is a function of position, it must remain inside the derivative operator; otherwise, it can be factored out, and the PDE reduces to Laplace's equation.

Another problem that is governed by a PDE of the same form is a static electric field in a dielectric. This case was considered in Section 4.3.5, where the PDE

$$\nabla \cdot \varepsilon \nabla \phi = 0 \quad \text{(dielectric medium, static fields)} \tag{8.41}$$

was obtained. Equations (8.40) and (8.41) are mathematically identical and are both the generalizations of Laplace's equation.

8.3.1 Functional Form of the Generalized Laplace Equation

These PDEs can be solved numerically using the finite element method, but must first be placed into functional form. We will consider the dielectric case, but the derivation for the conducting case is identical, since the only difference between the two cases is the physical meaning of the scalar quantity σ or ε appearing in the modified Laplacian operator. This highlights an important aspect of numerical methods: once we have a code that can solve a PDE of a given mathematical form, any other physical system that can be modeled by a PDE of the same form can also be analyzed.

The functional corresponding to the PDE (8.41) is

$$I(\phi) = \frac{1}{2} \int_V \varepsilon(\overline{r}) \left| \nabla \phi(\overline{r}) \right|^2 dV \tag{8.42}$$

which is similar to (8.14), except that it includes an additional permittivity factor in the integrand. The functional represents the energy stored in the electric field corresponding to the potential ϕ.

If the permittivity and boundary conditions vary only in one dimension, then the functional simplifies to

$$I(\phi) = \frac{1}{2} \int_a^b \varepsilon(x) \left| \frac{\partial \phi(x)}{\partial x} \right|^2 dx \quad \text{(1-D)} \tag{8.43}$$

This is a 1-D problem, since the unknown field is a function only of x. This BVP is easy to solve using a variety of techniques, and a full FEM code is generally not required for 1-D problems, but (8.43) does provide a simple way to introduce the FEM approach and to illustrate the connection between FEM and FD.

8.3.2 Mesh Representation

For a 1-D simulation domain consisting of the single interval $[a, b]$, the mesh consists of N subintervals that span the simulation domain. The endpoints are specified by the list

$$x_1, x_2, x_3, \ldots, x_{N_{\text{nodes}}}$$

where $N_{\text{nodes}} = N + 1$. The endpoints of the domain should be included as node points. If we choose to order the nodes from smallest to largest values of x (the node list does not necessarily have to be ordered in any special way), $x_1 = a$, and $x_{N_{\text{nodes}}} = b$.

We will denote the element array introduced in (6.126) as $t(i, e)$, where

$$e = \text{element number} \tag{8.44}$$

$$i = \text{local node number (for the 1-D case, } i = 1, 2) \tag{8.45}$$

$$t(i, e) = \text{global node number or the index into the list of node coordinates} \tag{8.46}$$

The array $t(i, e)$ together with the list of the x coordinates of each node defines the mesh.

8.3.3 Rayleigh–Ritz Procedure

The unknown potential is approximated as a linear combination of basis functions, so that

$$\phi(x) \simeq \hat{\phi}(x) = \sum_{n=1}^{N_{\text{nodes}}} \phi_n f_n(x) \tag{8.47}$$

where the ϕ_n are unknown samples of the approximate potential solution at the mesh node points. This notation assumes that the basis functions are interpolatory. If the basis functions were not interpolatory, the coefficients of the expansion would not represent samples of $\phi(x)$, and a different notation such as a_n would be used for the coefficients of the expansion (8.47).

From the approximate expansion of the unknown potential solution, we can derive a linear system using the Rayleigh–Ritz method. Using (8.43) and (8.47), the value of the functional on the entire mesh is

$$I(\hat{\phi}) = \frac{1}{2}\sum_{m,n=1}^{N_{\text{nodes}}} \underbrace{\left[\int \varepsilon(x)\frac{\partial f_m}{\partial x}\frac{\partial f_n}{\partial x}\,dx\right]}_{C_{mn}} \phi_m\phi_n \tag{8.48}$$

$$= \frac{1}{2}\overline{\phi}^T C\overline{\phi} \tag{8.49}$$

The matrix C is the Rayleigh–Ritz matrix, but it is also often called the stiffness matrix by analogy with the applications of FEM in structural mechanics. The vector $\overline{\phi}$ consists of sample values of the approximate potential at each of the nodes of the mesh:

$$\overline{\phi} = \begin{bmatrix} \hat{\phi}(x_1) \\ \hat{\phi}(x_2) \\ \vdots \\ \hat{\phi}(x_{N_{\text{nodes}}}) \end{bmatrix} \tag{8.50}$$

In this chapter, we use an overline on column vectors of coefficients to help distinguish between the continuous function $\phi(x)$ that is the solution to the boundary value problem we are solving and the coefficients in the vector $\overline{\phi}$ that represent samples or basis function weights in the approximation (8.47) to $\phi(x)$.

If we minimize the functional by setting partial derivatives with respect to each element of the vector $\overline{\phi}$ to zero, we obtain the linear system

$$C\overline{\phi} = 0 \tag{8.51}$$

If there were a source term in the original PDE, the right-hand side would become a nonzero vector. For nodes lying on the boundary of the simulation domain with a Dirichlet boundary condition, the value of the potential at those nodes is given, and the linear system can be rearranged to move these values to the right-hand side.

8.3.4 Element Stiffness Matrix

Because the integration in the functional (8.42) is linear, the value of the functional for the entire simulation domain is the sum of the functional evaluated on individual elements:

$$I(\phi) = \sum_{e=1}^{N_e} I_e(\phi) \tag{8.52}$$

where

$$I_e(\phi) = \frac{1}{2} \int_{\text{Element } e} \varepsilon(\bar{r})\, |\nabla\phi(\bar{r})|^2 \, dV \tag{8.53}$$

Based on this observation, instead of filling the stiffness matrix directly, it is common practice in implementing FEM to derive a smaller stiffness matrix for a single element and then to assemble the element matrices into the full matrix for the entire mesh. Since the stiffness matrix is sparse, filling it by adding contributions from each mesh element rather than by computing each matrix element avoids looping over many zero entries and improves the computational efficiency of the algorithm.

8.3.5 Basis Functions and Shape Functions

To implement the Rayleigh–Ritz procedure, the basis functions must be specified explicitly. The basis functions we will choose are triangle functions that change linearly from zero to one on each element. Each triangle basis function on the interior of the mesh is nonzero on two adjacent elements and is equal to unity at the shared mesh node. The basis functions at the two boundary nodes consist of half of a triangle function.

Triangle functions are handled slightly differently for FEM than for MoM. The triangle functions are divided into two halves on adjacent elements, are treated separately, and are recombined during the assembly process. On each element, the approximate potential solution is a linear combination of the overlapping halves (referred to as shape functions) of two triangle functions. This combination is linear over the element, which means that the global approximate solution $\hat{\phi}(x)$ is piecewise linear.

On the eth element of a 1-D mesh, a piecewise linear approximation for the potential has the form

$$\begin{aligned}
\phi_e(x) &= f_{t(1,e)}(x)\phi_{t(1,e)} + f_{t(2,e)}(x)\phi_{t(e,2)} \\
&= \frac{x_2 - x}{x_2 - x_1}\phi_{t(1,e)} + \frac{x - x_1}{x_2 - x_1}\phi_{t(2,e)} \\
&= \alpha_1(x)\phi_{t(1,e)} + \alpha_2(x)\phi_{t(2,e)}
\end{aligned} \tag{8.54}$$

where the indices 1 and 2 are local node numbers for the two nodes of the element, and $t(1, e)$ and $t(2, e)$ are the corresponding global node numbers.

The functions $\alpha_1(x)$ and $\alpha_2(x)$ are called shape functions. Each shape function is one on one node and zero on other nodes, as shown in Figure 8.2. The shape functions for a 1-D mesh are equivalent to the first-order Lagrange polynomials. For a given

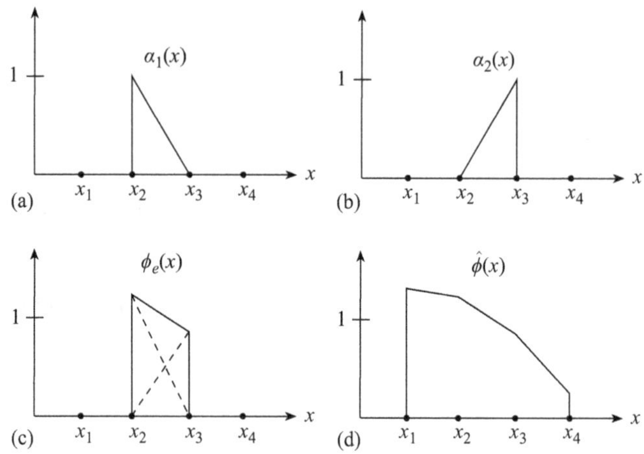

Figure 8.2 (a) Shape function $\alpha_1(x)$ on the mesh element $x_2 \le x \le x_3$. This shape function is equal to one at $x = x_2$ and zero at $x = x_3$. (b) Shape function $\alpha_2(x)$, which is zero at $x = x_2$ and one at $x = x_3$. (c) The linear combination $\phi_e(x) = \alpha_1(x)\phi_{t(1,e)} + \alpha_2(x)\phi_{t(2,e)}$ represents the approximate potential solution on the eth element. $\phi_e(x)$ interpolates linearly between the value $\phi_{t(1,e)}$ at node 1 to the value $\phi_{t(2,e)}$ at node 2. (d) The global approximate solution $\hat{\phi}(x)$ is the sum of the individual element linear approximations $\phi_e(x)$ over all elements $e = 1, 2, \ldots, N_e$ and is a piecewise linear function

node, two shape functions are nonzero at that node, one on each of the two adjacent mesh elements that share the node. The two adjacent shape functions join together to form a complete triangle function.

Since the piecewise linear shape functions $\alpha_1(x)$ and $\alpha_2(x)$ evaluate to one on one node and zero on the other, they are interpolatory basis functions. This means that the coefficients of the basis functions represent samples of the approximate solution at the mesh node points. This motivates the choice of notation $\phi_{t(1,e)}$ and $\phi_{t(2,e)}$ for the coefficients of the two shape functions.

The potential $\phi_e(x)$ on the eth element is a linear combination of the two shape functions, as shown in Figure 8.2(c). Combining the approximate potential solutions $\phi_e(x)$ for all the mesh elements leads to the piecewise linear potential shown in Figure 8.2(d). The goal of the FEM algorithm is to determine the coefficients $\phi_{t(i,e)}$, $i = 1, 2$, $e = 1, 2, \ldots, N_e$, so that the piecewise linear approximation $\phi(x)$ is as close as possible to the desired boundary value problem solution.

Since each triangle function is associated with a mesh node, these basis functions are referred to as nodal basis functions, whereas edge basis functions such as the Rao–Wilton–Glisson (RWG) basis discussed in Section 6.8.2 are associated with mesh element edges.

8.3.6 Evaluating the Element Stiffness Matrix

Using the expansion (8.54) for ϕ_e as a linear combination of shape functions, the value of the functional on the eth element is

$$
\begin{aligned}
I_e &= \frac{1}{2} \int\limits_{\text{Element } e} \varepsilon \left| \frac{\partial \phi_e}{\partial x} \right|^2 dx \\
&= \frac{1}{2} \int_{x_1}^{x_2} \varepsilon \left| \sum_{i=1}^{2} \phi_{t(i,e)} \frac{\partial \alpha_i}{\partial x} \right|^2 dx \\
&= \frac{1}{2} \varepsilon_e \sum_{i=1}^{2} \sum_{j=1}^{2} \underbrace{\left[\int_{x_1}^{x_2} \frac{\partial \alpha_i}{\partial x} \frac{\partial \alpha_j}{\partial x} dx \right]}_{C^e_{ij}} \phi_{t(i,e)} \phi_{t(j,e)}
\end{aligned}
\tag{8.55}
$$

where C^e is the element Rayleigh–Ritz or stiffness matrix and ε_e is the permittivity of the dielectric at the eth element. If the integrals are evaluated numerically, the permittivity can be allowed to vary over the element and left under the integral.

The element functional can be placed in matrix form, so that

$$
\begin{aligned}
I_e &= \frac{1}{2} \varepsilon_e \left[\phi_{e,1} \; \phi_{e,2} \right] \left[\; C^e \; \right] \underbrace{\begin{bmatrix} \phi_{e,1} \\ \phi_{e,2} \end{bmatrix}}_{\overline{\phi}_e} \\
&= \frac{1}{2} \varepsilon_e \overline{\phi}_e^T C^e \overline{\phi}_e
\end{aligned}
\tag{8.56}
$$

For piecewise linear 1-D basis functions, the matrix C^e can be computed in closed form. For example,

$$
\begin{aligned}
C^e_{11} &= \int_{x_1}^{x_2} \left(\frac{-1}{x_2 - x_1} \right) \left(\frac{-1}{x_2 - x_1} \right) dx \\
&= \frac{1}{(x_2 - x_1)^2} (x_2 - x_1) \\
&= \frac{1}{h_e}
\end{aligned}
\tag{8.57}
$$

where h_e is the width of the element. The other matrix elements are

$$
\begin{aligned}
C^e_{12} &= -\frac{1}{h_e} \\
C^e_{21} &= -\frac{1}{h_e} \\
C^e_{22} &= \frac{1}{h_e}
\end{aligned}
\tag{8.58}
$$

The complete element matrix is

$$C^e = \frac{1}{h_e}\begin{bmatrix} 1 & -1 \\ -1 & 1 \end{bmatrix} \tag{8.59}$$

In a typical FEM implementation, this matrix is computed for each element, and the values are assembled into a large, sparse stiffness matrix for the full mesh. The assembly process effectively joins pairs of shape functions into triangle basis functions.

8.3.7 Assembly of the Global Stiffness Matrix

So far, we have considered only the value of the functional on one element. To obtain the full stiffness matrix, we must evaluate the functional on all elements in the mesh. From (8.48), the global stiffness matrix is given by

$$C_{mn} = \sum_{e=1}^{N_e} \varepsilon_e \int_{\text{Element } e} \frac{\partial f_m}{\partial x}\frac{\partial f_n}{\partial x}\, dx \tag{8.60}$$

The summation means that the original integral over the entire mesh is evaluated element-by-element. Because the basis functions have support on only two adjacent elements, the integrand is zero unless nodes m and n are shared by an element or are the same node. The eth element makes a contribution to C_{mn} if nodes m and n are both local nodes of the element. This means that instead of computing C_{mn} for each m, n, we can loop over the elements and add in the contributions from each shape function on the element to the corresponding C_{mn}.

This element-by-element method of filling the stiffness matrix can be represented using the assembly equation

$$C_{t(i,e),t(j,e)} = C_{t(i,e),t(j,e)} + \varepsilon_e C^e_{ij} \tag{8.61}$$

This equation must be applied for each element and each pair of local node numbers i and j. The element stiffness matrix C^e is indexed by the local node number of the nodes for the eth element. The global stiffness matrix C is indexed by the global node numbers for the entire mesh, and its size is N_{nodes} by N_{nodes}.

Embedding the assembly equation in loops over the elements and local node numbers leads to the algorithm

Matrix Assembly for 1-D FEM

Initialization: $C = 0$
Loop over elements, $e = 1, 2, \ldots, N_e$
Loop over local node number $i = 1, 2$
Loop over local node number $j = 1, 2$

$$C_{t(i,e),t(j,e)} = C_{t(i,e),t(j,e)} + \varepsilon_e C^e_{ij}$$

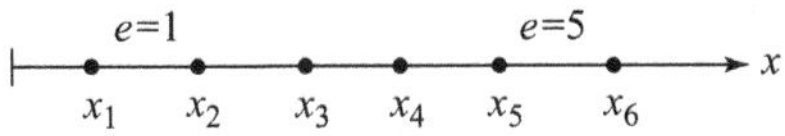

Figure 8.3 *One-dimensional mesh with six nodes and five elements. The boundary points are $x_1 = a$ and $x_6 = b$. The nodes need not be evenly spaced, but for this simple 1-D mesh, it is typically convenient to choose the nodes to be evenly spaced, so that the mesh is regular*

8.3.8 Example: Five-Element Mesh

We will consider the 1-D Laplace BVP on the interval $[a, b]$, with the Dirichlet boundary condition $\phi(a) = 0$, $\phi(b) = 10\,\text{V}$. The mesh consists of six nodes and $N_e = 5$ elements, with nodes and elements ordered from left to right, as shown in Figure 8.3.

Because there are six nodes, the global stiffness matrix is a 6×6 square matrix. The values of the entries in the global stiffness matrix are sums of elements of the element stiffness matrices, according to the assembly equation (8.61). Some of the entries in the assembled stiffness matrix are

Row 1

$$C_{11} = \varepsilon_1 C_{11}^1 = \varepsilon_1/h_1$$

$$C_{12} = \varepsilon_1 C_{12}^1 = -\varepsilon_1/h_1$$

$$C_{13} = C_{14} = C_{15} = C_{16} = 0 \text{ (these nodes are not common to any element)}$$

Row 2

$$C_{21} = \varepsilon_1 C_{21}^1 = -\varepsilon_1/h_1$$

$$C_{22} = \varepsilon_1 C_{22}^1 + \varepsilon_2 C_{11}^2 = \varepsilon_1/h_1 + \varepsilon_2/h_2$$

$$C_{23} = \varepsilon_2 C_{12}^2 = -\varepsilon_2/h_2$$

$$C_{24} = C_{25} = C_{26} = 0$$

The other rows can be obtained similarly. If all the elements have the same width, $h_e = h$, and if the permittivity is constant, $\varepsilon_e = \varepsilon$. The global stiffness matrix becomes

$$C = \frac{\varepsilon}{h}\begin{bmatrix} 1 & -1 & 0 & 0 & 0 & 0 \\ -1 & 2 & -1 & 0 & 0 & 0 \\ 0 & -1 & 2 & -1 & 0 & 0 \\ 0 & 0 & -1 & 2 & -1 & 0 \\ 0 & 0 & 0 & -1 & 2 & -1 \\ 0 & 0 & 0 & 0 & -1 & 1 \end{bmatrix} \tag{8.62}$$

Because of the boundary condition, $\phi_1 = 0$ and $\phi_6 = 10\,\text{V}$. This leads to the linear system

$$\begin{bmatrix} 1 & -1 & 0 & 0 & 0 & 0 \\ -1 & 2 & -1 & 0 & 0 & 0 \\ 0 & -1 & 2 & -1 & 0 & 0 \\ 0 & 0 & -1 & 2 & -1 & 0 \\ 0 & 0 & 0 & -1 & 2 & -1 \\ 0 & 0 & 0 & 0 & -1 & 1 \end{bmatrix} \begin{bmatrix} 0 \\ \phi_2 \\ \phi_3 \\ \phi_4 \\ \phi_5 \\ 10 \end{bmatrix} = \begin{bmatrix} 0 \\ 0 \\ 0 \\ 0 \\ 0 \\ 0 \end{bmatrix} \tag{8.63}$$

By dropping the first and last rows and reordering the unknowns, we obtain the linear system

$$\begin{bmatrix} 2 & -1 & 0 & 0 & -1 & 0 \\ -1 & 2 & -1 & 0 & 0 & 0 \\ 0 & -1 & 2 & -1 & 0 & 0 \\ 0 & 0 & -1 & 2 & 0 & -1 \end{bmatrix} \begin{bmatrix} \phi_2 \\ \phi_3 \\ \phi_4 \\ \phi_5 \\ 0 \\ 10 \end{bmatrix} = \begin{bmatrix} 0 \\ 0 \\ 0 \\ 0 \end{bmatrix} \tag{8.64}$$

Finally, if we move the last two columns to the right-hand side, we arrive at

$$\begin{bmatrix} 2 & -1 & 0 & 0 \\ -1 & 2 & -1 & 0 \\ 0 & -1 & 2 & -1 \\ 0 & 0 & -1 & 2 \end{bmatrix} \underbrace{\begin{bmatrix} \phi_2 \\ \phi_3 \\ \phi_4 \\ \phi_5 \end{bmatrix}}_{\substack{\text{Interior} \\ \text{nodes}}} = - \begin{bmatrix} -1 & 0 \\ 0 & 0 \\ 0 & 0 \\ 0 & -1 \end{bmatrix} \begin{bmatrix} 0 \\ 10 \end{bmatrix} = \begin{bmatrix} 0 \\ 0 \\ 0 \\ 10 \end{bmatrix} \tag{8.65}$$

This linear system can be solved to find the unknown potential at the interior nodes.

8.3.9 Comparison of FEM and FD

As observed already, there is a close relationship between FEM and FD. To illustrate this, we will compare the result (8.65) with the finite difference method for the same BVP, with the grid points chosen to be the six nodes of the FEM mesh. The stencil for the 1-D Laplacian is $(-\phi_{n-1} + 2\phi_n - \phi_{n+1})/h^2$. Applying the stencil at each interior node leads to the difference equations

$$\begin{array}{rrrrrrrrrrl} 0 & + & 2\phi_2 & - & \phi_3 & & & & & & = 0 \\ & - & \phi_2 & + & 2\phi_3 & - & \phi_4 & & & & = 0 \\ & & & - & \phi_3 & + & 2\phi_4 & - & \phi_5 & & = 0 \\ & & & & & - & \phi_4 & + & 2\phi_5 & - \;10 & = 0 \end{array} \tag{8.66}$$

which is identical to the linear system obtained from the finite element method. This shows that FEM and FD are closely related.

One important difference between the two methods is that it is quite difficult to apply FD to nonrectangular grids, whereas FEM is generally easy to apply on a nonuniform or curved mesh. Thus, FEM can be viewed as a systematic way to extend FD to arbitrary grids.

8.3.10 Sparse Matrix and Dense Matrix Methods

Because each basis function is nonzero on at most two elements, nearly all of the elements of the stiffness matrix C are zero. For this reason, FEM is a sparse matrix method. This is in contrast to numerical methods based on integral equations, which essentially replace the derivatives in (8.48) by convolutions with a kernel or Green's function. The convolution changes the functions with local support to functions with support over the entire mesh, producing a dense moment matrix.

The fact that FEM produces a sparse matrix, whereas MoM for an integral equation generally leads to a dense matrix, represents one of the key performance trade-offs between the two classes of numerical methods. Sparse matrices can be stored efficiently, since zero matrix elements need not be retained in memory, and large dense matrices require extremely high amounts of memory. Well-conditioned sparse linear systems can be solved with lower computational cost than dense linear systems. For surface integral equations, however, the matrix size may be smaller with MoM than with FEM, since only the surface of an object must be meshed and the moment matrix need not be filled explicitly if the fast multipole method algorithm is employed. The relative efficiencies of FEM and MoM, therefore, depend strongly on the type of problem to be solved and the details of the implementations.

8.4 Helmholtz Equation: 2-D FEM

As an illustration of the finite element method for 2-D problems, we will consider two types of boundary value problems. First, we will develop a mode solver for metallic waveguides. This is the same problem considered in Section 4.3.2 using the finite difference method. In this chapter, we will call this an "unknown k" or eigenvalue problem. This simple FEM algorithm will allow the local matrices and global matrix assembly process for the Laplacian operator to be implemented and tested.

Second, we will apply FEM to the 2-D electromagnetic scattering boundary value problem introduced in Section 2.13.6. This particular BVP has been a key example throughout the book and has been solved with a variety of numerical methods. The 2-D scattering BVP was solved with the FDTD algorithm in Section 4.4 and with the method of moments applied to an integral equation in Section 6.4. With the FDTD algorithm, we solved the coupled first-order Maxwell's equation form of the boundary value problem with the Yee cell construction. With the method of moments, we solved the same BVP but using an integral equation formulation. Here, we will solve the 2-D EM problem using a third approach, with the BVP formulated as a second-order Helmholtz equation. Like the 2-D FDTD algorithm, the FEM scattering problem solver will require an absorbing boundary condition. To distinguish this FEM application from the waveguide mode solver, we will refer to this BVP as a "known k" or scattering problem.

For both the eigenvalue solver and the scattering solver, the governing differential equation will be the 2-D Helmholtz equation,

$$(\nabla^2 + k^2)\phi(x, y) = 0 \tag{8.67}$$

To be consistent with the notation used earlier in this chapter, the field variable is denoted here by ϕ. Depending on the physical problem we are solving, ϕ may represent different quantities. For an FEM-based metallic waveguide mode solver, ϕ represents the transverse variation of the z component of the electric field associated with a waveguide mode for the TM polarization (Dirichlet boundary condition) or the z component of the magnetic field for the TE polarization (Neumann boundary condition). For scattering problems, ϕ represents E_z^s, the field scattered by a dielectric or conducting object when illuminated with an incident field. For the waveguide solver, k represents the unknown transverse wave number of a waveguide mode. For the scattering problem, k represents the material composition of the scatterer.

The associated Helmholtz functional is

$$I(\phi) = \frac{1}{2}\int [|\nabla\phi|^2 - k^2|\phi|^2]\, dS \tag{8.68}$$

For a given boundary condition on ϕ, this functional has a unique minimum solution. This basic functional is at the heart of both the waveguide mode (unknown k) and the scattering problem (known k) solvers that we will develop in this section.

8.4.1 Boundary Conditions for FEM

As with the finite difference time-domain method, the FEM algorithm requires constraints on the solution at the boundary of the simulation domain. Dirichlet and Neumann conditions are relatively easy to handle. Other boundary conditions, such as an absorbing boundary condition (ABC), can be more difficult to implement but are critical in practical applications of FEM to radiation and scattering problems.

Dirichlet condition

For the Dirichlet condition, we have a specified value of the unknown potential on the boundary

$$\phi(\bar{r})|_{\bar{r}\in\Gamma} = h(\bar{r}) \tag{8.69}$$

where Γ is the boundary of the domain. Implementing this boundary condition simply requires setting all nodal values of the potential on the boundary to the given value, and only interior nodes in the problem are unknowns.

Neumann condition

In this case, we have a specified value of the normal derivative of the potential on the boundary

$$\left.\frac{\partial\phi(\bar{r})}{\partial n}\right|_{\bar{r}\in\Gamma} = h(\bar{r}) \tag{8.70}$$

where n represents a coordinate that is normal to the boundary. For a triangular mesh, implementing this boundary condition leads to relationships between boundary nodes and adjacent nodes that constrain the solution to tend to a constant near the boundary.

Impedance condition

The Dirichlet and Neumann conditions can be generalized to

$$\frac{\partial \phi(\bar{r})}{\partial n} = h(\bar{r})\phi(\bar{r}) \tag{8.71}$$

Instead of specifying the potential or its normal derivative, we specify the ratio. This type of boundary condition can be used to approximate the boundary condition at a conducting layer or impedance surface.

Absorbing boundary condition

For radiation and scattering problems, an absorbing boundary condition is required. As with the FDTD method, a material layer can be added to the outside of the simulation domain with permittivity, permeability, and conductivity (loss) chosen to absorb incident radiation. This is the perfectly matched layer (PML) approach. A surface integral equation can also be used to implement an absorbing boundary condition. This leads to the finite element–boundary element method (FEM–BEM) treated in Section 8.5. Other ABCs for FEM are available, such as infinite elements and the Bayliss–Turkel radiation condition, but most commercial software packages use a PML.

8.4.1.1 Boundary Terms in the Functional

With the finite difference method, boundary conditions are usually implemented as additional equations involving boundary grid points. For FEM, boundary conditions can be included in a more rigorous way by modifying the functional used with the Rayleigh–Ritz procedure to take into account the behavior of the solution on the boundary of the region of interest.

To include boundary terms in the functional, we need to rederive the functional from the governing PDE. For the Helmholtz equation, using (8.19) with $L = \nabla^2 + k^2$ leads to

$$\begin{aligned} I &= \langle \phi, (\nabla^2 + k^2)\phi \rangle - 2\langle \phi, f \rangle \\ &= \int_V (\phi^* \nabla^2 \phi + k^2 |\phi|^2 - 2f\phi^*)\, dV \end{aligned} \tag{8.72}$$

where f is a given source or forcing function. Using the identity $f\nabla^2 g = \nabla \cdot (f\nabla g) - \nabla f \cdot \nabla g$, this becomes

$$I = \int_V (-|\nabla\phi|^2 + k^2|\phi|^2 - 2f\phi^*)\, dV + \int_V \nabla \cdot (\phi^* \nabla \phi)\, dV \tag{8.73}$$

This is a form of multidimensional integration by parts. Applying the divergence theorem to the second term leads to

$$I = \int_V (-|\nabla\phi|^2 + k^2|\phi|^2 - 2f\phi^*)\,dV + \oint_S (\phi^*\nabla\phi)\cdot d\overline{S} \tag{8.74}$$

To simplify the last term on the right, we can evaluate the gradient and dot product using

$$\begin{aligned}\nabla\phi\cdot d\overline{S} &= \nabla\phi\cdot\hat{n}\,dS \\ &= (\hat{n}\cdot\nabla\phi)dS \\ &= \frac{\partial\phi}{\partial n}dS\end{aligned} \tag{8.75}$$

This allows the functional to be placed in the form

$$I = \int_V (-|\nabla\phi|^2 + k^2|\phi|^2 - 2f\phi^*)\,dV + \oint_S \phi^*\frac{\partial\phi}{\partial n}\,dS \tag{8.76}$$

This more general form of the functional allows boundary conditions to be implemented naturally as part of the Rayleigh–Ritz procedure.

With the homogeneous Dirichlet or Neumann boundary conditions, the boundary term in (8.76) vanishes. For this reason, functionals are typically written without the boundary term. For other types of boundary conditions, when the Rayleigh–Ritz method is applied, the boundary term of the functional leads to an additional set of unknowns for the normal derivative of the field on the boundary. Because the number of unknowns increases, the FEM linear system does not provide enough linear equations to solve for the unknowns. The boundary condition must be applied to obtain an additional set of relationships between the potential and its normal derivative on the boundary. This approach is used to realize an absorbing boundary condition in Section 8.5.

8.4.2 Rayleigh–Ritz Method for the Helmholtz Functional

As with the 1-D FEM algorithm, the global stiffness matrix can be assembled from stiffness matrices for individual mesh elements. If we evaluate the Helmholtz functional on one mesh element and place the result in matrix form as in (8.56), we obtain

$$I_e(\overline{\phi}_e) = \frac{1}{2}\overline{\phi}_e^T C^e\overline{\phi}_e - \frac{k^2}{2}\overline{\phi}_e^T T^e\overline{\phi}_e \tag{8.77}$$

where $\overline{\phi}_e = [\phi_{e1}\ \phi_{e2}\ \phi_{e3}]^T$ are the samples of the numerical approximation to the unknown field $\phi(x,y)$ at the three nodes of the eth triangular mesh element. The element matrices are

$$C^e_{ij} = \int_e \nabla\alpha_i\cdot\nabla\alpha_j\,dS \tag{8.78}$$

and

$$T_{ij}^{e} = \int_{e} \alpha_i \alpha_j \, dS \tag{8.79}$$

where the α_i are shape functions on the mesh element.

If we number the nodes so that interior nodes are first and nodes on the boundary are last, the global functional for the mesh can be written in block form as

$$I = \frac{1}{2}\begin{bmatrix} \overline{\phi}_{\mathrm{i}}^{T} & \overline{\phi}_{\mathrm{b}}^{T} \end{bmatrix}\begin{bmatrix} C_{\mathrm{ii}} & C_{\mathrm{ib}} \\ C_{\mathrm{bi}} & C_{\mathrm{bb}} \end{bmatrix}\begin{bmatrix} \overline{\phi}_{\mathrm{i}} \\ \overline{\phi}_{\mathrm{b}} \end{bmatrix} - \frac{k^2}{2}\begin{bmatrix} \overline{\phi}_{\mathrm{i}} & \overline{\phi}_{\mathrm{b}} \end{bmatrix}\begin{bmatrix} T_{\mathrm{ii}} & T_{\mathrm{ib}} \\ T_{\mathrm{bi}} & T_{\mathrm{bb}} \end{bmatrix}\begin{bmatrix} \overline{\phi}_{\mathrm{i}} \\ \overline{\phi}_{\mathrm{b}} \end{bmatrix} \tag{8.80}$$

Since the boundary values of the solution will be determined by a boundary condition, only the interior coefficients are unknown.

Minimizing the functional over the interior values $\overline{\phi}_{\mathrm{i}}$ leads to the linear system

$$\begin{bmatrix} C_{\mathrm{ii}} & C_{ib} \end{bmatrix}\begin{bmatrix} \overline{\phi}_{\mathrm{i}} \\ \overline{\phi}_{\mathrm{b}} \end{bmatrix} - k^2\begin{bmatrix} T_{\mathrm{ii}} & T_{\mathrm{ib}} \end{bmatrix}\begin{bmatrix} \overline{\phi}_{\mathrm{i}} \\ \overline{\phi}_{\mathrm{b}} \end{bmatrix} - 0 \tag{8.81}$$

This result was derived for the homogeneous Helmholtz equation with no source term. With a nonzero source term on the right-hand side of the Helmholtz equation, (8.81) becomes

$$\begin{bmatrix} C_{\mathrm{ii}} & C_{\mathrm{ib}} \end{bmatrix}\begin{bmatrix} \overline{\phi}_{\mathrm{i}} \\ \overline{\phi}_{\mathrm{b}} \end{bmatrix} - k^2\begin{bmatrix} T_{\mathrm{ii}} & T_{\mathrm{ib}} \end{bmatrix}\begin{bmatrix} \overline{\phi}_{\mathrm{i}} \\ \overline{\phi}_{\mathrm{b}} \end{bmatrix} = -\begin{bmatrix} T_{\mathrm{ii}} & T_{\mathrm{ib}} \end{bmatrix}\begin{bmatrix} \overline{f}_{\mathrm{i}} \\ \overline{f}_{\mathrm{b}} \end{bmatrix} \tag{8.82}$$

where $\overline{f}_{\mathrm{i}}$ and $\overline{f}_{\mathrm{b}}$ are samples of the right-hand side of (8.85) at the interior and boundary mesh node points. To solve this linear system, a boundary condition is required. One of the boundary conditions discussed in Section 8.4.1 can be implemented, or the FEM–BEM method of Section 8.5 can be used. Depending on the nature of the physical problem to be solved, there are two ways to apply the FEM algorithm for the Helmholtz equation.

8.4.3 Eigenvalue Problems (Unknown k)

For a problem such as finding waveguide modes, k in the Helmholtz equation represents an unknown transverse wave number. If the unknown field in the Helmholtz equation is the z component of the electric field inside a PEC waveguide, the boundary samples $\overline{\phi}_{\mathrm{b}}$ vanish by the Dirichlet boundary condition at the waveguide walls. The linear system becomes

$$[C_{\mathrm{ii}} - k^2 T_{\mathrm{ii}}]\overline{\phi}_{\mathrm{i}} = 0 \tag{8.83}$$

where k represents the transverse wave number of a waveguide mode. This can be rearranged into a eigenvalue/eigenvector problem of the form

$$T_{\mathrm{ii}}^{-1} C_{\mathrm{ii}} \overline{\phi}_{\mathrm{i}} = k^2 \overline{\phi}_{\mathrm{i}} \tag{8.84}$$

which shows that the transverse mode wave numbers k can be found from the eigenvalues of $T_{\mathrm{ii}}^{-1} C_{\mathrm{ii}}$.

8.4.4 Scattering Problems (Known k)

Throughout this book, we have used as a key example one particular boundary value problem, electromagnetic scattering from a 2-D conducting or dielectric cylinder. 2-D EM problems have two possible polarizations, TM^z and TE^z. As boundary conditions are simpler, we have focused on the TM^z polarization. This BVP has multiple formulations that provide the same unique solution. It can be solved with Maxwell's equations directly, as was done with the 2-D FDTD algorithm based on the Yee cell construction in Section 4.4. It can be reformulated using an integral equation and solved using the method of moments as in Section 6.4. A third formulation is the second-order, scalar Helmholtz equation. This approach will be used here as an example application of the finite element method.

When the Helmholtz equation is used to solve a scattering problem, k is known and we wish to find a function $\phi(\overline{\rho})$ that is a solution to the PDE and satisfies a given boundary condition. It is convenient to use the scattered field formulation, so that the Helmholtz equation is modified to

$$\left[\nabla^2 + k^2(\overline{\rho})\right]\phi(\overline{\rho}) = \left[k_0^2 - k^2(\overline{\rho})\right]\phi^{\mathrm{i}}(\overline{\rho}) \tag{8.85}$$

where for the transverse magnetic (TM) polarization, $\phi(\overline{\rho})$ represents the scattered electric field E_z^{s} and $\phi^{\mathrm{i}}(\overline{\rho})$ represents the z component E_z^{i} of the incident electric field.

For the Helmholtz formulation of the 2-D scattering problem, we are solving the problem in the frequency domain. The field quantity ϕ represents the phasor electric field intensity. For plane wave scattering, or to find scattering widths and scattering amplitudes, the incident field $\phi^{\mathrm{i}}(\overline{\rho})$ in the right hand side is the phasor electric field given in (2.6.2).

In this section, we will continue to use the notation ϕ for the unknown field quantity, to emphasize the continuity with the solutions to Laplace's equation discussed earlier. It is important to remember that in this section ϕ represents the scattered electric field rather than electric potential or an angle.

8.4.5 FEM Formulation of the 2-D Scattering Problem

For Laplace eigenvalue problems, the wave number k is unknown and constant over the simulation region. For scattering problems, k is known but varies over the simulation region. We need to modify the previous derivation to allow k to be a function of position. This means that $k(\overline{\rho})$ must remain under the integral for the elements of T^e, and the definition of T^e changes to

$$T_{ij}^{\prime e} = \int_e k^2(\overline{\rho})\alpha_i\alpha_j\, dS \tag{8.86}$$

where $\overline{\rho}$ represents a point (x, y) in the simulation region. For a dielectric scatterer, the value of the squared wave number as a function of position is given by

$$k^2(\overline{\rho}) = \omega^2\mu\varepsilon = k_0^2\varepsilon_r(\overline{\rho}) \tag{8.87}$$

where $k_0^2 = \omega^2 \mu_0 \varepsilon_0$, and $\varepsilon_r(\overline{\rho})$ is the relative permittivity of the scatterer. For a conductive scatterer, in the frequency domain, the permittivity becomes complex, so that

$$\varepsilon = \varepsilon' - j\frac{\sigma}{\omega} \tag{8.88}$$

where ε' is the real part of the permittivity. For a very good conductor, the imaginary part of ε is typically -2 to -10, and a value such as -100 suffices to model a PEC scatterer.

To evaluate the integral in (8.86), we will approximate the wave number as constant on each element, so that the matrix element becomes

$$T_{ij}^{\prime e} = k_e^2 \int_e \alpha_i \alpha_j \, dS \tag{8.89}$$

where $k_e^2 = k^2(\overline{\rho}_c)$, and $\overline{\rho}_c$ is the center of the eth element. In terms of the element nodes, the center is given by

$$x_c = \frac{x_1 + x_2 + x_3}{3} \tag{8.90a}$$

$$y_c = \frac{y_1 + y_2 + y_3}{3} \tag{8.90b}$$

Since one value of the wave number is required for each mesh element, during the mesh generation process, an array of permittivity or wave number values can be created with one entry per mesh element. This material parameter array is used in the matrix assembly process to fill T'.

Since the wave number is absorbed into the matrix T', the FEM linear system becomes

$$\begin{bmatrix} C_{ii} & C_{ib} \end{bmatrix} \begin{bmatrix} \overline{\phi}_i \\ \overline{\phi}_b \end{bmatrix} - \begin{bmatrix} T'_{ii} & T'_{ib} \end{bmatrix} \begin{bmatrix} \overline{\phi}_i \\ \overline{\phi}_b \end{bmatrix} = -\begin{bmatrix} T'_{ii} & T'_{ib} \end{bmatrix} \begin{bmatrix} \overline{f}_i \\ \overline{f}_b \end{bmatrix} \tag{8.91}$$

It remains to determine the element matrices T^e and C^e, so that the global matrices can be assembled from the element matrices.

8.4.6 Triangular Mesh

For two-dimensional problems, the most common type of mesh is triangular. The mesh is defined by (1) a list of the coordinates of the node points and (2) a list of node numbers for each triangle. Many implementations of FEM also include an array that defines the boundary of the mesh to facilitate the application of boundary conditions. The triangular mesh used here is identical to the triangular mesh used with RWG basis functions in Section 6.8.2, except that basis functions are associated with nodes instead of element edges, so the edge list is not required here.

For the simple mesh shown in Figure 8.4, the node list is

$$p = \begin{bmatrix} x_1 & x_2 & x_3 & x_4 & x_5 \\ y_1 & y_2 & y_3 & y_4 & y_5 \end{bmatrix} \tag{8.92}$$

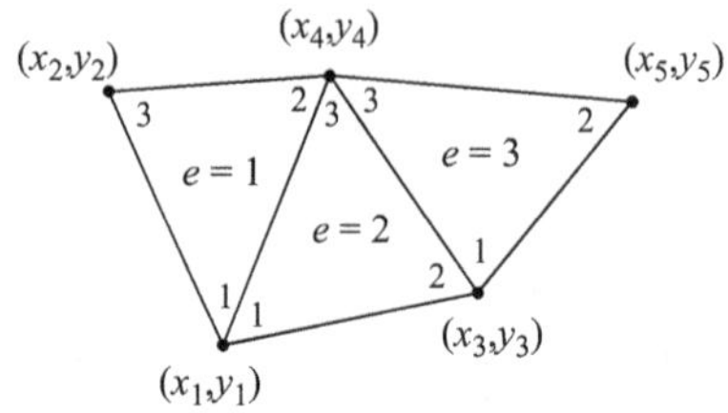

Figure 8.4 Three-element triangular mesh. The subscripts on coordinate papers are global node numbers. Local node numbers and element numbers are indicated on each mesh triangle

The column number in the array p is the global node number. The mesh triangles are defined by the array

$$t = \begin{bmatrix} 1 & 1 & 3 \\ 4 & 3 & 5 \\ 2 & 4 & 4 \end{bmatrix} \tag{8.93}$$

The row number of the array $t(i, e)$ is the local node number with respect to the eth triangle. All geometrical information is contained in the node list, and the triangle array $t(i, e)$ contains only topological information. In the MATLAB implementation of finite element mesh functions, the array t contain an additional row with the region number of each triangle.

An FEM code should be implemented so that the results of a calculation are independent of the order in which nodes and triangles are listed. Some codes assume that the local nodes are listed in clockwise or counterclockwise order with respect to a given choice of normal direction.

In the MATLAB implementation of the finite element method, an additional boundary or edge array `e` is also produced as part of the mesh-generation process. The first two rows of this array list the nodes of the edges of patches on the boundary of the mesh. For the example mesh considered here, the first two rows of the edge array are

$$\texttt{e(1:2,:)} = \begin{bmatrix} 1 & 3 & 5 & 4 & 2 \\ 3 & 5 & 4 & 2 & 1 \end{bmatrix}$$

Other rows of the edge array define the values of a parameter that ranges from one to zero on each continuous edge segment of the mesh and the domain number of the regions on the right and left of the boundary with respect to a given direction of traverse around the boundary.

8.4.7 Basis Functions and Shape Functions

Once a mesh is specified, basis functions must be defined. The most common basis functions for the 2-D FEM are 2-D generalizations of 1-D triangle functions. This basis is also referred to as piecewise linear, since the resulting approximation for the unknown field is piecewise linear, and the graph of the approximation is made up

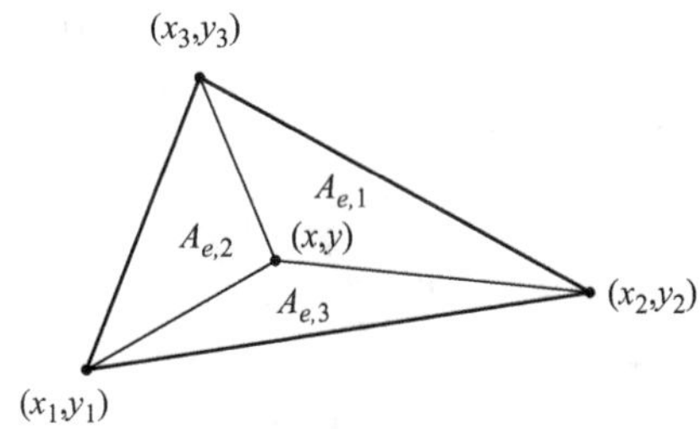

Figure 8.5 Barycentric coordinates on a mesh triangle

of triangular facets. For this basis type, each mesh node has an associated generalized triangle basis function. Each basis function is made up of several linear shape functions, one for each triangle that is adjacent to the node.

One way to define the shape functions that make up the piecewise linear basis functions is using the barycentric coordinates introduced in (6.140) and illustrated in Figure 8.5. For the eth triangle in the mesh,

$$\alpha_1 = \frac{A_{e,1}}{A_e} \tag{8.94a}$$

$$\alpha_2 = \frac{A_{e,2}}{A_e} \tag{8.94b}$$

$$\alpha_3 = \frac{A_{e,3}}{A_e} \tag{8.94c}$$

where the area of the triangle in terms of the node coordinates is

$$A_e = \frac{1}{2}\det\begin{bmatrix} 1 & x_1 & y_1 \\ 1 & x_2 & y_2 \\ 1 & x_3 & y_3 \end{bmatrix} \tag{8.95}$$

$A_{e,1}$ is the area of the triangle with nodes (x,y), (x_2,y_2), and (x_3,y_3), where (x,y) is an arbitrary point on the mesh element. $A_{e,2}$ is the area of the triangle with nodes (x,y), (x_1,y_1), and (x_3,y_3), and $A_{e,3}$ is the area of the triangle with nodes (x,y), (x_1,y_1), and (x_2,y_2). To obtain the areas $A_{e,1}$, $A_{e,2}$, and $A_{e,3}$, the determinant formula (8.95) can be used with one of the coordinates (x_i,y_i) replaced by (x,y).

From these definitions, it is easy to show that

$$\alpha_1 = \frac{1}{2A_e}[x(y_2-y_3)+y(x_3-x_2)+x_2y_3-x_3y_2] \tag{8.96a}$$

$$\alpha_2 = \frac{1}{2A_e}[x(y_3-y_1)+y(x_1-x_3)+x_3y_1-x_1y_3] \tag{8.96b}$$

$$\alpha_3 = \frac{1}{2A_e}[x(y_1-y_2)+y(x_2-x_1)+x_1y_2-x_2y_1] \tag{8.96c}$$

The function α_1 is equal to one at (x_1,y_1) and changes linearly to zero at the side of the triangle opposite to the first node. α_2 and α_3 are, respectively, equal to one at the second and third nodes, and zero at the opposite sides. As with 1-D triangle functions, if the shape functions are added together, the sum is one everywhere on the element (which can be easily seen from (8.94a)–(8.94c) and Figure 8.5).

8.4.8 Evaluating Element Matrices

For piecewise linear or triangle basis functions, we can explicitly evaluate the element matrices C^e and T^e. For the C^e matrix, the 1,1 element is

$$
\begin{aligned}
C_{11}^e &= \int_e \nabla\alpha_1 \cdot \nabla\alpha_1 \, dx\, dy \\
&= \int_e \left\| \frac{\partial\alpha_1}{\partial x}\hat{x} + \frac{\partial\alpha_1}{\partial y}\hat{y} \right\|^2 dx\, dy \\
&= \int_e \left\| \frac{y_2 - y_3}{2A_e}\hat{x} + \frac{x_3 - x_2}{2A_e}\hat{y} \right\|^2 dx\, dy \\
&= \frac{1}{4A_e^2} \int_e \left[(y_2 - y_3)^2 + (x_3 - x_2)^2 \right] dx\, dy \\
&= \frac{1}{4A_e} [(y_2 - y_3)^2 + (x_3 - x_2)^2]
\end{aligned}
$$

Since the shape functions are linear, the gradients of the shape functions are constant vectors. The integral over x and y evaluates to the element area A_e. Other matrix elements can be found similarly, including the following:

$$
\begin{aligned}
C_{12}^e &= \int_e \nabla\alpha_1 \cdot \nabla\alpha_2 \, dx\, dy \\
&= \frac{1}{4A_e^2} \int_e [(y_2 - y_3)(y_3 - y_1) + (x_3 - x_2)(x_1 - x_3)] \, dx\, dy \\
&= \frac{1}{4A_e} [(y_2 - y_3)(y_3 - y_1) + (x_3 - x_2)(x_1 - x_3)] \\
C_{13}^e &= \frac{1}{4A_e} [(y_2 - y_3)(y_1 - y_2) + (x_3 - x_2)(x_2 - x_1)] \\
C_{23}^e &= \frac{1}{4A_e} [(y_3 - y_1)(y_1 - y_2) + (x_1 - x_3)(x_2 - x_1)]
\end{aligned}
$$

Since the matrix is symmetric, $C_{ij}^e = C_{ji}^e$. Other elements of the matrix can be found by permuting the indices in these formulas.

For the T^e matrix, the integrations are more complicated, but we can compute them by performing a change of variables to the barycentric coordinates α_1 and α_2:

$$
\begin{aligned}
T^e_{ij} &= \int_e \alpha_i\alpha_j\, dx\, dy \\
&= \int_0^1 \int_0^{1-\alpha_2} \alpha_i\alpha_j\, d[\alpha_1 x_1 + \alpha_2 x_2 + (1-\alpha_1-\alpha_2)x_3] \\
&\qquad \wedge [\alpha_1 y_1 + \alpha_2 y_2 + (1-\alpha_1-\alpha_2)y_3] \\
&= \int_0^1 \int_0^{1-\alpha_2} \alpha_i\alpha_j\, [(x_1-x_3)d\alpha_1 + (x_2-x_3)d\alpha_2] \\
&\qquad \wedge [(y_1-y_3)d\alpha_1 + (y_2-y_3)d\alpha_2] \\
&= \int_0^1 \int_0^{1-\alpha_2} \alpha_i\alpha_j\, [(x_1-x_3)(y_2-y_3)-(x_2-x_3)(y_1-y_3)]d\alpha_1\, d\alpha_2 \\
&= 2A_e \int_0^1 \int_0^{1-\alpha_2} \alpha_i\alpha_j\, d\alpha_1\, d\alpha_2
\end{aligned}
$$

The factor $2A_e$ is the Jacobian of the coordinate transformation. This can be found using a determinant of partial derivatives or by inserting an exterior product $\wedge$ between the differentials. The exterior product has the property that $d\alpha_1 \wedge d\alpha_2 = -d\alpha_2 \wedge d\alpha_1$ and $d\alpha_1 \wedge d\alpha_1 = d\alpha_2 \wedge d\alpha_2 = 0$ and allows differentials to be manipulated algebraically. The first element of the matrix is

$$
\begin{aligned}
T^e_{11} &= 2A_e \int_0^1 \int_0^{1-\alpha_2} \alpha_1^2\, d\alpha_1\, d\alpha_2 \\
&= 2A_e \int_0^1 \frac{1}{3}(1-\alpha_2)^3\, d\alpha_2 \\
&= \frac{A_e}{6}
\end{aligned}
$$

The complete matrix is

$$
T^e = A_e \begin{bmatrix} 1/6 & 1/12 & 1/12 \\ 1/12 & 1/6 & 1/12 \\ 1/12 & 1/12 & 1/6 \end{bmatrix} \tag{8.97}
$$

The full mesh matrices C and T are assembled using these expressions for the element matrices C^e and T^e.

8.4.9 Matrix Assembly

Filling the global FEM matrices C and T for the 2-D FEM algorithm follows the same procedure as in Section 8.3.7. The assembly algorithm is

Matrix Assembly for 2-D FEM

Initialization: $C = 0$, $T = 0$
Loop over elements, $e = 1, 2, \ldots, N_e$
Double loop over local node numbers $i, j = 1, 2, 3$

$$C_{t(i,e),t(j,e)} = C_{t(i,e),t(j,e)} + C^e_{ij} \tag{8.98}$$

$$T_{t(i,e),t(j,e)} = T_{t(i,e),t(j,e)} + T^e_{ij} \tag{8.99}$$

where $t(i,e)$ is the triangle array in (8.93). If the wave number $k(\overline{\rho})$ depends on position, it is included in the T matrix as indicated in (8.86).

To illustrate the assembly process, we will fill the matrix T for the simple mesh shown in Figure 8.4. The assembly process for C is similar. For element $e = 1$, the following values are assembled into the global T matrix:

$$\begin{array}{lll} T_{11} = T_{11} + T^e_{11} & T_{14} = T_{14} + T^e_{12} & T_{12} = T_{12} + T^e_{13} \\ T_{41} = T_{41} + T^e_{21} & T_{44} = T_{44} + T^e_{22} & T_{42} = T_{42} + T^e_{23} \\ T_{21} = T_{21} + T^e_{31} & T_{24} = T_{24} + T^e_{32} & T_{22} = T_{22} + T^e_{33} \end{array}$$

For element $e = 2$, the assembled values are

$$\begin{array}{lll} T_{11} = T_{11} + T^e_{11} & T_{13} = T_{13} + T^e_{12} & T_{14} = T_{14} + T^e_{13} \\ T_{31} = T_{31} + T^e_{21} & T_{33} = T_{33} + T^e_{22} & T_{34} = T_{34} + T^e_{23} \\ T_{41} = T_{41} + T^e_{31} & T_{43} = T_{43} + T^e_{32} & T_{44} = T_{44} + T^e_{33} \end{array}$$

Because the basis functions associated with global nodes 1 and 2 are confined to elements 1 and 2, at this point in the assembly process, the global matrix elements T_{11} and T_{22} are fully assembled. The values of these matrix elements are

$$\begin{aligned} T_{11} &= T^1_{11} + T^2_{11} = A_1/6 + A_2/6 \\ T_{12} &= T^1_{13} = A_1/12 \\ T_{21} &= T^1_{31} = A_1/12 \\ T_{22} &= T^1_{33} = A_1/6 \end{aligned}$$

where A_e is the area of the eth mesh element.

The other elements of the T matrix are not fully assembled until contributions for the remaining mesh element ($e = 3$) are added. The basis function associated with node 4, for example, consists of shape functions on all three elements. Consequently, the matrix element T_{44} receives a contribution for $e = 1$, $e = 2$, and $e = 3$. Once the assembly loop has run through all mesh elements, the matrix T is complete.

8.5 Finite Element Method–Boundary Element Method

As we have already seen for the FDTD algorithm, absorbing boundary conditions are essential for radiation and scattering simulations. The absorbing boundary conditions we have considered so far have been discretizations of a one-way wave equation and perfectly matched layers. Another approach to developing an ABC or radiation boundary condition is to use a surface integral equation on the boundary of the domain.

From (8.76), it can be seen that a boundary condition can be viewed as a relationship between an unknown field or potential and its normal derivative. If the field and its normal derivative are related properly, the boundary condition can be set up to absorb incident waves without reflection. A surface integral equation can be derived for this relationship and discretized using the method of moments to provide a set of linear equations that relate the potential and its normal derivative. Since the method of moments for surface integral equations is also called the boundary element method (BEM), the combination of FEM and a surface integral equation on the boundary is generally referred to as FEM–BEM, and occasionally as the finite element method–boundary integral equation (FEM–BIE) method.

To derive the linear system for the FEM–BEM algorithm, several different approaches can be used, but here it is convenient to use the method of weighted residuals. If ϕ is an arbitrary function, the residual error for the Helmholtz equation $(\nabla^2 + k^2)\phi = f$ is

$$r = \nabla^2\phi + k^2\phi - f \tag{8.100}$$

If we form inner products with a set of testing functions $t_m(\bar{r})$ and set the result to zero, we have

$$\langle t_m, r\rangle = \int_V t_m \left(\nabla^2\phi + k^2\phi - f\right) dV = 0 \tag{8.101}$$

This is known as the weak form of the PDE. If we were to substitute a basis function expansion for ϕ, we would obtain the same linear system that would result from the method of weighted residuals.

By following a derivation similar to that of (8.76), we can express the weak form of the PDE as

$$\int_V \left(\nabla t_m \cdot \nabla\phi - t_m k^2\phi\right) dV - \oint_S t_m \frac{\partial\phi}{\partial n}\, dS = -\int_V t_m f\, dV \tag{8.102}$$

which is similar in form to the functional for the Helmholtz equation and includes a boundary term on the left-hand side. The first term on the left will give rise to an FEM-type linear system, and the second term on the left will produce a coupling matrix that links the FEM linear system to a BEM linear system for a surface integral equation.

To discretize this equation, we need unknowns for the solution in the interior of V and for the normal derivative of the field on the boundary. Inside V, we will expand the unknown field as a linear combination of basis functions, so that

$$\hat{\phi}(\overline{r}) \simeq \sum_{n=1}^{N} a_n f_n(\overline{r}) \tag{8.103}$$

For the normal derivative, we can use the same basis functions f_n restricted to the boundary, so that

$$\frac{\partial \phi}{\partial n} \simeq \sum_{S} b_n f_n(\overline{r}), \quad \overline{r} \in S \tag{8.104}$$

where the sum is over the basis functions that are nonzero on the boundary. Substituting (8.103) and (8.104) into the weak form of the PDE leads to

$$\int_V \left(\nabla t_m \cdot \nabla \sum a_n f_n - t_m k^2 \sum_{n=1}^{N} a_n f_n \right) dV - \oint_S t_m \sum_S b_n f_n \, dS = - \int_V t_m f \, dV \tag{8.105}$$

Rearranging this expression gives

$$\sum_{n=1}^{N} \underbrace{\int_V (\nabla t_m \cdot \nabla f_n - t_m k^2 f_n) \, dV}_{A_{mn}} a_n - \sum_S \underbrace{\oint_S t_m f_n \, dS}_{B_{mn}} b_n = - \underbrace{\int_V t_m f \, dV}_{c_m} \tag{8.106}$$

where the coefficients a_n represent a basis expansion of the unknown field solution $\phi(x, y)$. The coefficients b_n represent the normal derivative of ϕ on the boundary of the simulation domain.

The boundary value problem we are solving involves an incident field and a dielectric or conducting object. The goal is to find the scattered field, represented by the unknown function ϕ in the Helmholtz equation. The forcing function f in the Helmholtz equation is given by the right-hand side of (8.85). The elements of the vector $\overline{c}$ are zero for testing functions outside the scatterer and nonzero for testing functions on or inside the scatterer. For plane wave scattering and calculation of scattering widths and scattering amplitudes, $\overline{c}$ includes a phasor incident plane wave as discussed in Section 8.4.4.

In matrix form, (8.106) can be written as

FEM Linear System

$$A\overline{a} - B\overline{b} = -\overline{c} \tag{8.107}$$

where A is an $N \times N$ matrix, and N is the total number of basis functions. B is an $N \times N_b$ matrix, where N_b is the number of basis functions on the boundary. In this linear system, there are $N + N_b$ unknowns but only N equations. To complete the linear system, N_b additional equations are required. This will provide an absorbing boundary condition for the model. The additional equations can be obtained using the boundary element method.

8.5.1 Boundary Element Method

The radiation integral in (4.137) can be used to obtain a relationship between the unknown ϕ and its normal derivative on the boundary of the finite element mesh. By the equivalence theorem, we can think of ϕ and $\partial_n\phi$, where ∂_n is the normal derivative, as surface currents on the boundary S. We want those surface currents to be such that they radiate energy to the exterior of the surface S, with no energy radiated back into the interior V. This condition can be enforced using a radiation integral.

The radiation integral for equivalent surface currents on the boundary S is

$$\begin{aligned} E_z(\overline{\rho}) &= -\frac{k\eta}{4}\int_S d\overline{\rho}'\, H_0^{(2)}(k|\overline{\rho}-\overline{\rho}'|)J_z(\overline{\rho}') + \hat{z}\cdot\nabla \\ &\quad \times \frac{j}{4}\int_S d\overline{\rho}'\, H_0^{(2)}(k|\overline{\rho}-\overline{\rho}'|)\hat{t}' M_t(\overline{\rho}') \end{aligned} \tag{8.108}$$

where $\hat{t} = \hat{z}\times\hat{n}$, and the prime means that the unit tangent vector is evaluated at the point $\overline{\rho}'$. We need to relate J_z and M_t to the unknown ϕ and its normal derivative at the boundary, using the fact that for a 2-D scattering problem with the TM^z polarization, ϕ represents E_z^s.

The magnetic surface current is

$$\overline{M}_s = -\hat{n}\times(\hat{z}E_z) = \hat{t}E_z \tag{8.109}$$

which shows that $M_t = E_z = \phi$. The electric surface current is

$$\overline{J}_s = \hat{n}\times(\hat{t}H_t) = \hat{z}H_t$$

so that $J_z = H_t$. Using Ampere's law,

$$H_t = \hat{t}\cdot\frac{1}{jk\eta}\nabla\times(\hat{z}E_z)$$

Using the identity $\overline{A}\cdot(\nabla\times\overline{B}) = \overline{B}\cdot(\nabla\times\overline{A}) - \nabla\cdot(\overline{A}\times\overline{B})$,

$$\begin{aligned} H_t &= \frac{1}{jk\eta}\left[\hat{z}E_z\cdot\underbrace{(\nabla\times\hat{t})}_{0} - \nabla\cdot(\hat{t}\times\hat{z}E_z)\right] \\ &= \frac{1}{jk\eta}\nabla\cdot(\hat{n}E_z) \\ &= \frac{1}{jk\eta}\frac{\partial E_z}{\partial n} \end{aligned} \tag{8.110}$$

from which we can see that the normal derivative of ϕ is $jk\eta J_z$.

We will now simplify the second term of (8.108) to remove the curl operator. The integrand of this term can be written as

$$\begin{aligned}
\hat{z} \cdot \nabla \times (g\hat{t}'M_t) &= \hat{z} \cdot \left[g \underbrace{(\nabla \times \hat{t}'M_t)}_{0} - \hat{t}'M_t \times \nabla g \right] \\
&= M_t \hat{z} \cdot [\nabla g \times (\hat{z} \times \hat{n}')] \\
&= M_t \hat{z} \cdot [(\nabla g \cdot \hat{n}')\hat{z} - (\nabla g \cdot \hat{z})\hat{n}'] \\
&= M_t \nabla g \cdot \hat{n}'
\end{aligned} \tag{8.111}$$

where g is the Green's function. Expanding the gradient,

$$\begin{aligned}
\nabla g \cdot \hat{n}' &= kH_0^{(2)'}(k|\overline{\rho} - \overline{\rho}'|) \left[\hat{x} \frac{\partial |\overline{\rho} - \overline{\rho}'|}{\partial x} + \hat{y} \frac{\partial |\overline{\rho} - \overline{\rho}'|}{\partial y} \right] \cdot \hat{n}' \\
&= kH_0^{(2)'}(k|\overline{\rho} - \overline{\rho}'|) \left[n'_x \frac{x - x'}{|\overline{\rho} - \overline{\rho}'|} + n'_y \frac{y - y'}{|\overline{\rho} - \overline{\rho}'|} \right] \\
&= kH_0^{(2)'}(k|\overline{\rho} - \overline{\rho}'|) \cos \psi' \qquad (8.112) \\
&= -kH_1^{(2)}(k|\overline{\rho} - \overline{\rho}'|) \cos \psi' \qquad (8.113)
\end{aligned}$$

where ψ' is the angle between the vector $\overline{\rho} - \overline{\rho}'$ and $\hat{n}'$. The last step makes use of the recursion relation $Z'_p(x) = -Z_{p+1}(x) + pZ_p(x)/x$, where Z_p is a Bessel function. This simplifies the radiation integral to

$$\begin{aligned}
E_z(\overline{\rho}) &= -\frac{k\eta}{4} \int_S d\overline{\rho}' \, H_0^{(2)}(k|\overline{\rho} - \overline{\rho}'|) J_z(\overline{\rho}') \\
&\quad - \frac{jk}{4} \int_S d\overline{\rho}' \, H_1^{(2)}(k|\overline{\rho} - \overline{\rho}'|) \cos \psi' M_t(\overline{\rho}')
\end{aligned} \tag{8.114}$$

This derivation has assumed that the point $\overline{\rho}$ is outside V. To change the radiation integral into a boundary integral equation, we must take the limit of the radiation integral (8.114) as the observation point goes to the boundary from inside V. For the first term, this limit is straightforward—we just evaluate the integrand for $\overline{\rho}$ on S. For the second term, the limit is nontrivial, because the kernel singularity is less well behaved. It can be shown that

$$\begin{aligned}
\lim_{\overline{\rho} \to S} \frac{jk}{4} \int_S d\overline{\rho}' \, H_1^{(2)}(k|\overline{\rho} - \overline{\rho}'|) \cos \psi' M_t(\overline{\rho}') &= -\frac{M_t(\overline{\rho})}{2} \\
&\quad + \frac{jk}{4} \int_S d\overline{\rho}' \, H_1^{(2)}(k|\overline{\rho} - \overline{\rho}'|) \cos \psi' M_t(\overline{\rho}')
\end{aligned} \tag{8.115}$$

For this integral operator, when the observation point moves to the boundary S, the limit results in the appearance of a new term, $-M_t(\overline{\rho})/2$.

Using (8.115) in the radiation integral (8.114) leads to

$$E_z(\overline{\rho}) = -\frac{k\eta}{4}\int_S d\overline{\rho}'\, H_0^{(2)}(k|\overline{\rho}-\overline{\rho}'|)J_z(\overline{\rho}') + \frac{M_t(\overline{\rho})}{2} - \frac{jk}{4}\int_S d\overline{\rho}'\, H_1^{(2)}(k|\overline{\rho}-\overline{\rho}'|)\cos\psi' M_t(\overline{\rho}') \tag{8.116}$$

where $\overline{\rho}$ lies on the boundary S. This is an integral equation that relates the electric field on a 2-D surface S to the z-directed electric current and magnetic current flowing in the x–y plane on the surface. Since J_z and M_t are related to ϕ and its normal derivative, this provides a connection between ϕ and $\partial_n\phi$ on S.

8.5.1.1 Extinction Theorem

If the observation point is outside the boundary surface S, the field $E_z(\overline{\rho})$ represents fields radiated by the scatterer outside V. Inside the boundary, the field $E_z(\overline{\rho})$ should vanish, since the equivalent sources on S represent sources inside V and there are no sources outside the region of interest to radiate into V. Enforcing this condition on the field and its normal derivative will ensure that no energy is radiated toward the inside of the boundary S, resulting in the desired absorbing boundary condition.

In the integral equation (8.116), we have already taken the limit as the observation point $\overline{\rho}$ moved to the boundary from outside V. If the observation point had moved to the boundary from the inside, the extinction theorem would result. The only difference between (8.116) and the extinction theorem is the direction of the surface normal vector $\hat{n}$. If the observation point $\overline{\rho}$ is outside V, the surface normal vector points to the outside, but if $\overline{\rho}$ is inside V, the normal vector points to the inside of V. From (8.113), it can be seen that the effect of the switch in the surface normal direction is to change the sign of $\cos\psi'$. With this change in the direction of the surface normal, the extinction theorem is

$$0 = -\frac{k\eta}{4}\int_S d\overline{\rho}'\, H_0^{(2)}(k|\overline{\rho}-\overline{\rho}'|)J_z(\overline{\rho}') + \frac{M_t(\overline{\rho})}{2} - \frac{jk}{4}\int_S d\overline{\rho}'\, H_1^{(2)}(k|\overline{\rho}-\overline{\rho}'|)\cos\psi' M_t(\overline{\rho}') \tag{8.117}$$

Using the relationships $M_t = E_z$ and $J_z = H_t$, this becomes

$$0 = \frac{k\eta}{4}\int_S d\overline{\rho}'\, H_0^{(2)}(k|\overline{\rho}-\overline{\rho}'|)H_t^{s}(\overline{\rho}') - \frac{E_z^{s}(\overline{\rho})}{2} + \frac{jk}{4}\int_S d\overline{\rho}'\, H_1^{(2)}(k|\overline{\rho}-\overline{\rho}'|)\cos\psi' E_z^{s}(\overline{\rho}') \tag{8.118}$$

where we have multiplied by -1 so that the first integral operator has the same sign as the electric field integral equation (EFIE) operator in (6.11).

Finally, replacing E_z^s and H_t^s with ϕ and $\partial_n\phi/(jk\eta)$ leads to the surface integral equation

$$\frac{j}{4}\int_S d\overline{\rho}'\, H_0^{(2)}(k|\overline{\rho}-\overline{\rho}'|)\frac{\partial\phi(\overline{\rho}')}{\partial n} - \frac{\phi(\overline{\rho})}{2}$$
$$+\frac{jk}{4}\int_S d\overline{\rho}'\, H_1^{(2)}(k|\overline{\rho}-\overline{\rho}'|)\cos\psi'\phi(\overline{\rho}') = 0 \tag{8.119}$$

In this integral equation, we can identify the first term as the EFIE operator $\mathscr{L}$, and the second term is the magnetic field integral equation (MFIE) operator $1/2 - \mathscr{M}_{\mathrm{TE}}$. Using operator notation, the integral equation can be written as

$$\mathscr{L}\frac{\partial\phi}{\partial n} - \left(\tfrac{1}{2} - \mathscr{M}_{\mathrm{TE}}\right)\phi = 0 \tag{8.120}$$

Discretizing this equation using the method of weighted residuals or the method of moments leads to the linear system

BEM Linear System

$$A_E\overline{b} - A_M\overline{a}_{\mathrm{b}} = 0 \tag{8.121}$$

which provides the additional equations needed to solve (8.107).

8.5.1.2 FEM–BEM Linear System

As with the finite element method in the previous section, we will partition the unknowns into interior and boundary nodes. This induces a block partitioning of the FEM matrices, and facilitates combining (8.107) and (8.121). Combining the two sets of equations leads to the block linear system

FEM–BEM Linear System

$$\left[\begin{array}{c|c|c} A_{\mathrm{ii}} & A_{\mathrm{ib}} & 0 \\ \hline A_{\mathrm{bi}} & A_{\mathrm{bb}} & -B_{\mathrm{bb}} \\ \hline 0 & -A_M & A_E \end{array}\right]\left[\begin{array}{c} \overline{a}_{\mathrm{i}} \\ \hline \overline{a}_{\mathrm{b}} \\ \hline \overline{b} \end{array}\right] = \left[\begin{array}{c} -\overline{c}_{\mathrm{i}} \\ \hline -\overline{c}_{\mathrm{b}} \\ \hline 0 \end{array}\right] \tag{8.122}$$

The matrix on the left-hand side is sparse except for the two BEM matrices, which are dense because they are associated with integral operators rather than differential operators. Solving this linear system leads to fields on the interior of the simulation domain as well as boundary values of the field and its normal derivative. The boundary values can be used conveniently in postprocessing to accomplish a near-to-far transformation.

8.5.2 Implementation

The FEM–BEM algorithm is a fairly complex combination of several core computational routines, and it is critical that each section of the code be debugged individually by checking results by hand for simple cases.

The following is a code outline for the 2-D FEM–BEM algorithm:

FEM–BEM Code Outline

1. *Problem parameters:* Specify the frequency and angle of arrival of the incident plane wave. Define the scatterer geometry and material parameters.
2. *Preprocessing:* Create a mesh of the simulation domain, defined by a node array `p`, triangle array `t`, and edge array `e`. Using the scatterer geometry information, fill an array of the wave number $k(\overline{\rho})$ using the relative dielectric constant $\varepsilon_r(\overline{\rho})$ at each mesh element barycenter. $k(\overline{\rho})$ can be approximated as constant on each element, evaluated at element barycenters, and stored in an array for use in assembling the matrix T'.
3. Assemble the FEM matrix $A = C - T'$, taking into account the value of $k^2(\overline{\rho})$ on each element. Use `spalloc` to allocate memory for the sparse matrix data type to avoid storing many zeros.
4. Assemble the right-hand side vector $\overline{c}$. This can be done with a loop over one local node number nested inside a loop over elements. The source function in the integral for c_n in (8.106) can be approximated as constant over each element. For each element and local node, the contribution to c_n is then given by evaluating the source function at the current node point and multiplying by the integral of the associated shape function over the element. For linear shape functions, this integral is given by

$$\int_e \alpha_i \, dx \, dy = 2A \int_0^1 \int_0^{1-\alpha_2} \alpha_i \, d\alpha_1 \, d\alpha_2 = \frac{A}{3}$$

   where A is the element area.
5. Partition a list of global node numbers into interior and boundary nodes. The array `boundary_nodes` can be taken to be the first row of the boundary array `e` in MATLAB format, and `interior_nodes` is a list of the global node numbers that are not boundary nodes. Use this to find A_{ii} using

```
Aii = A(interior_nodes, interior_nodes)
```

   A_{ib} can be found using

```
Aib = A(interior_nodes, boundary_nodes)
```

   A_{bi}, A_{bb}, $\overline{c}_{\text{i}}$, and $\overline{c}_{\text{b}}$ can be found similarly.

6. Assemble the overlap matrix B from FEM basis functions and BEM basis functions. Typically, the BEM basis functions on the boundary are chosen to be the boundary functions induced by the FEM basis functions. For linear shape functions, this would lead to triangle functions on the boundary. To simplify the computation of BEM matrix elements, we will instead use pulse functions centered at the boundary nodes. With this approximation, the contribution to B for a given element is

$$
\begin{aligned}
B_{ij}^{e} &= \int_{e} \alpha_i(s) f_j(s)\, ds \\
&= \int_{0}^{h_e/2} \alpha_i(s)\, ds \\
&= \begin{cases} \frac{3h_e}{8} & i = j \\ \frac{h_e}{8} & i \neq j \end{cases}
\end{aligned}
$$

where h_e is the length of the boundary element. B can be assembled by looping over all elements with a side on the boundary, with a nested double loop over local nodes $i = 1, 2$ and $j = 1, 2$. Within this loop, the first index of B is the global node number corresponding to the current node, which can be found in the first row of the MATLAB mesh edge array `e`. The second index is the global number of the current boundary element, which can be taken to be the column index of the node in the array `e`. In pseudo-code, the assembly algorithm for B is

```
B = zeros(N, N_b)
Loop over edge elements, n_e = 1, 2, ..., N_b
Loop over local element numbers i = 1, 2, j = 1, 2
  B(e(i,n_e), n_e + j - 1) = B(e(i,n_e), n_e + j - 1) + B^e(i,j)
```

For $n_e = N_b$, the second index of B needs to wrap around to 1. After filling B, the global node numbers corresponding to the first index need to be sorted to match A_{bb} and A_{bi}. This can be done by forming $B_{bb} =$ `B(boundary_nodes,:)`. Alternately, B_{bb} can be filled directly by restricting the assembly loop to the boundary nodes.

7. Create a boundary mesh description from the FEM mesh. In general, this would be similar to the node and element lists of the FEM mesh. For the case of pulse functions, a simpler description suffices. All that is required is a list of boundary element center points, element widths, and the angle of the normal vector for each element. The boundary element center points can be taken to be the first node of each FEM edge element on the boundary S. The angle of the inward normal direction can be found by computing the angle of the edge element and subtracting $\pi/2$. At corner nodes, the normal direction

can be taken to be the average of the normal directions of the two adjacent boundary elements or the angle of one of the two adjacent elements, which avoids having to treat the corners specially. These simplifications reduce the convergence rate of the numerical method but make implementation easier, since the matrix A_E can be filled with the same algorithm used in Problem 6.3.

8. Fill the BEM matrices A_E and A_M. The FEM assembly process could actually be used to fill the BEM matrices, but because these matrices are dense, the "loop over basis functions" approach of Chapter 6 is just as efficient. Matrix elements of A_E for the EFIE operator can be evaluated using the point matching and single-point integration approach of Section 6.4. The matrix elements of A_M can also be computed with a single-point integration rule for both the source and observation integrals. For the matrix element $A_{M,mn}$, the angle ψ' can be computed using

```
psi_prime = atan2(ny(n),nx(n)) - ...
            atan2(y(n) - y(m), x(n) - x(m));
```

where x and y are arrays of edge node coordinates, and nx and ny are arrays of components of the inward-pointing surface normal vector at each edge node. Or, since the cosine of the angle is needed, we can also use

```
cospsi_prime = (nx(n)*(x(m)-x(n)) + ...
                ny(n)*(y(m)-y(n)))/rho
```

where rho is the length of the vector $\overline{\rho} - \overline{\rho}'$.
The diagonal matrix elements of A_M are particularly easy to compute, because the identity term dominates the integral term and the diagonal elements are approximately $A_{M,mm} = 1/2$.

9. Combine all the matrix blocks into the full FEM–BEM linear system according to (8.122). This can be done in MATLAB using

```
M = [[Aii, Aib, Z];
     [Aib.', Abb, -Bbb];
     [Z.', -AM, AE]];
```

where Z is an $N_i \times N_b$ matrix of zeros. The right-hand side vector can be filled using

```
c = [-ci; -cb; zeros(N_EDGE,1)];
```

The MATLAB backslash operator can then be used to solve for the interior and boundary unknowns, so that `x = M\c`.

10. *Postprocessing:* Boundary values of ϕ and $\partial_n\phi$ are available directly as part of the FEM–BEM solution. These can be used in the far-field radiation integral

to perform a near-to-far transformation. The far-field limit of the radiation integral (8.114) is

$$E_z(\overline{\rho}) = \frac{k}{4}\sqrt{\frac{2j}{\pi k \rho}}e^{-jk\rho}\int d\overline{\rho}'\, e^{jk\hat{\rho}\cdot\overline{\rho}'}\left[-\eta J_z(\overline{\rho}') + \cos\psi' M_t(\overline{\rho}')\right] \qquad (8.123)$$

where ψ' is the angle between $\hat{\rho}$ and the outward normal vector to S at $\overline{\rho}'$. M_t and J_z are proportional to ϕ and the normal derivative of ϕ, and the unknowns in $\overline{a}_b$ and $\overline{b}$ are samples of ϕ and $\partial_n\phi$ at the boundary element centers. If the integral is evaluated using the boundary element centers as quadrature points, the integral becomes a sum over a linear combination of $\overline{a}_b$ and $\overline{b}$ weighted by the appropriate factors according to the integrand of (8.123). The required boundary values can be extracted from the vector x using

```
Mt = x((Ni + 1):(Ni + Nb));
Jz = x((Ni + Nb + 1):(Ni + 2*Nb))/(j*k*eta0);
```

The far fields can be used to find scattering amplitudes and scattering widths.

8.6 Numerical Results

To illustrate the performance of the FEM–BEM algorithm, we will give results for an implementation of this algorithm with a rectangular boundary geometry. Figure 8.6

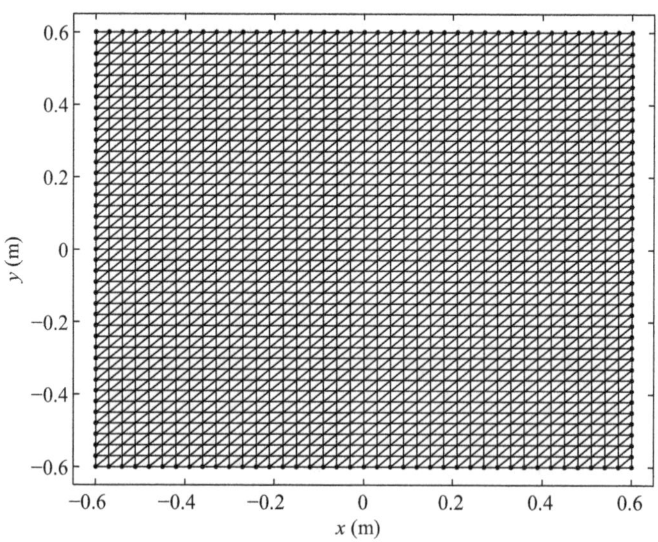

Figure 8.6 FEM mesh for rectangular region. Dots mark boundary mesh nodes

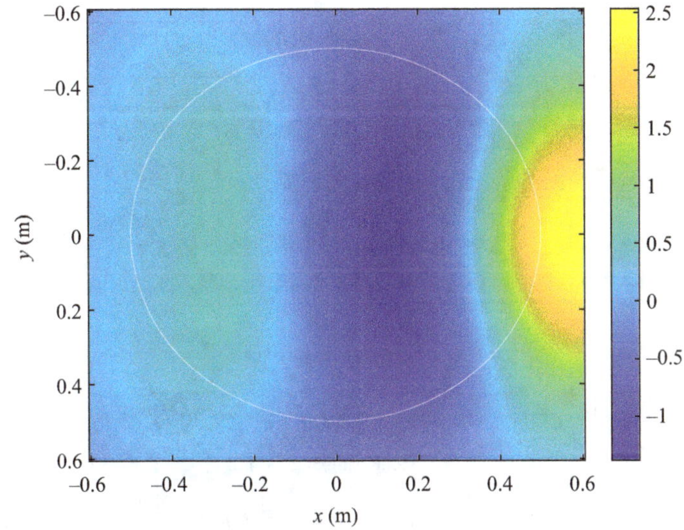

Figure 8.7 Real part of the electric field intensity E_z computed with the FEM–BEM algorithm for a circular dielectric cylinder ($\varepsilon_r = 2$) with half wavelength radius. The scatterer boundary is indicated for reference as a white line

shows FEM and boundary meshes for a 1.2 m by 1.2 m simulation domain. The FEM mesh consists of $N_{\text{nodes}} = 1,681$ nodes and $N_e = 3,200$ triangles and was plotted using the MATLAB PDE Toolbox command `pdemesh`. The BEM mesh consists of 160 edge segments. The BEM matrices are discretized using a single-point integration rule at each boundary node.

Near fields computed using the FEM–BEM algorithm with these meshes are shown in Figure 8.7 for a dielectric circular cylinder of radius $a = \lambda/2$ with relative permittivity $\varepsilon_r = 2$. The incident field is a TM-polarized plane wave at the angle of incidence $\phi^{\text{i}} = \pi$. The frequency of the incident field is 300 MHz. To generate the field magnitude density plot, the MATLAB command `tri2grid` was used to interpolate the field unknown values from mesh node points to points on a rectangular grid, which allows the field values to be plotted using `imagesc`.

Computed and analytical scattering width results for the dielectric cylinder are shown in Figure 8.8. It can be seen that accuracy is reasonably good. Since the boundary mesh has 40 nodes on each side and the simulation domain is a 1.2 λ square, the mesh density is $n_\lambda = 33$. For this scatterer, the surface method of moments would provide more accurate results than FEM–BEM with a smaller matrix size. The FEM–BEM method can accommodate inhomogeneous scatterers with variable permittivity, permeability, or conductivity; however, so FEM–BEM is more flexible than MoM for surface integral equations.

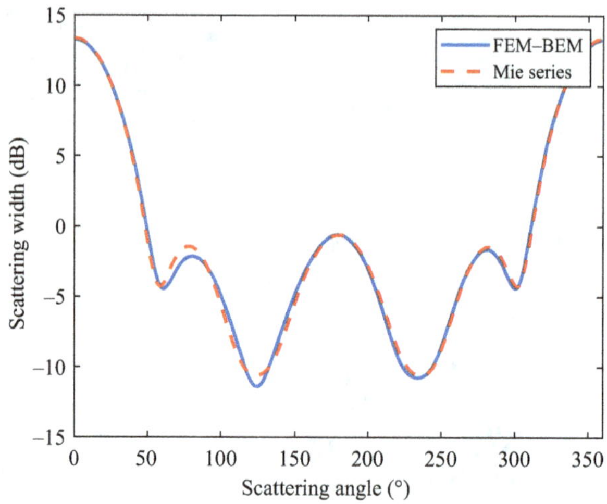

Figure 8.8 Computed and exact scattering widths for the circular dielectric cylinder

Since the volume MoM of Section 6.6.1 also can be applied to inhomogeneous scatterers, FEM–BEM and the volume MoM offer comparable capabilities. Numerical accuracy for the volume moment method is better than FEM–BEM for a given discretization density. For the half wavelength dielectric cylinder, FEM–BEM requires two dense matrices with size 160 × 160, whereas the matrix size for volume MoM is 861 × 861, so the memory storage requirement and matrix solution time are larger for the volume MoM.

Problems

8.1 (a) Write the functional for the arc length of a path in the plane between the points (x_1, y_1) and (x_2, y_2). (b) Show that the functional is stationary for a straight line.

8.2 Derive the Helmholtz equation from the functional

$$I(\phi) = \frac{1}{2} \int_V dV \left[|\nabla \phi|^2 - k^2 \phi^2 \right]$$

8.3 Derive (8.34) from (8.32).

8.4 The method of moments can be derived using the bivariate functional $I(u, w) = \langle \mathscr{L}u, w \rangle - \langle f, w \rangle - \langle g, u \rangle$, where f and g are given source functions. Substitute basis expansions for u in terms of expansion functions f_n and w in terms of testing functions t_n. Impose the condition that derivatives of the functional with respect to the expansion coefficients of w vanish and show that the method of moments with linear system (6.50) and right-hand side (6.51) is obtained.

8.5 (a) Find the analytical solution to the boundary value problem posed in Section 8.3.8. (b) Solve the linear system (8.65) numerically and compare the resulting samples of the potential with the exact solution. What is the numerical solution error for this problem? Explain in terms of the behavior of the potential solution and the choice of piecewise linear basis functions.

8.6 (a) Determine by hand the linear system (8.65) for a six-element mesh. (b) Find the modified linear system in the case that the region $[a, b]$ is filled with two dielectrics, $\varepsilon_{r1} = 1$ for $a \leq x \leq c$, and $\varepsilon_{r2} = 2$ for $c < x < b$, with $c = (a + b)/2$.

8.7 Implement the 1-D FEM algorithm for Laplace's equation for the static electric potential in a material with an inhomogeneous dielectric constant.

(a) Create a vector of node coordinates `x` on the interval $[a, b]$ for an arbitrary number of nodes. In general, mesh nodes need not be evenly spaced, but in this case, it is easiest to generate a regular mesh.

(b) Create a $2 \times N_{\text{el}}$ array `t(i,e)` that has two entries, `i = 1,2`, for element `e`, such that `t(1,e)` is the index in the vector `x` of the left node of the element and `t(2,e)` is the index of the right node. The element and node lists do not need to be in any particular order, but if the nodes and elements are ordered from left to right, `t(1,e) = e` and `t(2,e) = e+1`.

(c) Use the assembly equation in a loop over the element index `e` and a double loop over local node numbers `i` and `j` to fill the global stiffness matrix `C` from the 2×2 element stiffness matrix. Within the loop, use logic based on the location of the mesh element center

```
xc = (x(t(1,e)) + x(t(2,e)))/2
```

to determine the correct value of the permittivity to use in the assembly equation.

(d) Since the boundary node potential values are fixed by the Dirichlet boundary condition, the linear system in (8.65) should include unknowns only for the interior node sample values. The matrix `C` must be reduced to only the entries corresponding to interior nodes. If the nodes are ordered from left to right, this can be done using `Ci = C(2:N-1,2:N-1)`, where `N` is the number of nodes.

(e) Verify the stiffness matrix by comparing with the hand-assembled matrices from Problem 8.6.

(f) Fill the right-hand side vector `bi` of the linear system by initializing all array entries to zero and setting the entry corresponding to the right-most node to the boundary condition $V = 10$ multiplied by `C(n,n)`, where n is the global node number of the right boundary node.

(g) Find the potential solution using `phi = Ci\bi`. Check the solution using the exact result from Problem 8.5.

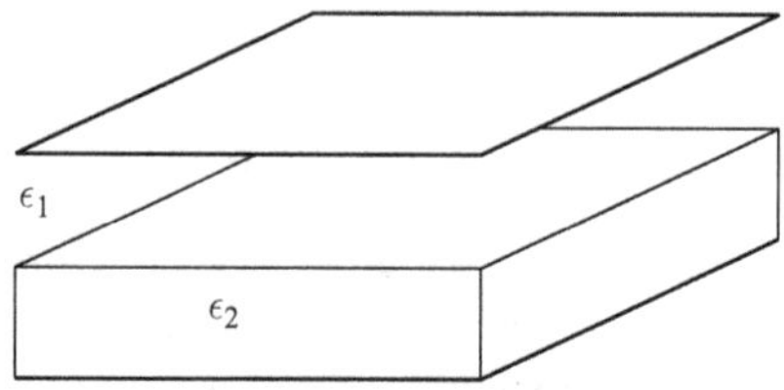

Figure 8.9 Parallel plate capacitor partially filled with dielectric

8.8 Develop postprocessing for the 1-D FEM algorithm of Problem 8.7 to compute the capacitance of an inhomogeneous dielectric-filled parallel plate capacitor, with one plate at $x = a$ and the other at $x = b$ (see Figure 8.9). Neglecting fringing fields at the plate edges effectively makes the problem one-dimensional.

(a) Convert the potential to electric field intensity, using a finite difference approximation for the derivative in (4.65).

(b) Convert the electric field intensity to electric flux density using the constitutive relation (2.2a).

(c) By Gauss's law, the electric charge stored on one plate is equal to the integral of the electric flux density over a closed surface enclosing the plate. Since the electric field is small outside the region between the plates, this integral can be evaluated by multiplying the flux density by an arbitrary plate area A. From the stored charge, use the relationship $C = Q/V$ to find the capacitance.

(d) Check the code by verifying the capacitance computed for a parallel plate capacitor filled with a homogeneous dielectric (ε = constant).

(e) Compare the capacitance obtained using this method for the two-layer dielectric-filled parallel plate capacitor of Problem 8.6(b) with the exact result obtained from the series combination of two parallel plate capacitors with homogeneous dielectrics, one for each dielectric layer.

8.9 Derive the formulas in (8.96a)–(8.96c).

8.10 (a) Draw a picture of a simple two-dimensional mesh with two or three triangles. Give each mesh element a global element number, each node a global node number, and the three nodes in each mesh element a local node number. (b) Write down the full triangle array `t` for the mesh. (c) Assemble the 2-D FEM T matrix by hand. Leave the element area A as a variable, but otherwise give numerical values for the matrix elements.

8.11 Use the MATLAB command `[p,e,t] = poimesh('rect',N);` to create a finite element mesh for a rectangular simulation domain. `rect.m` is a geometry function that defines a rectangular boundary. The MATLAB PDE Toolbox function `squareg.m` defines a square geometry with unit side lengths. This function can be modified to produce a rectangular boundary with arbitrary side lengths. (If the `poimesh` function is not available, the code of Problem 6.12 can be used to generate a mesh for rectangular domain.)

8.12 Implement the mesh generation method of [8]. Test the algorithm by creating a mesh for one of the example cases in that paper.

8.13 Apply the 2-D FEM algorithm to the rectangular waveguide of Problem 4.8. Begin with a mesh generated by the code from Problem 8.11 (or Problem 6.12) for a rectangular domain with the same dimensions as the waveguide. (a) Write a code to assemble the finite element matrices for the Laplace eigenvalue problem on this domain. (b) From the eigenvalues of the resulting matrix discretization of the Laplacian operator, estimate the cutoff frequencies for the waveguide. Plot the cutoff frequencies of the first 20 modes from FEM on a semilogy scale overlaid with the exact solution.

8.14 Develop a TM waveguide mode solver for a waveguide with arbitrary polygonal cross section. Create a Rayleigh–Ritz matrix for a rectangular domain that is larger than the waveguide cross section. In the assembly equation, the `inpolygon` function can be used to fill only the matrix entries for which all three nodes lie inside the waveguide (this effectively implements the Dirichlet boundary condition). Use the `inpolygon` function again to find the vector `interior_nodes` of indices of nodes that lie inside the waveguide and form the reduced matrix `A(interior_nodes,interior_nodes)`. Compute the eigenvalues and the corresponding waveguide mode cutoff frequencies. Verify the code by comparing with the result of Problem 4.9.

8.15 Implement the FEM–BEM method and use it to compute the bistatic scattering width of a circular PEC cylinder of radius one half wavelength. Model the PEC by setting the imaginary part of the complex permittivity to a large negative value such as -100. Compare with the exact solution. Hint: Plot the near fields around the cylinder using

```
a = [];
a([interior_nodes,boundary_nodes]) = x(1:N);
pdesurf(p,t,real(a))
```

and compare with the fields obtained with the FDTD algorithm to ensure that the fields are correct before debugging the postprocessing.

8.16 Use the FEM–BEM algorithm to compute the bistatic scattering width of a circular dielectric cylinder of half-wavelength radius and relative permittivity $\varepsilon_r = 2$. Compare with results obtained using the FDTD algorithm.

8.17 Modify the FEM–BEM algorithm using the `inpolygon` function to allow for scatterers of arbitrary polygonal cross section. Compute the bistatic scattering width of a square dielectric cylinder one half wavelength on a side and compare with results obtained with the FDTD algorithm and the volume MoM algorithm of Problem 6.8. Which method is the most accurate for a given number of degrees of freedom? For a given run time?

References

[1] M. O. Sadiku, "A simple introduction to finite element analysis of electromagnetic problems," IEEE Trans. Educ., vol. 32, no. 2, pp. 85–93, 1989.

[2] M. N. O. Sadiku, *Numerical Techniques in Electromagnetics With MATLAB*. Boca Raton, FL: CRC Press, 2009.

[3] C. S. Desai and J. F. Abel, *Introduction to the Finite Element Method: A Numerical Approach for Engineering Analysis*. New York, NY: Van Nostrand Reinhold, 1972.

[4] M. V. K. Chari and P. P. Silvester, *Finite Elements for Electrical and Magnetic Field Problems*. Chichester: John Wiley, 1978.

[5] A. F. Peterson, D. R. Wilton, and R. E. Jorgenson, "Variational nature of Galerkin and non-Galerkin moment method solutions," IEEE Trans. Antennas Propag., vol. 44, no. 4, pp. 500–503, 1996.

[6] C. Geuzaine and J. Remacle, "GMSH: A three-dimensional finite element mesh generator with built-in pre-and post-processing facilities," Int. J. Numer. Methods Eng., vol. 79, no. 11, pp. 1309–1331, 2009.

[7] K. Ho-Le, "Finite element mesh generation methods: A review and classification," Comput.-Aided Des., vol. 20, no. 1, pp. 27–38, 1988.

[8] P. O. Persson and G. Strang, "A simple mesh generator in MATLAB," SIAM Rev., vol. 46, no. 2, pp. 329–345, 2004.

[9] J. C. Nedelec, "Mixed finite elements in R3," Numer. Math., vol. 35, pp. 315– 341, 1980.

Chapter 9
Optimization Methods

9.1 Introduction

So far, we have considered analysis problems, in the sense that a structure is given, and the goal is to solve Maxwell's equations, an integral equation, or a partial differential equation to find an unknown field or current solution. In engineering, analysis is typically only a preparatory step or a building block for tools that can be used to solve design synthesis problems. A synthesis problem typically involves one or more performance goals and constraints, which must be met by the designed structure, circuit, or system. Synthesis problems can be solved using a variety of techniques, including analytical formulas, trial and error, and empirical methods. As the cost of computational power has decreased, it has become feasible to solve synthesis problems using algorithms for numerical optimization.

An example of a synthesis problem is the design of an antenna for a specified gain. The goal is to determine the geometry and composition of an antenna structure that has the required gain, while meeting constraints on physical size, cost, manufacturability, or other aspects of the antenna. One approach to solving the problem is to consider an antenna structure for which an analytical formula for the gain in terms of physical parameters and dimensions is available. Analytical techniques have been used for many decades to solve antenna design applications, but in recent years wireless devices and microwave systems have increased in complexity, and more sophisticated design tools are required. System design is now often done using computational electromagnetics codes coupled with optimization algorithms. A numerical model that computes the performance of a given design is embedded in an optimization algorithm that updates geometrical parameters of the structure until the required performance specifications are met. Optimization methods have become ubiquitous in engineering and can be used for the design of radio frequency circuits, microwave components, filters, optical structures, and many other types of devices and systems.

9.1.1 Optimization Problems

The general optimization problem can be posed in terms of a real-valued objective function or cost function $f(x_1, x_2, x_3, \ldots, x_N)$. The values $x_1, x_2, x_3, \ldots, x_N$ are adjustable parameters of a design or structure, and the value of f represents a quantity or combination of quantities that characterize the performance of the design.

The objective function is typically defined such that the smaller the value of f, the better the performance of the design. The goal is to find the vector $\mathbf{x}$ consisting of values of the parameters x_n that achieve a global minimum of the objective function.

Mathematically, an optimization problem can be expressed in the form

Optimization Problem

$$\mathbf{x}_{\text{opt}} = \operatorname{argmin} f(\mathbf{x}) \tag{9.1}$$

where we have assembled the parameters $x_1, x_2, x_3, \ldots, x_N$ into a vector $\mathbf{x}$. The optimization problem may also include constraints on the parameters. Constraints include physical size limits on a structure, minimum or maximum thickness of materials used in fabrication, or restrictions on overlap between components. For a constrained optimization problem, solutions $\mathbf{x}$ that satisfy the constraints are said to be in the feasible region.

For an antenna synthesis problem, f might represent the inverse or negative antenna gain, and the vector $\mathbf{x}$ is a set of dimensions for parts of the antenna. An optimization algorithm should return the set of dimensions that minimizes the objective function and maximizes the antenna gain. A more complex design problem may also require low sidelobe levels, in which case the cost function would be obtained from a linear combination of gain and sidelobe level. Alternately, the sidelobe level could be included as a constraint, so that the feasible region consists of all designs with sidelobe level above a given value, and the constrained optimization algorithm finds the antenna design with maximum gain subject to the sidelobe level constraint.

Figure 9.1 shows a functional diagram for a software-based design optimization process. The structure, component, or system to be designed is represented through a set of geometrical parameters. These parameters form the vector $\mathbf{x}$ in (9.1). A forward numerical model such as a finite difference, method of moments, or finite element algorithm is used to model the field solution for the structure or component. For an

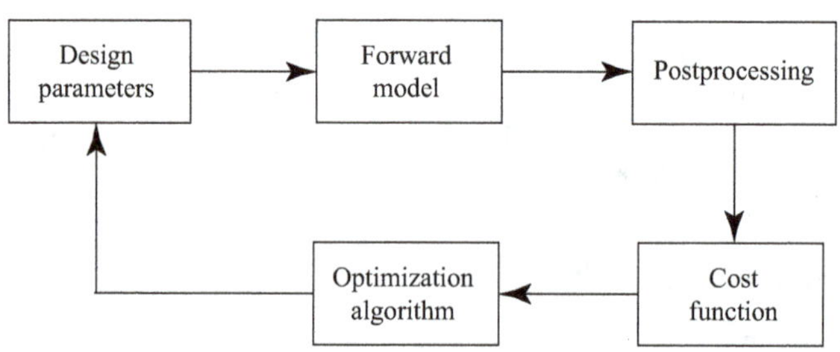

Figure 9.1 Diagram of a software design optimization process

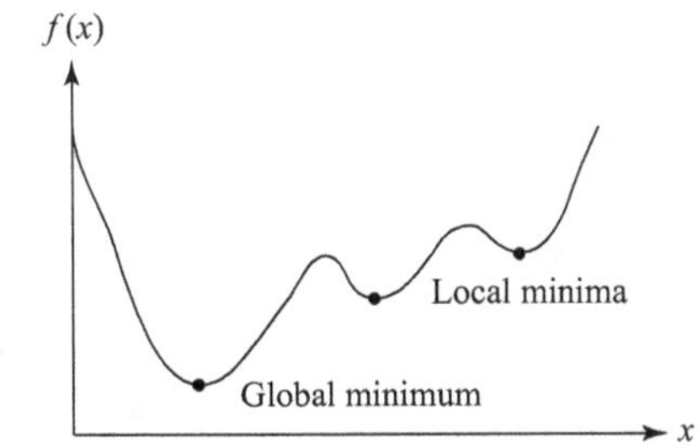

Figure 9.2 Local and global minima

antenna design problem, this would be the electric field radiated by the antenna for a given input current excitation. In postprocessing, figures of merit such as antenna gain are computed. The figures of merit are assembled into a cost function value. These blocks evaluate the function $f(\mathbf{x})$ in (9.1). The optimization algorithm generates a new set of design parameter values with the goal of moving toward the global minimum of the cost function.

9.1.2 Local and Global Optimization

For functions with a single local minimum, a variety of methods can be used to update a guess $\mathbf{x}_k$ to produce a new set of parameters $\mathbf{x}_{k+1}$ such that the sequence converges to the minimum of f. The conjugate gradient method considered in Chapter 7 is a method of this type. Algorithms that find a minimum by updating the vector $\mathbf{x}_k$ in such a way that the objective function decreases are local optimization methods.

For complex design problems, it is common for the objective function to have many local minima in addition to a global minimum, as illustrated in Figure 9.2. This type of optimization problem is more challenging than local optimization. A local optimization algorithm will converge to the nearest local minimum and halt without finding the global optimum. To find the global minimum, a local optimization algorithm can be used with many randomly chosen initial starting values, and the solution with the smallest value of the objective function is selected as the global minimum. Another approach is to find a local minimum, perturb the solution to move away from the local minimum, and rerun the local optimization to find a different minimum. Alternately, the global minimization problem can be solved with search-type methods such as genetic algorithms and simulated annealing.

9.2 Classes of Optimization Problems

Because of the difficulty of global optimization for an arbitrary objective function with many local minima, the development of methods for global optimization is a vast field of research in mathematics, numerical analysis, and scientific computation.

The goal is to create robust algorithms that can identify the global minimum for a wide range of objective functions with as few evaluations of the objective function as possible. No single algorithm performs well for all objective functions. The best approach for a given application is often selected from the wide range of available optimization methods by trial and error. General treatments of numerical optimization include [1–3].

There are several ways to organize the broad spectrum of available optimization problems into a taxonomy. The theory and development of optimization methods is sometimes referred to as mathematical programing, and in this framework methods are grouped according to increasing complexity of the objective function:

Classes of Mathematical Programing

Linear programing: The objective function is linear in the variables $x_1, x_2, x_3, \ldots, x_N$.

Quadratic programing: The objective function is quadratic in the variables.

Nonlinear programing: The objective function is nonlinear (i.e., an arbitrary objective function).

Dynamic programing: The objective function is complex enough that it must be broken up into a combination of smaller, interacting optimization problems that are solved using linear, quadratic, or nonlinear programing techniques.

The conjugate gradient algorithm for linear system solution discussed in Section 7.5.1 minimizes a quadratic cost function and so can be considered to be a quadratic programing method. The quadratic form minimized by the conjugate gradient algorithm is a convex optimization problem (see Section 9.2.1).

Optimization problems with linear or quadratic cost functions are much easier to solve than problems with nonlinear objective functions. Most optimization problems that arise in engineering design fall into the category of nonlinear programing. Nonlinear problems are extremely challenging in general, as there can be many local minima and the behavior of the objective function can be extremely complex.

9.2.1 Convex Optimization

If the objective function is a convex function and if constraints on the solution are convex, then a special class of algorithms can be applied, known as convex optimization algorithms. Convex optimization problems arise in control theory, signal processing, communications, circuit design, mathematical finance, structures, and other fields.

Convex optimization problems can be viewed as an easier subclass of the general nonlinear optimization problem. An example of a convex function is a paraboloid. A paraboloid has only one local and global minimum and finding the minimum is relatively easy. Convex problems become challenging when they have many degrees of freedom and complex constraints on the feasible region. The conjugate gradient method for quadratic optimization can be extended to objective functions that are not quadratic but still convex using the Fletcher–Reeves algorithm.

A general nonconvex and nonlinear optimization problem might be intractable with only four or five unknowns, whereas a convex optimization problem of that dimensionality can be trivial. Sophisticated algorithms for convex problems with millions of degrees of freedom have been developed [4]. Some nonconvex problems can be recast into convex optimization problems, which often dramatically speeds up the solution of the original problem.

9.2.2 Types of Optimization Algorithms

Optimization methods can also be classified according to the general characteristics of the algorithm. Some important properties of optimization algorithms are as follows:

Types of Optimization Algorithms

Local algorithms find a local minimum. This may be the minimum that is nearest to the starting point.

Global algorithms can locate the global minimum, at least for a restricted class of well-behaved objective functions.

Iterative algorithms progressively update a candidate solution $\mathbf{x}_k$ to yield a hopefully better solution $\mathbf{x}_{k+1}$. Iterative algorithms tend to converge quickly but can miss global minima.

Search methods divide the feasible region into subsets and progressively move through them until a minimum is found. Search methods can require many more evaluations of the objective functions and are more computationally intensive than iterative methods but work well with careful tuning of the algorithm for certain classes of problems with many local minima.

Gradient methods use derivatives of the objective function in the computation of $\mathbf{x}_{k+1}$ from $\mathbf{x}_k$ to move more rapidly to minima of the objective function.

Gradient-free methods require only evaluations of the objective functions. Gradient-free methods are typically less efficient than methods that make use of derivatives of the objective function, but it can be difficult or impossible to evaluate derivatives for complex objective functions.

Deterministic methods yield the same solution each time they are run for the same cost function and initial guess $\mathbf{x}_0$.

Stochastic methods introduce randomness when updating a collection of candidate solutions to produce new solutions that move toward the global minimum.

9.2.3 Common Optimization Algorithms

Nonlinear or global optimization methods that are used in engineering design include the following:

Common Optimization Algorithms

Local methods
- One-dimensional minimization
  - Golden section search
  - Brent's method (parabolic interpolation)
- Conjugate gradient methods
  - Fletcher–Reeves
  - Polak–Ribiere
- Variable metric or quasi-Newton methods
  - Davidon–Fletcher–Powell (DFP)
  - Broyden–Fletcher–Goldfarb–Shanno (BFGS)
  - Levenberg–Marquardt
- Direction-set methods
  - Powell's method
- Simplex methods
  - Nelder–Mead
  - Flexible polyhedron

Global methods
- Random search
- Simulated annealing
- Genetic algorithms
- Particle swarm optimization
- Ant colony optimization

Optimization is a broad and rich field, so these are just a sampling of the many available algorithms. Optimization algorithms come in so many types and flavors that it is difficult to survey all of them. Within specialized applications of optimization techniques, comparisons can be made of the merits and relative advantages of different algorithms (see, e.g., [5] for a comparison of various algorithms for antenna design).

Some of the features of important classes of optimization algorithms are described next.

Quasi-Newton methods are variants of Newton's method. Newton's method was introduced in Section 3.6 as a technique for finding zeros of a function, but when applied to the derivative, it can be used to find local minima. When Newton's method is applied to the derivative of an objective function, the second derivative or Hessian matrix is required. For some objective functions, computing the second derivative is tedious or impossible, so various approaches have been developed for estimating the second derivative have been developed. These are collectively referred to as quasi-Newton methods.

Conjugate gradient methods solve a one-dimensional optimization problem at each iteration to find the minimum of the objective function along a line in the search space. Algorithms are distinguished by the way new search directions are determined.

Simplex methods update a small number of points in the search space in such a way that the geometrical shape spanned by the points moves toward a minimum. The Nelder–Mead simplex method is treated in detail in Section 9.4.

Random search is a brute-force method based on checking the cost function at random points in the search space. Since the volume of the search space grows exponentially with the dimension of the optimization problem, this approach is feasible only in limited circumstances. Random search is sometimes used as a coarse optimization step to initialize another algorithm.

Particle swarm optimization is a heuristic for developing optimization algorithms based on a group of independent agents (commonly thought of as "birds") seeking a desirable location ("food"). The algorithm is initiated with a randomly chosen set of solutions, and at the next algorithm step the solutions are updated according to a dynamical model that moves solutions toward the solutions with best values of the objective function. The group of solutions tends to converge on minima. A parameter representing inertia keeps candidate solutions from moving too quickly toward local minima and allows the algorithm to locate global minima.

Simulated annealing is a class of methods that are based on a model for the formation of crystal domains in a cooling metal. As temperature decreases, domains join and coalesce until the material is dominated by a few large domains. Heating the material pushes the structure out of local minima in the internal energy, ultimately allowing larger domains to be formed in the cooled material. As a numerical optimization method, candidate solutions are updated to move toward nearby minima, but at each step in the algorithm some solutions are randomly replaced by perturbed solutions with a probability that depends on a parameter referred to as temperature. At high temperatures, candidate solutions are likely to move out of local minima. At later stages in the simulated annealing algorithm with a lower temperature parameter, the solutions are less likely to be perturbed, to allow the algorithm to locate the global minimum.

Genetic algorithms (GAs) are a class of search methods based on concepts from evolutionary biology [6,7]. A GA is initialized with an initial population of randomly selected values of $\mathbf{x}$, and the population is successively updated by combining pairs of solutions using the evolutionary principles of natural selection based on fitness. Fitness for each member of the population is determined by the smallness of the objective function, so candidate solutions with smaller objective function values are more likely to reproduce and propagate to the next generation. When parent solutions are combined to form child solutions, random mutations are introduced to allow the population to cover the full search space and hopefully move toward a global minimum. Genetic algorithms can require a significant amount of tuning to work well for a given problem, but successful applications to a wide variety of design synthesis problems have been reported in the literature.

9.2.4 Gradient and Gradient-Free Methods

A key practical consideration for optimization methods is the availability of derivatives of the cost function. Optimization algorithms that use derivatives are typically more efficient than gradient-free methods, since derivative information can allow the search point to move rapidly toward a minimum at each iteration. For electromagnetic structures, however, the cost function is typically evaluated using a forward scattering algorithm such as the finite difference time-domain (FDTD) method or method of moments (MoM), and derivatives are not available. For these reasons, we will focus in this chapter on algorithms that do not require derivatives.

There are two approaches for dealing with the lack of analytical derivatives. One is to estimate the derivative using samples of the cost function as is done by quasi-Newton methods. Another approach is to use a gradient-free algorithm such as the Nelder–Mead simplex method, conjugate gradient algorithms, or global methods such as simulated annealing.

9.3 One-Dimensional Optimization

One-dimensional optimization is not as computationally challenging as multidimensional optimization, but algorithms for the one-dimensional problem are used as subroutines for conjugate gradient and direction set algorithms. There are two basic gradient-free algorithms for one-dimensional optimization: golden section search and parabolic interpolation.

9.3.1 Golden Section Search

If a, b, and c are points such that $f(b) < f(a)$ and $f(b) < f(c)$, then there must be at least one minimum of $f(x)$ in the interval $[a, b]$. This is illustrated in Figure 9.3. The triplet of points (a, b, c) is said to bracket a minimum. The golden section search method progressively replaces a bracketing interval with a smaller bracketing interval until a convergence condition is reached.

The basis for the golden section search is a method of choosing the next search point within a given triplet in an optimal way. It is most efficient to choose the new point x in the wider interval. Assuming that $c - b > b - a$, then x will be between b and c. For the original triplet,

$$\frac{c-b}{b-a} = \phi \tag{9.2}$$

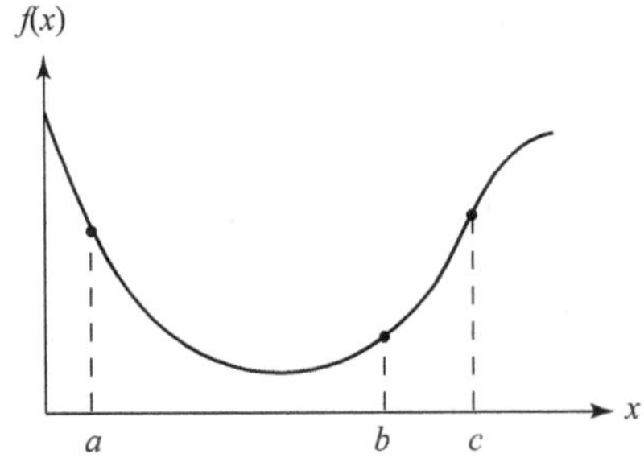

Figure 9.3 A triplet of points (a, b, c) that bracket a minimum of $f(x)$

where ϕ is the ratio of the larger interval to the smaller. If the objective function at the new point x is greater than $f(b)$, the new triplet will be (a, b, x). We will choose x so the ratio of the largest to smallest interval remains the same:

$$\frac{b-a}{x-b} = \phi \tag{9.3}$$

Alternately, if $f(x) < f(b)$, the new triplet is (b, x, c), and

$$\frac{c-x}{x-b} = \phi \tag{9.4}$$

Solving this system of linear equations for ϕ leads to

$$\phi^2 - \phi - 1 = 0 \tag{9.5}$$

The positive solution is

$$\phi = \frac{1+\sqrt{5}}{2} \simeq 1.618 \tag{9.6}$$

where the constant ϕ is known as the golden ratio.

All optimization algorithms require a termination condition. For the golden section search, the algorithm can be terminated when the interval becomes sufficiently small. If the condition

$$|c-b| < \tau(|b| + |x|) \tag{9.7}$$

is met, the algorithm stops and returns whichever of x or b has the smallest value of the objective functions as the minimum. The tolerance parameter τ remains to be specified.

9.3.2 Tolerance Parameter

The accuracy with which a minimum can be determined depends on the resolution of the finite precision numbers used in the algorithm software implementation [8]. The smallest representable difference between two computed numbers in a relative sense is the machine precision ε. The machine precision is given by the function `eps` in MATLAB®. Optimization algorithms operate by comparing values of the objective function at different points. Near a minimum, the objective function is approximately quadratic:

$$f(x) \simeq f(x_0) + c(x - x_0)^2 \tag{9.8}$$

This means that if two values for the objective function $f(x_1)$ and $f(x_2)$ differ by ε, the two points x_1 and x_2 are on the order of $\sqrt{\varepsilon}$ apart. Consequently, the finest possible relative resolution for the minimum point x is $\sqrt{\varepsilon}$, and the tolerance parameter should be $\tau = \sqrt{\varepsilon}$ or larger.

9.3.3 Inverse Quadratic Interpolation

The golden section search is robust, but for well-behaved, smooth objective functions, faster methods are available for locating the minimum. Near a minimum, a smooth function is approximately quadratic. In this case, a parabola can be fit to a bracket a, b, c and used to locate a good approximation to a minimum in the interval $[a, b]$. For a parabola that crosses the points $[a, f(a)]$, $[b, f(b)]$, and $[c, f(c)]$, the minimum is located at the point

$$x = b - \frac{1}{2}\frac{(b-a)^2[f(b)-f(c)] - (b-c)^2[f(b)-f(c)]}{(b-a)[f(b)-f(c)] - (b-c)[f(b)-f(c)]} \tag{9.9}$$

For some problems, inverse quadratic interpolation converges rapidly to a minimum, but for others, convergence can be slow. To overcome the disadvantages of either algorithm when used alone, the golden section and quadratic interpolation methods can be combined using Brent's method.

9.3.4 Brent's Method

Brent's method is a combination of the golden section search and parabolic interpolation. At each step of the algorithm, a parabolic interpolation is first attempted. If the extrapolated minimum point satisfies a reasonableness criterion in relation to previous search points, the parabolic minimum is accepted. If not, the cost function is assumed to be poorly behaved, and a golden section step is made instead.

9.4 Nelder–Mead Simplex Method

The bracketing method cannot be generalized in a simple way to higher dimensions, but the simplex search method of Nelder and Mead [9] is a robust, easily implemented method with some similarities to the golden section method.

For an N-dimensional optimization problem, the algorithm is initialized with $N + 1$ points $\mathbf{x}_1, \mathbf{x}_2, \ldots, \mathbf{x}_{N+1}$ in the N-dimensional parameter space. The points are vertices of an N-dimensional simplex. In two dimensions, the simplex is a triangle, and in three dimensions it is a tetrahedron. The points are sorted according to the value of the objective function at each point, and points are successively replaced in a way that moves the simplex spanned by the points toward a minimum.

The simplex method is quite old, and more sophisticated algorithms have been developed, but it converges reasonably rapidly for many practical classes of objective functions and is simpler to code than most other multidimensional optimization algorithms.

9.4.1 Initial Simplex

The initial simplex can be generated in various ways, but a common approach is to begin with a point $\mathbf{x}_1$ and generate N additional points by incrementing the nth coordinate of $\mathbf{x}_1$ by a given step size h_n. The initial simplex can also be chosen to be a polyhedron with given side lengths, or the vertices can be chosen randomly.

The algorithm begins by evaluating the objective function for each of the points to yield the list of numbers $f_n = f(\mathbf{x}_n)$, $n = 1, 2, N + 1$. The vertices are ordered by decreasing value of the objective function so that $f_1 \geq f_2 \geq \cdots \geq f_{N+1}$.

9.4.2 Simplex Transformations

At each iteration of the Nelder–Mead algorithm, one or more transformations to the simplex vertices are attempted. The goal is to replace the worst vertex (i.e., the vertex with the largest objective function value) with a new vertex that moves the simplex toward a minimum. If the new point satisfies a goodness condition, it is accepted; otherwise, another transformation is attempted. The algorithm ends when the simplex is small enough in size or the values of the objective function at each vertex are sufficiently close.

The best face of the simplex is the one opposite the worst vertex $\mathbf{x}_1$ with the largest objective function value. The centroid of the best face is

$$\mathbf{x}_b = \frac{1}{N}\sum_{n=2}^{N+1} \mathbf{x}_n \tag{9.10}$$

The centroid is a base point that is used to perform one of the following simplex transformations.

Reflect

The reflection point is

$$\mathbf{x}_r = \mathbf{x}_b + \alpha(\mathbf{x}_b - \mathbf{x}_1) \tag{9.11}$$

The reflection parameter is commonly chosen to be $\alpha = 1$. This point lies on the line from the worst point $\mathbf{x}_1$ to the centroid of the best face. If the value of the objective function $f_r = f(\mathbf{x}_r)$ satisfies the conditions $f_r < f_2$ and $f_r \geq f_{N+1}$, then the reflection point objective function value is better than the second-worst value f_2 but larger than the best value f_{N+1}, and the reflection point is accepted and the current iteration step ends. If the condition is not met, an expansion or contraction is attempted.

Expand
If the reflection point is better than the best current vertex ($f_r < f_{N+1}$), the expansion point

$$\mathbf{x}_e = \mathbf{x}_b + \gamma(\mathbf{x}_r - \mathbf{x}_b) \tag{9.12}$$

is evaluated, where the expansion parameter is $\gamma = 2$. The objective function value $f_e = f(\mathbf{x}_e)$ at the expansion point is computed. If $f_e < f_r$, then the expansion point is accepted and the iteration ends. If $f_e \geq f_r$, then the expansion point is not helpful, so the reflection point $\mathbf{x}_r$ is accepted and the iteration step ends.

Outside contraction
If the reflection point has a larger objective function value than the second-worst vertex ($f_r \geq f_2$) but is better than the worst vertex ($f_r < f_1$), the outside contraction point

$$\mathbf{x}_c = \mathbf{x}_b + \beta(\mathbf{x}_r - \mathbf{x}_b) \tag{9.13}$$

is evaluated, where the contraction parameter is $\beta = 1/2$. The objective function value $f_c = f(\mathbf{x}_c)$ is computed. If $f_c \leq f_r$, then the outside contraction point is better than the reflection point, $\mathbf{x}_c$ is accepted, and the iteration step ends.

Inside contraction
If $f_r \geq f_1$, the inside contraction point

$$\mathbf{x}_c = \mathbf{x}_b + \beta(\mathbf{x}_1 - \mathbf{x}_b) \tag{9.14}$$

is evaluated and the objective function value $f_c = f(\mathbf{x}_c)$ is computed. If $f_c \leq f_1$, then the inside contraction point is better than the worst point, $\mathbf{x}_c$ is accepted, and the iteration step ends.

Shrink
For certain choices of the initial simplex, it is possible for the simplex to become "stuck" and the algorithm fails to converge without an additional transformation. The shrink transformation can overcome this problem. Or, if the initial simplex is chosen randomly, the algorithm can be rerun for a different initial simplex in order to achieve convergence. If needed, the shrink transformation is accomplished by replacing all points except the best with $\mathbf{x}_n = \mathbf{x}_{N+1} + \delta(\mathbf{x}_n - \mathbf{x}_{N+1})$, where $\delta = 1/2$.

Once the new vertex from one of the transformations is accepted, the worst vertex is replaced by the new vertex, and the vertices are resorted by objective function value before the next iteration of the algorithm.

There are several variants of the basic Nelder–Mead simplex method. In some implementations, the shrink transformation is used as an additional transformation step, and in others, only one of the contractions is used. There are also methods for reducing the cost of sorting the objective function values and other steps associated with the Nelder–Mead algorithm. For engineering design the objective function evaluation takes far more time than the other calculations required for each step of the optimization algorithm itself, so these modifications are not important for CEM applications. The values for the reflection, expansion, contraction, and shrink parameters α, γ, β, and δ given above have been found empirically to work well for many problems. More guidance on choosing these parameters can be found in the literature on the Nelder–Mead algorithm.

9.4.3 Termination

The algorithm terminates when the simplex is sufficiently small or the objective function values at the vertices are close together. A simple criterion is to check if the difference between the objective function values at the best and worst vertices is close. If the condition

$$|f_1 - f_{N+1}| \leq \tau \tag{9.15}$$

is met, the algorithm terminates and returns the best vertex as the result of the optimization. The tolerance parameter τ can be chosen according to the considerations in Section 9.3.2. This termination criterion fails if the simplex lies in a very flat region of the objective function and in that case can be combined with a check of the distances between the vertices.

9.4.4 Implementation Details

The following outline may help in implementing the Nelder–Mead optimization method. To initialize the algorithm $N + 1$ vectors of parameter values $\mathbf{x}_1, \mathbf{x}_2, \ldots, \mathbf{x}_{N+1}$ are chosen. The parameter values are usually selected randomly within a known bounding box. These represent the vertices of a simplex in N dimensions.

The cost function is evaluated for each vertex of the initial simplex. The vertices are ordered so that the cost function values $f(\mathbf{x}_1) \geq f(\mathbf{x}_2) \geq \cdots \geq f(\mathbf{x}_{N+1})$ range from highest to lowest. From the sorted list, the worst, second worst, and best vertices (e.g., with the highest, second highest, and lowest cost function values) are identified.

The initial simplex is then transformed by applying the following steps within a loop until a termination condition is reached:

Nelder–Mead Optimization Algorithm

1. Compute the centroid of the best simplex face $\mathbf{x}_b$ using (9.10). Check the cost function values to see if the termination criterion (9.15) is met, and if so, stop the algorithm and return the best vertex.
2. **Reflect:** Compute $\mathbf{x}_r = \mathbf{x}_b + \alpha(\mathbf{x}_b - \mathbf{x}_1)$ and $f(\mathbf{x}_r)$. If the reflection point has a cost value $f(\mathbf{x}_r)$ that is between the best and second worst vertices of the current simplex, replace the worst vertex with the reflection point and return to step 1.
3. **Expand:** If the reflection point is better than the best vertex, compute the expansion point $\mathbf{x}_e = \mathbf{x}_b + \gamma(\mathbf{x}_r - \mathbf{x}_b)$. If the expansion point is better than the reflection point, replace the worst vertex with the expansion point and

return to step 1. If the expansion point is worse than the reflection point, replace the worst vertex with the reflection point and return to step 1.

4. **Outside contraction:** If the cost value of the reflection point is between the second worst and worst vertices, compute the outside contraction point $\mathbf{x}_c = \mathbf{x}_b + \beta(\mathbf{x}_r - \mathbf{x}_b)$. If the outside contraction point is better than the reflection point, replace the worst vertex with the outside contraction point and return to step 1.
5. **Inside contraction:** If the reflection point is worse than all the vertices of the current simplex, compute the inside contraction point $\mathbf{x}_c = \mathbf{x}_b + \beta(\mathbf{x}_1 - \mathbf{x}_b)$. If the inside contraction point is better than the worst point, replace the worst vertex with the outside contraction point and return to step 1.
6. **Shrink:** If needed, shrink the simplex by replacing all vertices except the best with $\mathbf{x}_n = \mathbf{x}_{N+1} + \delta(\mathbf{x}_n - \mathbf{x}_{N+1})$ and return to step 1.

To debug optimization codes, it is convenient to define a simple cost function as a MATLAB function handle using

```
f = @(x) x(1)^2 + x(2)^2
```

The function handle can be passed to an optimization routine as an argument:

```
x = simplex(f, xm);
```

where the vector `xm` defines an initial simplex for the Nelder–Mead method. An objective function stored as a MATLAB *m*-file can also be passed to the optimization routine as an argument. For more information about function handles, use the MATLAB command `help function_handle`.

9.4.5 Numerical Example

For a two-dimensional minimization problem, the simplex produced by the Nelder–Mead method has three nodes and is a triangle. The objective function is Rosenbrock's function

$$f(x_1,x_2) = 100(x_2 - x_1^2)^2 + (1 - x_1)^2 \tag{9.16}$$

which is a common test case for optimization algorithms. The simplices for nine iterations of the algorithm are shown in Figure 9.4. Rosenbrock's function has a narrow trough that makes finding the global minimum challenging for many algorithms. At the last iteration of the Nelder–Mead method shown in Figure 9.4, the simplex is moving along the trough toward the minimum at (1, 1).

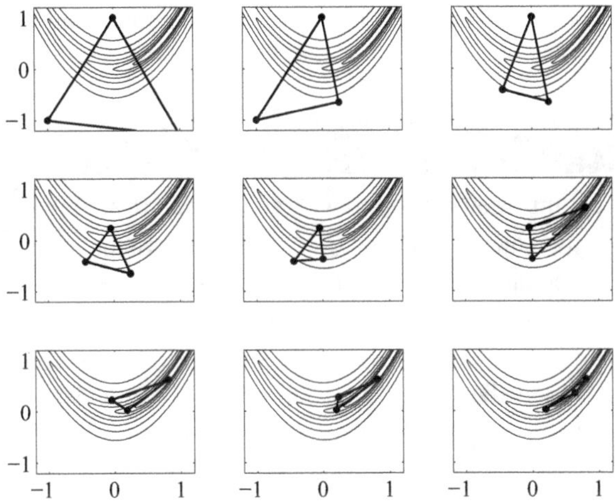

Figure 9.4 Nine iterations of the simplex method for a two-dimensional minimization problem. The level sets of the objective function are shown along with the simplex at each iteration. The minimum of the objective function is at the point (1,1)

9.5 Practical Considerations for Optimization

The general N-dimensional global optimization problem is extremely challenging, to say the least. An objective function may converge steeply in a very small region to a global minimum in a way that an optimizer must find trial values of the parameters that are extremely close to the global minimum before it can be located. There are an infinite variety of objective functions that could be designed to defeat any given algorithm. It has been proved that over a wide range of possible objective functions, all optimization algorithms perform equally poorly in an average sense [10]. This implies that there is no "magic" algorithm that converge rapidly for all objective functions.

For many physics-based optimization problems, there is a natural smoothness of the objective function with respect to the parameters. This implies that global minima are surrounded by regions of relatively slow decrease of the objective function and can therefore be found by optimization algorithms within a reasonable number of iterations or evaluations of the objective function. In Chapter 10, we will consider classes of problems that do not have well-behaved objective functions and are particularly difficult to solve.

Even when engineering optimization problems have smooth, well-behaved objective functions, evaluating the objective function can require a large amount of computation time. Adding just one more parameter to the objective function can significantly increase the number of iterations or objective function evaluations required for convergence, and the total time for the optimization blows up as the number of parameters increases. To deal with this in practice, the following recommendations can help:

Tips for Practical Optimization Problems

- Debug an optimization algorithm with a fast, analytical objective function.
- To get a feel for convergence rates and the behavior of the optimization, before attempting a complex engineering optimization problem with many design parameters start with a simplified problem with just one or two parameters.
- Find ways to simplify the forward model to speed up the objection function evaluation. Be sure the optimization converges for a faster version of the objective function with fewer parameters before attempting the full optimization problem.
- Intuition from early optimization rest runs can be used to limit the range of each parameter, guide the choice of initial values of parameters, or select new parameters that lead to meaningful progress toward the ultimate optimization goal.
- Gradually add parameters and retest the convergence of the optimization to lead up to the full optimization.

These issues are explored in the problems at the end of the chapter, particularly Problem 9.4.

Problems

9.1 Implement the golden section 1-D search method as a MATLAB function that accepts a cost function and a bracketing triplet of points as arguments and returns the location of a minimum of the cost function. Debug the algorithm using $f(x) = x^2$.

9.2 Implement the Nelder–Mead simplex algorithm. Test the algorithm using (a) a simple cost function such as $f(x_1, x_2) = x_1^2 + x_2^2$ and (b) Rosenbrock's function.

9.3 Compare the performance of the golden section and Nelder–Mead algorithms for a one-dimensional objective function. Overlay plots of the convergence history (cost function value as a function of iteration count) for the two algorithms.

9.4 Using the optimization algorithm of Problem 9.2, find a stealth or low observable scatterer shape with as low as possible backscattering width over a 20° range of angles from 170° to 190° at $f = 300$ MHz. The scatterer must be such that a circular cylinder of radius 0.5λ can fit inside it without touching the scatterer surface, and its longest dimension must be less than 10 λ. Model the scatterer as a PEC or a conductor with high conductivity. Earlier codes have computed bistatic scattering widths. For this problem, since we are interested in the backscattering direction the MoM postprocessing must be changed to compute monostatic scattering widths ($\phi^s = \phi^i$). Use a mesh density of at least $n_\lambda = 10$ (although a courser mesh can be used in initial testing of the algorithm to save time).

For your scatterer, prepare the following: (a) a plot of the scatterer shape; (b) the polyarc description; (c) a plot of the backscattering width from 0° to 360° (it is helpful to mark the 20° range with vertical lines); and (d) a plot of the convergence history of the optimization algorithm for your design (cost function value as a function of Nelder–Mead iteration number). (e) Give the maximum value of the backscattering width over the range $170° \leq \phi \leq 190°$.

References

[1] R. Fletcher, *Practical Methods of Optimization*, 2nd edn. New York, NY: Wiley, 1987.

[2] J. Nocedal and S. J. Wright, *Numerical Optimization*. New York, NY: Springer, 1999.

[3] A. Antoniou and W. sheng Lu, *Practical Optimization Algorithms and Engineering Applications*. New York, NY: Springer, 2007.

[4] S. Boyd and L. Vandenberghe, *Convex Optimization*. Cambridge: Cambridge University Press, 2004.

[5] V. Grout, M. O. Akinsolu, B. Liu, *et al.*, "Software solutions for antenna design exploration: A comparison of packages, tools, techniques, and algorithms for various design challenges," IEEE Antennas Propag. Mag., vol. 61, no. 3, pp. 48–59, 2019.

[6] D. Goldberg, *Genetic Algorithms in Search, Optimization and Machine Learning*. Reading, MA: Addison-Wesley, 1989.

[7] Y. Rahmat-Samii and E. Michielssen, *Electromagnetic Optimization by Genetic Algorithms*. New York, NY: Wiley, 1999.

[8] W. Press, S. Teukolsky, W. Vetterling, and B. Flannery, *Numerical Recipes in FORTRAN: The Art of Scientific Computing*. Cambridge: Cambridge University Press, 1992.

[9] J. A. Nelder and R. Mead, "A simplex method for function minimization," Comput. J., vol. 7. pp. 308–313, 1965.

[10] D. H. Wolpert and W. G. Macready, "No free lunch theorems for optimization," IEEE Trans. Evol. Comput., vol. 1, no. 1, pp. 67–82, 1997.

Chapter 10

Inverse Problems

10.1 Introduction

The numerical methods that we have considered so far solve forward problems, in the sense that the scatterer or material structures are given and we want to find the fields around those objects. An inverse problem is one for which the scattered or radiated fields are given, and the goal is to reconstruct the location, geometry, or composition of the object that produced those fields.

The most basic feature of an inverse problem is that the parameters of a structure, model, or system must be determined from observed data. For the purposes of this chapter, the inverse problem deals with an electromagnetic structure, such as an antenna, microwave component, or underground geophysical layer. We will refer to the structure to be identified or imaged as a scatterer. The forward problem is to compute scattered fields for a known scatterer, whereas the inverse problem considers the scatterer to be unknown and seeks to determine characteristics of the object from forward scattered field data. For reasons that we will explore in this chapter, inverse problems are generally more difficult to solve than forward problems. Surveys of inverse scattering methods include [1,2].

10.2 Types of Inverse Problems

Inverse problems in electromagnetics are highly diverse. Radar, geophysical exploration, medical imaging, antenna synthesis, and microwave component design are just a few examples of applications that can be unified under the broad heading of inverse problems. A few classes of inverse problems that are common in electromagnetics are surveyed in the following sections.

10.2.1 Inverse Scattering

For an inverse scattering problem, an object is illuminated by a known set of incident fields, and the measured scattered fields for each incident field provide the data from which we seek to reconstruct the object. Radar is a very simple inverse scattering problem. The goal is to determine the target location and perhaps its velocity and a rough estimate of size. Since only a restricted set of parameters about the object is sought, the radar problem can be solved with fewer scattered field measurements than would be required to create an image of the object.

With knowledge of the scattered fields for many incident fields, or with continuous knowledge of the scattered field for a single incident field over a range of scattering directions, under certain conditions it is theoretically possible in the noise-free case to determine uniquely the scatterer shape and composition [3,4]. These mathematical results are known as perfect or exact reconstruction theorems.

Theorems on inverse scattering that hold in the noise-free case have limited practical value. In the theory of information, without noise it is possible to communicate at an infinite rate over a finite signal bandwidth. Similarly, without noise in the scattering data and numerical error in the inversion algorithm, it is possible to reconstruct a scatterer shape and composition completely from scattered fields. In real-world problems, noise in the scattering data prevents perfect reconstruction and often limits images to a rough estimate of the scatterer shape without fine details. Consequently, theoretical results about inverse scattering in the noise-free case can fail to have practical value. The issue of noise sensitivity in inverse problems is discussed further in Section 10.3 on ill-posed problems.

10.2.2 Imaging

When the goal is to determine the shape and material properties of an unknown object from scattered fields, inverse scattering problems are often referred to as inverse imaging problems. Inverse imaging encompasses a broad range of technologies, including X-ray imaging, tomography [5], imaging radar, ultrasound, microwave imaging, and ground-penetrating radar. Some applications of imaging systems are geophysical exploration, medical diagnostics, radar target recognition, and nondestructive evaluation and testing.

In standard X-ray imaging, only amplitude information in the signal is used to reconstruct an image of the object under test. This is commonly referred to as tomography. With more advanced imaging technologies, phase information is added to extract more detail about an object from the way it scatters electromagnetic waves.

10.2.3 Inverse Source Problems

Another type of inverse problem in electromagnetics is the inverse source problem. For an inverse source problem, the data consists of radiated fields, and the goal is to determine the source distribution that produced those fields. Antenna synthesis is a classical example of an inverse source problem.

10.2.4 Design Synthesis

The design or synthesis of microwave systems, components, and electromagnetic structures can be viewed as an inverse problem, since characteristics of the waves and fields produced by the structure are given and from that the physical parameters of the structure must be determined. Designing an antenna with a specified radiation pattern or a microwave component with given S-parameters can be viewed as inverse problems. Other examples include design of low observable and low radar cross-section aircraft, frequency-selective surfaces, and filters.

There is a direct connection between the scattered fields on transmission lines connected to an electronic device and the scattered fields from an object in space. For a microwave network, there are a finite number of scattering directions, whereas in free space there are an infinite number, but otherwise the mathematics is actually quite similar. When designing an electronic device, we want to come up with a device configuration that scatters incident energy in prescribed ways over frequency and from one port to another. In microwave imaging, we want to determine the shape and composition of an object from the way it scatters incident waves arriving from different directions. This helps us to understand the mathematical connections between microwave imaging and design synthesis.

Due to the inherent mathematical and computational difficulties associated with inverse problems, the design of electromagnetic structures and systems has traditionally been accomplished using a model for the structure with an analytical or approximate solution to the forward problem. This provides a mathematical relationship between the object and specified field data that is relatively easily inverted to solve for the design parameters that meet performance requirements.

Design synthesis using the forward analysis is limited because the solution space is limited to a small set of parametric designs. Closed-form or analytical formulas are only available for a relatively small number of problems with simple, canonical geometrical shapes. For applications with stricter performance requirements, a more geometrically rich family of structures with more design parameters is needed. Due to the complexity of the analysis or forward problem for more complex structures, analytical solutions are not available and design parameters meeting the requirements cannot be directly calculated. The design must be done using numerical optimization.

A simple example of the relationship between analytical and optimization-based design procedures is antenna design. An analytical model for a dipole antenna can be used to determine the length of the dipole arms required to achieve a matched antenna input impedance. If a broadband input impedance match is desired, a simple dipole is inadequate, and a larger solution space of potential antenna geometries must be explored. Analytical models for more complex antenna geometries are not available, and numerical optimization methods must be used.

As discussed in Chapter 9, increasing computational power makes it feasible to use numerical optimization algorithms to solve a wider range of complex design synthesis problems. This approach requires significant computational time, complex algorithms, and the optimization can fail to converge, so that it is not always robust and human intervention is often required to adjust the optimization approach so that it reaches a useful solution. Despite the difficulty of optimization-based design synthesis, these methods are now in routine industrial use in many applications, particularly in optimizing antennas and other microwave and RF system designs to tune the response to better fit a desired set of characteristics.

10.2.5 Applications

Applications that require solution of inverse problems in electromagnetics and acoustics include the following:

Applications of Inverse Scattering

X-ray tomography: A standard X-ray image can be viewed as a simple inverse scattering method. By taking images from multiple look directions, a map of a target's optical density can be constructed, as is done with computerized axial tomography (CAT) scans. Tomographic images can be improved by adding phase information about the transmitted signal through an object. Image quality improves, but at the cost of increased system complexity.

Ultrasonic imaging: Medical ultrasound is typically based on results from a 1-D inversion algorithm for multiple look directions pieced together to form a 2-D image. More advanced ultrasound systems using transducer arrays have also been developed.

Geophysical exploration: Acoustic or electromagnetic field measurements can be used to map underground geological layers and deposits. Imaging of underground features can be done using an array of transducers at the surface of the earth.

Impedance tomography: The conductivity of ground layers or other inhomogeneous structures can be imaged using low-frequency field measurements. Since the field penetration depth is large for low frequencies, deeply buried objects can be imaged. A simple type of impedance tomography borehole logging, with a transceiver that is moved through a hole to detect the characteristics of rock layers based on changes in the electrical impedance as a function of depth.

Microwave imaging: An array of microwave transceivers can be used to image tissues or materials.

Nondestructive evaluation: Electromagnetic fields generally do not damage an object under test, so defects or other internal properties can be found using inverse scattering without cutting into and destroying the structure.

Imaging radar: Synthetic aperture radar, ground-penetrating radar, target identification, and other technologies use multiple scattered field measurements from an object to reconstruct its shape and other characteristics.

Microwave and optical systems: Design of a microwave component or electromagnetic structure given a desired response is an important inverse problem. Examples include antennas, radomes, antireflective coatings, frequency-selective surfaces, filters, combiners and splitters, passive interconnects on a printed circuit board, matching networks, couplers, and microwave circuits.

10.3 Ill-Posed Problems

The most common mathematical characteristic of inverse problems is that they are ill-posed. A well-posed problem is one for which a unique solution exists and the solution depends in a continuous way on the data. Exact reconstruction theorems guarantee that some inverse problems have unique solutions, but the solution may not depend continuously on the forward scattering data [6]. Ill-posedness has far-reaching ramifications on practical applications of solution methods for inverse problems.

One reason for this ill-posedness is that the space of possible solutions to an inverse problem is far larger than the space of possible functions representing forward data. Considering inverse scattering and imaging problems, the scattering data is generally a smooth function of the incident and scattering angles whereas the image itself can be discontinuous. Large changes in the object composition or shape may produce only very small changes in the forward data, so there are many shapes that are mapped to nearly the same forward scattering data.

An ill-posed problem is generally difficult to solve numerically. In the noise-free case, for many classes of scattering problems the map from object shape to scattering data is one to one. There is only one object that corresponds to each set of forward data and hence perfect imaging is theoretically possible. In practice there are always degree of noise in the data and numerical error in the solution process. Because a wide variety of possible image shapes can be distinguished only by tiny changes in the forward data, for an ill-posed problem in the presence of noise, the mapping from object to data is effectively many to one. Many possible object shapes become indistinguishable on the basis of small differences in their forward scattering data that are below the noise floor, and a unique solution to the inverse problem cannot be found.

Due to the ill-posed nature of inverse problems, there is a significant gap between theorems that hold in the noise-free case and the performance of inverse solution methods in practical applications. Based on exact reconstruction theorems, an object can be reconstructed with forward data over a small range of incident and scattering angles in the noise-free case. In principle forward data over a small parameter range can be extrapolated to data over a full circle or sphere of angles or a wide frequency band. A smooth function is uniquely determined by its variation over any interval, no matter how small. But the process of extending information over a limited range of observations is related to analytic continuation, or extrapolating a function from samples of the function within an interval to values outside the internal. This is extremely sensitive to noise and generally fails to be useful in practice.

If an inverse problem can be characterized by an operator, the operator is generally ill-conditioned. The degree of ill-conditioning is typically extreme in comparison to operators associated with forward problems. For the electric field integral equation (EFIE) applied to a 2-D perfect electric conductor (PEC) forward scattering problem, the moment matrix condition number is proportional to the number of unknowns as the fineness of the mesh is increased. The moment matrix is poorly conditioned for a very fine mesh, but the condition number only grows linearly with the number of mesh elements. This is a relatively weak case of ill-conditioning. For an inverse problem, the condition number of an associated matrix operator may grow exponentially with the size of the matrix. This is far worse than the condition number growth with a typical forward problem like the EFIE. Numerically, the degree of ill-conditioning associated with inverse problems is challenging to overcome, to say the least.

10.3.1 Regularization

Despite the inherent difficulty of solving ill-posed problems, numerical solutions can be feasibly obtained for a variety of important applications by transforming the ill-posed problem into a well-posed problem using regularization. To obtain a meaningful numerical solution for an inverse problem, the problem must be changed so that the relationship between the object and forward data becomes one to one and invertible. In the presence of noise, this implies that regularization provides a way to choose a unique solution out of the space of possible solutions for a given noise level.

There are many different methods for regularizing an inverse problem. Some are ad hoc and are used in practice because they work well in an empirical sense for a given class of inverse problems. Others are based on an optimality proof of some kind for a given class of problems and noise model.

An ill-posed problem can be regularized by restricting the space of possible scatterer or object shapes, typically by considering only smooth geometries or a parametric geometry with a small number of degrees of freedom. Many regularization schemes amount to choosing spatially smooth image reconstructions over nonsmooth images. In fact, the word "regular" in a broader sense means a shape or function that is geometrically smooth, so there is a fundamental connection between regularization of an inverse problem and spatial smoothness.

When an ill-posed problem can be characterized by an operator, small singular values of the operator typically represent rapidly varying modes or functions in the solution space. In this case, the problem can be regularized by ignoring small singular values of the operator and choosing a smooth member of the space of possible solutions to the inverse problem.

In numerical implementations of inverse scattering methods, the operator associated with such an ill-posed problem is often transformed into a linear system of the form

$$Ax = b \tag{10.1}$$

where x represents the scatterer shape or material composition, and b is the scattered field data. Since the scatterer typically is represented by many more degrees of freedom than the number of scattered field measurements, A will have fewer rows than columns, and the linear system is underdetermined. To transform the linear system into new system with a square matrix, we can multiply by A^H to obtain the following normal equation:

$$A^H Ax = A^H b \tag{10.2}$$

Since the original system was underdetermined, the matrix $A^H A$ is highly ill-conditioned, and the linear system cannot be solved directly.

A simple approach to regularizing the inverse problem is to add a scaled identify matrix and change the linear system to

Regularization by Singular Value Shifting

$$(A^H A + \varepsilon I)x = A^H b \tag{10.3}$$

where ε is a regularization parameter. The value of the regularization parameter is typically small in relation to the largest eigenvalues of $A^H A$. With the additive regularization term in (10.3), small eigenvalues of $A^H A$ are increased to approximately ε and no longer dominate the solution x as they would in $(A^H A)^{-1}b$. Since the small eigenvalues of many physical operators represent eigenfunctions with high spatial frequencies, components of the reconstructed object represented by the vector x with rapid spatial variation are suppressed, and the reconstructed object is spatially smooth.

An example of the eigenvalues associated with a poorly conditioned operator from an inverse algorithm is shown in Figure 10.1. The regularization term leaves the large eigenvalues associated with a matrix of forward scattering amplitude data unchanged but raises the small eigenvalues to the level of the regularization parameter. This reduces the matrix condition number and numerically stabilizes the solution of the linear system.

The regularization parameter ε can be chosen empirically by adjusting its value until a good reconstruction is obtained for a known object. Because the quality of

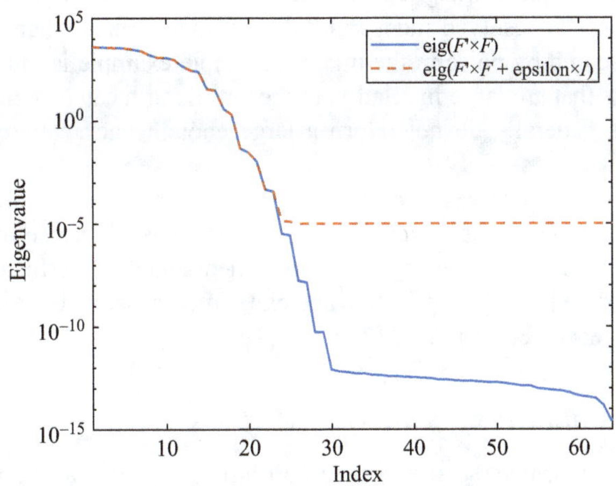

Figure 10.1 Eigenvalues associated with a matrix of scattering amplitudes before and after regularization. The forward scattering data corresponds to the example shown in Figure 10.7. The regularization parameter is $\varepsilon = 10^{-5}$

a reconstructed scatterer image or other type of inverse problem solution is hard to assess quantitatively, this ad hoc approach to regularization can be challenging. Despite its drawbacks, the empirical method is often used in practice.

There are more rigorous approaches to specifying the magnitude of the regularization parameter. One such method for choosing ε based on the noise level in the forward data is Tikhonov regularization. With this technique, the matrix singular values that represent scatterer characteristics for which changes in the scattering data are above the noise floor are only slightly perturbed, and singular values that represent information about the scatterer for which changes in the scattering data are below the noise floor are increased to near ε in value, to reduce the effect of those singular values on the reconstructed object solution x as given by (10.3). Increased rigor in the selection of the regularization parameter provides a theoretical basis for the regularization process, but in practice the range of reasonably good values for ε can be quite wide and a carefully chosen value does not lead to significantly better reconstruction than can be obtained with an empirical choice of the parameter.

10.3.2 Resolution

The ill-posed nature of the inverse scattering problem means that for an image computed from measured scattered fields, the degree of geometrical resolution of fine image details is inherently limited. Many regularization techniques suppress fine features in an image. Noise in measured forward scattering data also limits the resolution of a reconstructed object shape or composition map.

For narrowband, high-frequency inverse scattering applications based on acoustic or electromagnetic interrogation of an unknown object, spatial image resolution is typically close to the electromagnetic wavelength. Features that are large in relation to the wavelength can be resolved and features that are much smaller than the wavelength are usually obscured by noise in the image. A simple example is radar detection of objects. Objects that are large in relation to the wavelength can be detected, whereas smaller objects generally do not return a large enough backscattered pulse to be observed.

It is possible in some cases to reconstruct object features that are smaller than the wavelength in size. This is commonly referred to as superresolution. Methods based on superresolution can be tricky to implement and their performance depends strongly on the details of the application and experimental setup, but many successful demonstrations have been reported [7–9].

10.3.3 Types of Inverse Scattering Methods

Two major classes of inverse scattering algorithms are optimization-based methods and linearizing approximation methods. Methods of the first type use optimization algorithms to progressively update a trial image until its forward scattering matches the given data. The second class of methods is based on an approximation to the forward scattering problem, which is used to transform the inverse problem into a linear problem that can be readily solved numerically. There are other methods that do not fit into this classification scheme, including time reversal algorithms [10] and the regularized sampling method.

10.3.3.1 Optimization-Based Inverse Scattering Methods

Inverse problems and optimization methods are closely related, since any inverse problem for which a forward model is available can be formulated as an optimization problem. The cost function is the difference between the given measured scattering field data and the scattered fields computed by the forward model for a given scatterer. The optimization algorithm successively updates the scatterer shape or composition until the computed forward data converges to the measured forward data.

The key computational elements of an optimization-type inverse scattering algorithm are a fast forward solver and a method for updating the current object to provide a new guess at the next iteration in such a way that the search converges to a good solution. Since optimization-based methods can be implemented with little insight into the details of the inverse problem, such a method can be considered a "brute force" approach to solving inverse problems. Brute force methods can be straightforward to implement, at least in terms of the logic of the algorithm.

The drawback of optimization methods is that they are computationally expensive. For a complex scatterer, the number of degrees of freedom required to characterize its shape and composition is extremely large. The large number of degrees of freedom makes the optimization problem highly challenging, and the brute force approach of coupling a forward model with an optimizer code can require so many evaluations of the forward model that for many inverse problems the computation time is impractically large.

Optimization-based inverse scattering methods can be augmented in various ways to improve computational efficiency. Constraints or *a priori* information about the scatterer geometry or composition are helpful in restricting the solution space. One important class of optimization methods is based on taking a generalized derivative of the cost function with respect to changes in the object properties. This derivative is used to determine the next object update. These types of methods tend to be more difficult to implement. Another family of optimization-based methods relies on multiscale representations of the object. A first pass of the algorithm is used to compute the rough shape and location of the object, and successive passes recover progressively finer geometrical details of the object shape and composition.

10.3.3.2 Approximation-Based Methods

Many inverse scattering methods are based on approximations to the forward problem. These include high-frequency asymptotic limits, low-frequency or quasistatic limits, and various other approximations. The approximation often linearizes the forward problem in such a way that it can be inverted to provide a direct formula for a reconstructed image in terms of forward data.

The most important method of this type is the weak scatterer or single scattering approximation, commonly referred to as the Born approximation. With this approach, the scattered field on the scatterer is approximated by the incident field. This approximation can be used in conjunction with an integral equation for the forward scattering problem to develop several related Born-type inverse scattering algorithms. Other imaging methods that can be viewed as arising from approximations to the forward problem include microscopy and synthetic aperture radar.

10.3.3.3 Other Inverse Scattering Methods

Other classes of inverse scattering methods do not rely on optimization or approximations to the forward problem. Time reversal methods rely on the electromagnetic principle of reciprocity, which implies that if a scattered field distribution is reversed in time and propagates back toward the scatterer, it will focus on the location of the scatterer. This is sometimes referred to as backpropagation [11]. Most inverse scattering algorithms assume that the background medium is free space or is at least a known medium. Time reversal can be used as a way to image an object that is immersed in an unknown and possibly random medium such as ground or turbulent air. The regularized sampling method considered in Section 10.5 is another approach not based on a linearizing approximation.

10.4 Born Approximation Methods

To develop the Born approximation, we will consider a cylindrical dielectric object with transverse magnetic (TM) polarized incident fields. The scatterer is surrounded by N_T transmitters and N_R receivers. Often, the transmitters and receivers are collocated, so that each antenna is a transceiver and $N_T = N_R$. A generic inverse scattering configuration is shown in Figure 10.2. It is also common to assume that the transceivers are in the far field of the scatterer.

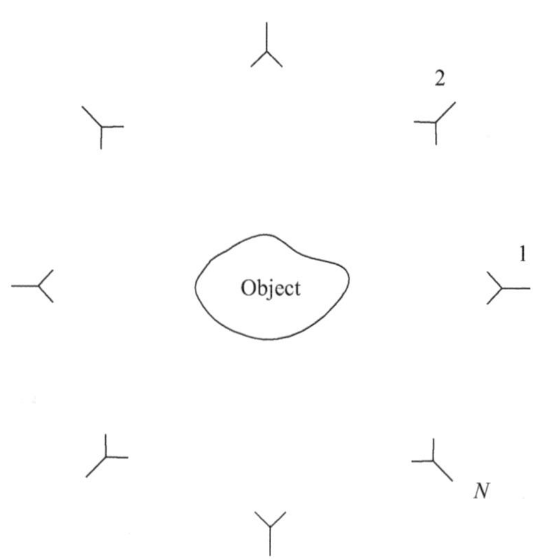

Figure 10.2 Scatterer surrounded by $N_T = N_R = N$ transceivers

We will derive a Born approximation-based algorithm from the forward integral equation for the scatterer. The 2-D volume EFIE for the forward scattering problem is

$$E_z^{\mathrm{i}}(\overline{\rho}) = E_z(\overline{\rho}) + \frac{k_0\eta}{4}\int d\overline{\rho}' H_0^{(2)}(k_0|\overline{\rho} - \overline{\rho}'|)J_z(\overline{\rho}') \tag{10.4}$$

where $E_z = E_z^{\mathrm{i}} + E_z^{\mathrm{s}}$ and

$$J_z(\overline{\rho}) = j\omega\varepsilon_0[\varepsilon_r(\overline{\rho}) - 1]E_z(\overline{\rho}) \tag{10.5}$$

represents an equivalent current on the object that radiates the scattered field E_z^{s}. This is the EFIE for a 2-D dielectric scatterer for the TM^z polarization that was solved in Section 6.6.1 using the volume moment method. Here, we are interested in an inverse scattering method, rather than a forward solver, so the integral equation will be applied differently.

We can write the current in the form

$$J_z(\overline{\rho}) = j\omega\varepsilon_0 O(\overline{\rho})E_z(\overline{\rho}) \tag{10.6}$$

where

$$O(\overline{\rho}) = \varepsilon_r(\overline{\rho}) - 1 \tag{10.7}$$

This quantity is called the object function, since it defines both the scatterer geometry and dielectric properties. The object function is zero outside the scatterer and nonzero inside. The goal of an inverse scattering method is to compute or estimate the object function given measurements of the scattered field E_z^{s} for a set of known incident fields E_z^{i}.

Using the object function in the EFIE leads to

$$E_z^{\mathrm{i}}(\overline{\rho}) = E_z(\overline{\rho}) + k_0^2\frac{j}{4}\int d\overline{\rho}' H_0^{(2)}(k_0|\overline{\rho} - \overline{\rho}'|)O(\overline{\rho}')E_z(\overline{\rho}') \tag{10.8}$$

where the $\overline{\rho}$ dependence is evaluated on the scatterer. In operator form,

$$(\mathscr{I} + \mathscr{L}\mathscr{O})E_z = E_z^{\mathrm{i}} \tag{10.9}$$

The derivation of the integral equation (10.4) in Section 6.6.1 made use of a radiation integral. As part of the derivation, the observation point $\overline{\rho}$ was restricted to the scatterer, so that both the source point $\overline{\rho}'$ and the observation point $\overline{\rho}$ range over the support of the scatterer. If we instead move the observation point $\overline{\rho}$ to the N_R receiver locations, (10.4) becomes

$$-\mathscr{L}_D\mathscr{O}E_z = E_z^{\mathrm{s}} \tag{10.10}$$

This is the radiation integral for the electric field at the receivers in terms of the equivalent source on the scatterer.

The integral equations (10.9) and (10.10) can be used to develop an inverse scattering algorithm. If we surround the object with N_T transmitters and N_R receivers, then the data available for each transmitter consists of the incident field $E_z^{\mathrm{i}}(\overline{\rho})$ on the scatterer and measurements of the scattered field $E_z^{\mathrm{s}}(\rho_n^D)$ where ρ_n^D are the receiver locations.

We must first discretize these integral equations and transform them into matrix relationships. If we discretize a region of space around the object using N pulse basis functions located at a grid of points ρ_n, so that

$$E_z = \sum_{n=1}^{N} a_n f_n(\overline{\rho}) \tag{10.11}$$

and apply the method of moments with point matching to the integral equations (10.9) and (10.10), we obtain linear systems of the form

$$a_n + \sum_{n=1}^{N} A_{mn} O_n a_n = E_z^{\mathrm{i}}(\rho_m) \tag{10.12}$$

$$\sum_{n=1}^{N} B_{mn} O_n a_n = E_z^{\mathrm{s}}(\rho_m^D) \tag{10.13}$$

where $O_n = O(\rho_n)$. For the first equation, we have tested the $\overline{\rho}$ dependence of (10.9) at N testing points ρ_n on a reconstruction region around the location of the object. For the second equation, we have tested the $\overline{\rho}$ dependence of (10.10) at the N_R receiver locations ρ_n^D. In matrix form,

$$(I + AO)E = E^{\mathrm{i}} \tag{10.14}$$

$$BOE = E^{\mathrm{s}} \tag{10.15}$$

where $O = \mathrm{diag}[O(\rho_1), \ldots, O(\rho_N)]$ and E^{i}, E^{s}, and E are column vectors. If multiple incident fields are used, then the linear system (10.14) can be repeated for each incident field to yield a larger linear system.

In these equations, both O and E are unknown. What we need to do is find a way to guess E so that we can solve the second equation for O. To do this, we will use the Born approximation. Assuming that the object contrast is low and $O(\overline{\rho}) = \varepsilon_r(\overline{\rho}) - 1$ is small, then in (10.9) the integral term is small, and we have

$$E_z \simeq E_z^{\mathrm{i}} \tag{10.16}$$

If we use this approximation in the operator term of (10.14) and solve for the initial E term, we obtain

$$E^{(1)} = E^{\mathrm{i}} - AOE^{\mathrm{i}} \tag{10.17}$$

This provides a slightly better approximation for E than (10.16). Using this in (10.15) leads to

$$BOE^{(1)} = E^{\mathrm{s}} \tag{10.18}$$

Since O is diagonal, we can think of the vector $OE^{(1)}$ as the unknown vector in this expression.

It is helpful to visualize the relative matrix sizes for the terms in (10.18). In block form, this relationship is

$$\begin{bmatrix} B_{11} & \cdots & B_{1N} \\ \vdots & \ddots & \vdots \\ B_{N_R 1} & \cdots & B_{N_R N} \end{bmatrix} \begin{bmatrix} O_{11}E_1^{(1)} \\ \\ \vdots \\ \\ O_{NN}E_N^{(1)} \end{bmatrix} = \begin{bmatrix} E_1^s \\ \vdots \\ E_{NR}^s \end{bmatrix} \tag{10.19}$$

Because B is an N_R-by-N matrix and $N_R \ll N$, the linear system is underdetermined. To solve the linear system, we use the least squares approach, so that

$$B^H Bx = B^H E^s \tag{10.20}$$

$B^H B$ is now a square matrix but is ill-conditioned, so we must regularize the linear system before it can be solved. The ill-conditioning of the linear system is a manifestation of the ill-posedness of the underlying inverse scattering problem. Using a scaled identity to regularize the linear system leads to

Inverse Scattering in the First-order Born Approximation

$$\begin{aligned} (B^H B + \varepsilon I)x &= B^H E^s \\ x &= OE^{(1)} \end{aligned} \tag{10.21}$$

where ε is a small regularization parameter. The resulting solution for the object function O is the first-order Born approximation. Since the scattering operator A is applied only once in obtaining $E^{(1)}$, this is also called the single scattering approximation.

10.4.1 Iterated Born Approximation

To improve the solution, we can iterate (10.14) to improve upon the Born approximation, so that

$$E^{(2)} = E^i - AO^{(1)}E^{(1)} \tag{10.22}$$

and then resolve (10.20) for $O^{(2)}$. This process can be repeated to obtain a series of higher order Born approximations for O. This series may not always converge, which means that a higher order Born approximation can fail to improve upon a lower order solution. The first-order Born approximation is almost always used in practice, although iterated Born approximations have been demonstrated in the literature [12].

10.4.2 Diffraction Tomography

Tomography is a general term for reconstructing a multidimensional object from lower dimensional images based on illumination or observation from different angles. A tomographic technique used routinely is X-ray imaging. Basic tomographic techniques rely only on intensity or magnitude information in the reconstruction process.

When modified to account for diffraction and wave scattering effects, tomographic methods are referred to as diffraction tomography. Diffraction tomography has many variants and applications. This family of methods is mathematically related to algorithms used in magnetic resonance imaging, synthesis imaging with sparse antenna arrays in radio astronomy, and other image reconstruction problems.

We can derive a simple diffraction tomography reconstruction method from the Born approximation used in the previous section. The algorithm developed in Section 10.4 makes no assumptions about the locations of the transmitters and receivers. If the transceivers are in the far field of the scatterer, in the Born approximation the relationship between the object function and forward scattering data becomes a Fourier transform. The imaging algorithm based on this observation leads to a type of diffraction tomography.

Using the Born approximation (10.16) in the integral equation (10.10) leads to

$$E_z^{\mathrm{s}}(\overline{\rho}) = -k_0^2 \int d\overline{\rho}'\, g(\overline{\rho} - \overline{\rho}')O(\overline{\rho}')E_z^{\mathrm{i}}(\overline{\rho}') \tag{10.23}$$

where $g(\overline{\rho} - \overline{\rho}') = (j/4)H_0^{(2)}(k_0|\overline{\rho} - \overline{\rho}'|)$. If the transmitters are far from the scatterer, the incident field can be approximated as a plane wave, so that $E_z^{\mathrm{i}}(\overline{\rho}) = e^{-j\overline{k}^{\mathrm{i}}\cdot\overline{\rho}}$. In the far field, the kernel can be simplified using the large argument asymptotic limit of the Green's function,

$$g(\overline{\rho} - \overline{\rho}') \sim \frac{e^{-jk\rho}}{\sqrt{\rho}} e^{j\overline{k}^{\mathrm{s}}\cdot\overline{\rho}'} \tag{10.24}$$

where $\overline{k}^{\mathrm{s}} = k_0\hat{\rho}$. Since the scattering amplitude is proportional to the scattered field, these results combine to yield the final relationship

Diffraction Tomography

$$S(\phi_i, \phi_s) \sim \int e^{j(\overline{k}^{\mathrm{s}} - \overline{k}^{\mathrm{i}})\cdot\overline{\rho}} O(\overline{\rho})\, d\overline{\rho} \tag{10.25}$$

where $\overline{k}^{\mathrm{i}}$ is the wave vector of the incident plane wave arriving from the angle ϕ_i and $\overline{k}^{\mathrm{s}}$ is a wave vector with angle ϕ_s.

This relationship shows that the object function can be obtained from the scattering amplitude using a two-dimensional Fourier transform (the algorithm can be generalized to three dimensions as well). If the image reconstruction points form a rectangular grid, the scattering amplitude data must be interpolated to a rectangular grid to apply the inverse fast Fourier transform to the data as part of a practical implementation of this diffraction tomography algorithm.

10.4.3 Holographic Backpropagation Tomography

One difficulty with diffraction tomography is that interpolation of the scattering data to a rectangular grid can lead to poor image quality in the presence of noise. To avoid interpolation, the method of holographic backpropagation tomography (HBT) can be used. This section provides a rough sketch of the derivation of the HBT method given in [13].

We begin with what is called the holographic field for the scatterer illuminated from the direction $\hat{k}^{\mathrm{i}}$:

$$E_z^{\mathrm{s}}(\overline{\rho};\hat{k}^{\mathrm{i}}) = \int d\hat{k}^{\mathrm{s}}\, S(\hat{k}^{\mathrm{s}},\hat{k}^{\mathrm{i}})e^{-j\overline{k}^{\mathrm{s}}\cdot\overline{\rho}} \tag{10.26}$$

This is the scattered field reconstructed using the scattering amplitude to weight a combination of plane waves integrated over the propagation direction. Using (6.72), this becomes

$$E_z^{\mathrm{s}}(\overline{\rho};\hat{k}^{\mathrm{i}}) = -\frac{k_0\eta}{4}\int d\hat{k}^{\mathrm{s}}\int d\overline{\rho}'\, e^{j\overline{k}^{\mathrm{s}}\cdot\overline{\rho}'}J_z(\overline{\rho}')e^{-j\overline{k}^{\mathrm{s}}\cdot\overline{\rho}} \tag{10.27}$$

where $J_z(\overline{\rho}')$ is the equivalent volume current on the scatterer that radiates $\overline{E}^{\mathrm{s}}$. Rearranging the integrals leads to

$$E_z^{\mathrm{s}}(\overline{\rho};\hat{k}^{\mathrm{i}}) = -\frac{k_0\eta}{4}\int d\overline{\rho}'\, J_z(\overline{\rho}')\int d\overline{k}^{\mathrm{s}}\, e^{-j\overline{k}^{\mathrm{s}}\cdot(\overline{\rho}-\overline{\rho}')} \tag{10.28}$$

By changing variables to the angle ψ between $\hat{k}^{\mathrm{s}}$ and $\overline{\rho}-\overline{\rho}'$, the last integral on the right-hand side of this expression becomes

$$\int d\psi\, e^{-jk_0|\overline{\rho}-\overline{\rho}'|\cos\psi} = 2\pi J_0(k_0|\overline{\rho}-\overline{\rho}'|) \tag{10.29}$$

With this result and (10.6), the scattered field can be written as

$$E_z^{\mathrm{s}}(\overline{\rho};\hat{k}^{\mathrm{i}}) = a_1\int d\overline{\rho}'\, O(\overline{\rho}')E_z(\overline{\rho})J_0(k_0|\overline{\rho}-\overline{\rho}'|) \tag{10.30}$$

where all the constants are combined in a_1. We next insert the Born approximation $E_z \simeq E_z^{\mathrm{i}} = e^{-j\overline{k}^{\mathrm{i}}\cdot\overline{\rho}}$, to obtain

$$E_z^{\mathrm{s}}(\overline{\rho};\hat{k}^{\mathrm{i}}) = a_1\int d\overline{\rho}'\, O(\overline{\rho}')e^{-j\overline{k}^{\mathrm{i}}\cdot\overline{\rho}'}J_0(k_0|\overline{\rho}-\overline{\rho}'|) \tag{10.31}$$

This relationship is identical to (10.8), except that the kernel in (10.31) is the nonsingular part of the 2-D Green's function.

We now perform the Fourier transform of this result with respect to position $\overline{\rho}$. Using the Fourier shift and convolution theorems, this yields

$$\tilde{E}_z^{\mathrm{s}}(\overline{k}+\overline{k}^{\mathrm{i}};\hat{k}^{\mathrm{i}}) = a_1\tilde{O}(\overline{k})\tilde{G}_i(\overline{k}+\overline{k}^{\mathrm{i}}) \tag{10.32}$$

where $\tilde{G}_i(\overline{k})$ is the Fourier transform of $J_0(k_0\rho)$. Next, we integrate both sides with respect to the incident field direction $\hat{k}^{\text{i}}$ to obtain

$$\int d\hat{k}^{\text{i}}\, \tilde{E}_z^{\text{s}}(\overline{k}+\overline{k}^{\text{i}};\hat{k}^{\text{i}}) = a_1\tilde{O}(\overline{k}) \underbrace{\int d\hat{k}^{\text{i}}\, \tilde{G}_i(\overline{k}+\overline{k}^{\text{i}})}_{f(\overline{k})} \tag{10.33}$$

The Fourier transform of the Green's function is certainly not constant, but to a very rough approximation, we can ignore its variation to simplify the relationship.

If we approximate $f(\overline{k})$ as a constant, we can solve for the Fourier transform of the object function to obtain

$$\tilde{O}(\overline{k}) \simeq a_2 \int d\hat{k}^{\text{i}}\, \tilde{E}_z^{\text{s}}(\overline{k}+\overline{k}^{\text{i}};\hat{k}^{\text{i}}) \tag{10.34}$$

Performing the inverse Fourier transform and using the Fourier shift theorem yields

$$O(\overline{\rho}) \simeq a_3 \int d\hat{k}^{\text{i}}\, e^{j\overline{k}^{\text{i}}\cdot\overline{\rho}} E_z^{\text{s}}(\overline{\rho};\hat{k}^{\text{i}}) \tag{10.35}$$

Finally, we insert the holographic reconstruction for the scattered field to obtain

$$O(\overline{\rho}) \simeq a_3 \int d\hat{k}^{\text{i}} \int d\hat{k}^{\text{s}}\, e^{j\overline{k}^{\text{i}}\cdot\overline{\rho}} S(\hat{k}^{\text{s}},\hat{k}^{\text{i}}) e^{-j\overline{k}^{\text{s}}\cdot\overline{\rho}} \tag{10.36}$$

If the function $f(\overline{k})$ is not approximated as a constant, the derivation is considerably more complicated. In this case, the resulting expression for the object function is [13]

$$O(\overline{\rho}) = a_4 \int d\hat{k}^{\text{i}} \int d\hat{k}^{\text{s}}\, |\sin\theta| e^{-j\overline{k}^{\text{s}}\cdot\overline{\rho}} S(\hat{k}^{\text{s}},\hat{k}^{\text{i}}) e^{j\overline{k}^{\text{i}}\cdot\overline{\rho}} \tag{10.37}$$

where θ is the angle between $\hat{k}^{\text{s}}$ and $\hat{k}^{\text{i}}$. This expression is the basis for the holographic backpropagation tomography inverse scattering method.

With a derived value for the constant a_4, the HBT imaging method can in principle be absolutely calibrated, in the sense that the double integral yields an approximation to the object function $\varepsilon_r(\overline{\rho}) - 1$. To avoid the complexity of calibrating the image to obtain the object dielectric properties, we will use HBT only to create an image of the shape of the reconstructed object.

Implementing the HBT method numerically requires evaluation of the integrals in (10.37). Because the integrand is a smooth function, simple quadrature rules can be used. If we evaluate the integrals using the midpoint rule, with integration points located at the angles corresponding to the receiver and transmitter locations, the scattering amplitude function in the integrand becomes an $N_R \times N_T$ matrix F with elements given by

$$F_{pq} = S\left(\hat{k}_p^{\text{s}}, \hat{k}_q^{\text{i}}\right) \tag{10.38}$$

Including the $\sin\theta$ factor from the integrand of (10.37) modifies this matrix to

$$F'_{pq} = |\sin(\phi_p - \phi_q)| S\left(\hat{k}^{\mathrm{s}}_p, \hat{k}^{\mathrm{i}}_q\right) \tag{10.39}$$

The plane wave terms become vectors b and c, which depend on the reconstruction point $\overline{\rho}$. Assuming that the receivers and transmitters are collocated, the components of the column vectors b and c are given by

$$b_p = e^{-jk_0\hat{\rho}_p\cdot\overline{\rho}}, \quad p = 1, 2, \ldots, N_T \tag{10.40a}$$

$$c_p = e^{jk_0\hat{\rho}_p\cdot\overline{\rho}}, \quad p = 1, 2, \ldots, N_R \tag{10.40b}$$

where $\hat{\rho}_p = \cos\phi_p\hat{x} + \sin\phi_p\hat{y}$ is a unit vector in the direction of the pth transceiver at angle ϕ_p and $\overline{\rho}$ is the current pixel point.

The numerically integrated value of (10.37) for the object function at the pixel with coordinates $\overline{\rho} = x\hat{x} + y\hat{y}$ is then given by the matrix–vector product

Holographic Backpropagation Tomography

$$O(\overline{\rho}) = a_5 c^H F' b \tag{10.41}$$

where a_5 is a constant scale factor that includes a_4 and the midpoint rule integration weights. With a value for the scale factor a_5, the object function could in principle be related to the material permittivity by (10.7). If only an image of the object shape is desired then the constant a_5 is not needed. The best choice of phase assumed for the image can be determined using test cases. Good results can be obtained by taking the image to be the absolute value of the object function in (10.41).

10.4.4 Simulating Forward Scattering Data

To test an inverse scattering algorithm, forward scattering data is required. For experimental validation or real-world applications, the bistatic scattered fields or scattering amplitudes are measured, but for algorithm development they can be simulated using any of the forward scattering algorithms developed in earlier chapters.

If the scattering amplitudes are obtained using the method of moments, a linear system must be solved for each incident plane wave. To speed up the simulation, it is helpful to find the LU decomposition of the moment matrix before postprocessing as suggested in Section 6.4.11. Within a double loop over N_T incidence angles and N_R scattering angles, the scattering amplitude can be computed for each incidence angle and scattering angle as described in Section 6.4.11. The result of the forward simulation is a matrix F of scattering amplitudes as defined in (10.38).

10.4.5 Implementing the Holographic Backpropagation Tomography Method

To implement the HBT algorithm, the following steps can be used:

Implementation Details for HBT

1. Assemble measured or simulated bistatic scattering amplitudes for a scatterer into an $N_R \times N_T$ matrix F modified according to (10.39).
2. Create a rectangular grid of pixel points $\overline{\rho}_{mn} = x_{mn}\hat{x} + y_{mn}\hat{y}$ over which the reconstructed image will be computed. To do this, it is helpful to know in advance a bounding box in which the scatterer lies.
3. For each pixel or reconstruction point, compute the value of the object function in (10.41). For the current pixel point $\overline{\rho} = \overline{\rho}_{mn}$, the column arrays `b` and `c` are computed according to (10.40a) and (10.40b) with the index p scanned over the angles corresponding to the transceiver locations. To implement the matrix–vector product, the MATLAB® expression

   ```
   O(m,n) = c'*F*b;
   ```

   can be used. In this expression, the indices `m` and `n` represent the x and y positions of the current image pixel point in a rectangular reconstruction grid.
4. After the object function has been computed for all pixel points, visualize the reconstructed image using `imagesc(x,y,abs(O.'))` where `x` and `y` are vectors of the x and y points used in the reconstruction grid.

10.4.6 Numerical Examples of Holographic Backpropagation Tomography

We will apply the holographic backpropagation inverse scattering algorithm first to a circular PEC cylinder. The radius of the cylinder is one wavelength at 100 MHz. The reconstructed image of the cylinder is shown in Figure 10.3. It can be seen that the image is hollow, in the sense that the reconstructed pixel values are large near the scatterer edge and small inside the support of the scatterer.

Figure 10.4 shows a reconstructed image of a rectangular PEC cylinder at 300 MHz. The shape is visible in the image, but the internal structure in the image inside the boundary of the object can make it difficult in practice to identify the object shape accurately.

An example of an aircraft profile with a more complex geometry is shown in Figure 10.5. The frequency of the illuminating field is 100 MHz. Details of the object shape near the wings and tail are fairly well resolved.

For PEC objects, it might be expected that the holographic backpropagation method would provide poor results. A PEC object is the strongest possible scatterer, since a perfect conductor has an effectively infinite magnitude for the permittivity ε in

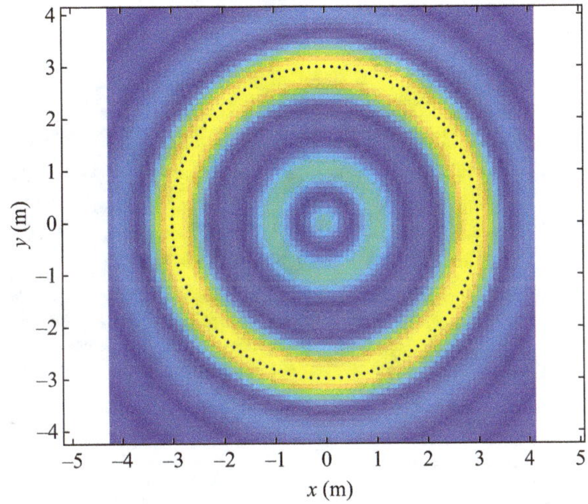

Figure 10.3 *Reconstruction of a circular PEC cylinder with radius 1 λ using the holographic backpropagation method. To compress the dynamic range of the image, pixel values are shown on a logarithmic scale. The boundary of the object is shown in black*

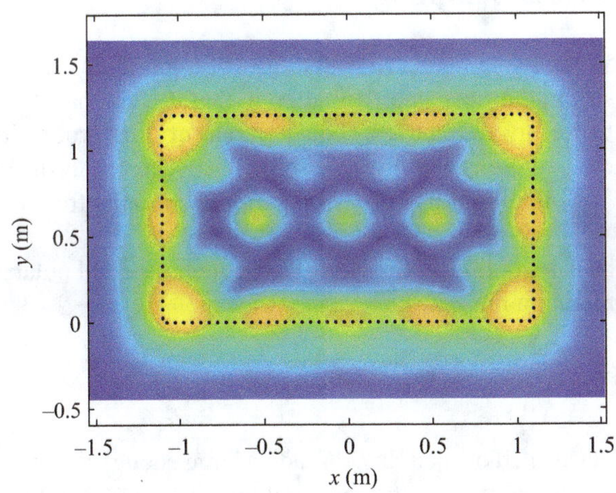

Figure 10.4 *Reconstruction of a 2.2 m-by-1.2 m PEC rectangle at 300 MHz using the holographic backpropagation method*

the object function (10.7). This implies that the weak scattering approximation does not apply. Even though the holographic backpropagation method is based on a weak scattering approximation; however, the reconstructed images for these examples are reasonably good.

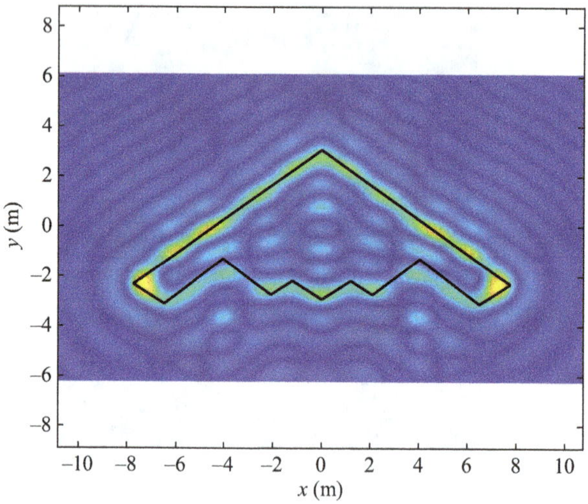

Figure 10.5 Reconstruction of an aircraft profile using the holographic backpropagation method

10.5 Regularized Sampling

The regularized sample method, also called linear sampling or the "simple method" [14–19], differs from the Born approximation in that it is not based on a linearizing approximation to the forward scattering equation. The original proof of the regularized sampling method is rigorous and involves some fairly deep analysis, but the method can be derived in a more intuitive way using simple concepts from electromagnetic theory [20].

Consider the case of a 2-D cylindrical PEC object with TM polarized fields. We begin with the radiation integral

$$E_z^s(\overline{\rho}) = -\frac{k\eta}{4}\int_S d\overline{\rho}' H_0^{(2)}(k|\overline{\rho}-\overline{\rho}'|)J_z(\overline{\rho}') \tag{10.42}$$

with the point $\overline{\rho}$ evaluated on a circle D of radius large enough that it is outside of the scatterer. We will refer to the resulting integral operator as $\mathscr{L}_D$, so that

$$\mathscr{L}_D J_z = E_z^s \tag{10.43}$$

We now choose E_Z^s to be the field radiated by a 2-D point source (3-D line source) located at ρ_0, so that

$$\mathscr{L}_D J_z = H_0^{(2)}(k|\overline{\rho}-\overline{\rho}_0|) \tag{10.44}$$

According to this integral relationship, J_z is the current that when impressed on the scatterer surface S radiates the same field as a point source located at ρ_0. This is similar to the EFIE, except the field is evaluated on the circle D rather than on the same surface S on which the current is impressed.

We now consider two cases:

1. ρ_0 *inside* S. In this case, by Huygens' principle there must exist a well-defined, finite current J_z that radiates the same field as the point source.
2. ρ_0 *outside* S. The right-hand side of (10.44) is singular at the location ρ_0 of the point source. By the uniqueness theorem, E_z on D uniquely determines the fields inside D, so there is no other (possibly nonsingular) field distribution that has the same fields on D as the point source. The right-hand side of (10.44) therefore must correspond to a field with a singularity outside of S. The field radiated by J_z cannot be singular outside the support of the scatterer, however, since the field around radiated by any finite source must be continuous and cannot have any singularities away from the source.

Physically, in the second case, there is no way that fields radiated by J_z on S can focus and appear as if they emanate from a point, if the point is outside of S. As a consequence, in the second case, the right-hand side of (10.44) is not in the domain of the integral operator $\mathscr{L}_D$. If we try to solve (10.44) for the second case, $\|J_z\|$ is theoretically infinite. In practice, regularization limits the magnitude of $\|J_z\|$, but the value is still large when computed numerically. Therefore, if we let the point ρ_0 scan across pixels on a reconstruction domain and numerically solve for J_z, the quantity $\|J_z\|^{-1}$ is large if ρ_0 is inside the support of the scatterer and is small if ρ_0 is outside the support. This provides a way to produce an image of the shape of the scatterer, except that we do not know what the operator $\mathscr{L}_D$ is without knowing the shape of the scatterer S in the first place. We need to further modify (10.44) to create a practical inverse scattering method.

We can do this by taking D into the far field. In this limit, we can replace the Hankel functions in the kernel of $\mathscr{L}_D$ and on the right-hand side of (10.44) with the far-field approximation. After canceling common factors on both sides of the equation, we have

$$\frac{k\eta}{4}\int_S d\overline{\rho}' e^{j\overline{k}^{\rm s}\cdot\overline{\rho}'} J_z(\overline{\rho}') = e^{j\overline{k}^{\rm s}\cdot\overline{\rho}_0} \tag{10.45}$$

where $\overline{k}^{\rm s} = k\hat{\rho} = \hat{x}k\cos\phi^{\rm s} + \hat{y}k\sin\phi^{\rm s}$. We define $E_z^{\rm i}$ to be the incident field that induces J_z on the PEC scatterer surface S, so that

$$\mathscr{L}J_z = E_z^{\rm i} \tag{10.46}$$

where $\mathscr{L}$ is the EFIE operator. We also expand $E_z^{\rm i}$ as an angular spectrum of incoming plane waves, so that

$$E_z^{\rm i}(\overline{\rho}) = \int_0^{2\pi} g(\phi^{\rm i}) e^{-j\overline{k}^{\rm i}\cdot\overline{\rho}}\, d\phi^{\rm i} \tag{10.47}$$

where $\overline{k}^{\rm i} = -\hat{x}k\cos\phi^{\rm i} - \hat{y}k\sin\phi^{\rm i}$.

Combining (10.45), (10.46), and (10.47) and rearranging the order of the integrations,

$$\int_0^{2\pi} g(\phi^{\mathrm{i}}) \underbrace{\left[\frac{k\eta}{4} \int_S d\overline{\rho}'\, e^{j\overline{k}^{\mathrm{s}}\cdot\overline{\rho}'} \mathscr{L}^{-1} e^{-j\overline{k}^{\mathrm{i}}\cdot\overline{\rho}} \right]}_{S(\phi^{\mathrm{s}},\phi^{\mathrm{i}})} d\phi^{\mathrm{i}} = e^{j\overline{k}^{\mathrm{s}}\cdot\overline{\rho}_0} \tag{10.48}$$

By comparing this expression to (6.8.2), the quantity in square brackets can be recognized as the scattering amplitude $S(\phi^{\mathrm{s}}, \phi^{\mathrm{i}})$ in the direction $\overline{k}^{\mathrm{s}}$ for a plane wave incident in the direction $\overline{k}^{\mathrm{i}}$. This leads to the defining integral equation of the regularized sampling method,

Regularized Sampling Equation

$$\int_0^{2\pi} S(\phi^{\mathrm{s}}, \phi^{\mathrm{i}}) g(\phi^{\mathrm{i}})\, d\phi^{\mathrm{i}} = e^{j\overline{k}^{\mathrm{s}}\cdot\overline{\rho}_0} \tag{10.49}$$

If the point $\overline{\rho}_0$ is inside the scatterer, then the solution g is finite. If not, then the right-hand side is not in the domain of the operator in (10.44), and the solution to that integral equation blows up. In view of (10.46) and (10.47), this implies that $\|g\| \to \infty$ as well. Consequently, the inverse of the norm of g provides an indicator function that can be used to determine the shape or region of support of the scatterer.

10.5.1 Discretization of the Regularized Sampling Equation

The regularized sampling integral equation is easy to discretize, because the kernel $S(\phi^{\mathrm{s}}, \phi^{\mathrm{i}})$ is smooth for a finite size scatterer. This means that we can use delta testing functions, pulse expansion functions, and a single-point integration rule to transform this into a linear system. We choose the testing points to be at the angular locations of N_R receivers, and the expansion points to be at the angular locations of N_T transmitters. We typically assume that the transmitters and receivers are far enough from the scatterer that the incident fields are plane waves, and the scattered field values with a suitable normalization become scattering amplitudes. The matrix corresponding to the discretized integral operator is the $N_R \times N_T$ matrix F of bistatic scattering amplitudes defined in (10.38). Typically, the transmitters and receivers are collocated, so that $N_T = N_R$.

The discretized linear system corresponding to (10.49) for a given pixel point is

$$Fx = b \tag{10.50}$$

where x represents samples of the function g at N_T angles, and b is a vector of samples of the right-hand side of (10.49) at N_R scattering angles. The linear system is highly ill-conditioned, so to implement the method numerically we must solve the regularized normal or least squares form of the equation,

$$(F^H F + \varepsilon I)x = F^H b \tag{10.51}$$

If $N_T = N_R$, the linear system can be solved directly rather than in normal form, although regularization is still required. The vector b takes on a different value at each pixel or reconstruction point $\overline{\rho}_0$ in an image of the object. The norm of g can be approximated by the norm of the solution x, so the image value at each pixel is given by $\|x\|^{-1}$.

10.5.2 Implementing the Regularized Sampling Method

The basic steps in an implementation of the regularized sampling method are as follows:

Implementation Details for Regularized Sampling

1. Assemble measured or simulated bistatic scattering amplitudes for a scatterer into an $N_R \times N_T$ matrix F.
2. Choose a value for the regularization parameter ε. This can be done empirically by using a known test scatterer to determine a reasonable value for ε that leads to good image quality.
3. For efficiency, find the LU decomposition of the matrix $F^H F + \varepsilon I$, since the linear system (10.51) must be solved with a different right-hand side at each image pixel point.
4. Create a grid of pixel points over which the reconstructed image will be computed. To do this, it is helpful to know in advance a bounding box in which the scatterer lies.
5. For each pixel or reconstruction point, compute the right-hand side of the linear system (10.51), and solve the linear system for x. This can be done using the MATLAB code

   ```
   x = U\(L\(P*(F'*b)));
   ```

   where `U`, `L`, and `P` represent the LU decomposition of the matrix $F^H F + \varepsilon I$, and `b` is the right-hand side of (10.51).
6. Store the values of $\|x\|^{-1}$ for the reconstruction points in an array with the same dimension as the reconstruction grid. The pixel value is computed using

   ```
   A(m,n) = 1/norm(x);
   ```

   where `A` is an array of image pixel values on the rectangular reconstruction domain. This array can be used to produce an image of the scatterer, since $\|x\|^{-1}$ will be large inside the scatterer and small outside. When plotting the image, it is common to show the logarithm of the image function to reduce the dynamic range of the pixel values in the image.

In practice, $\|x\|$ is not actually infinite for pixel points outside the scatterer, so the algorithm can be "calibrated" for a known scatterer to determine which value corresponds to the scatterer edge.

Images produced by the regularized sampling often have internal structure. In some cases, this corresponds to internal resonances of the scatterer shape for which the EFIE operator has a zero eigenvalue [20]. If the reconstruction point is at a location where an internally resonant mode is zero, the pixel value produced by the regularized sampling algorithm behaves as if the point were outside the scatterer, and the object indicator function becomes small, even though the point is inside the scatterer support.

10.5.3 Numerical Examples of the Regularized Sampling Method

Numerical examples using the regularized sampling inverse scattering method are shown in Figures 10.6–10.8. The first image is a circular PEC cylinder with one wavelength radius. The frequency of the illuminating field is 100 MHz. Since the image pixel values produces by the regularized sampling method have a high dynamic range, it is common to plot the logarithm of the pixel values. Qualitatively, the image shows the support of the scatter quite well, but the pixel value decreases near the edge of the scatterer, so a precise location of a scatterer edge using regularized sampling is difficult.

Figure 10.7 shows a reconstructed image of a rectangular PEC cylinder at 300 MHz. As with the HBT reconstruction of the same shape the internal structure in the image inside the boundary of the object makes it challenging to determine its shape accurately.

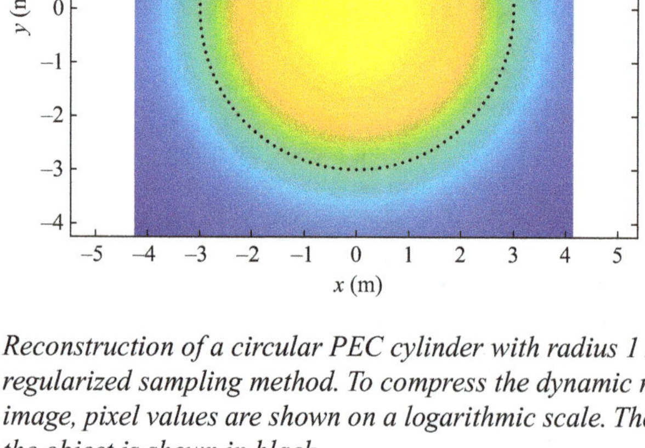

Figure 10.6 Reconstruction of a circular PEC cylinder with radius 1 λ using the regularized sampling method. To compress the dynamic range of the image, pixel values are shown on a logarithmic scale. The boundary of the object is shown in black

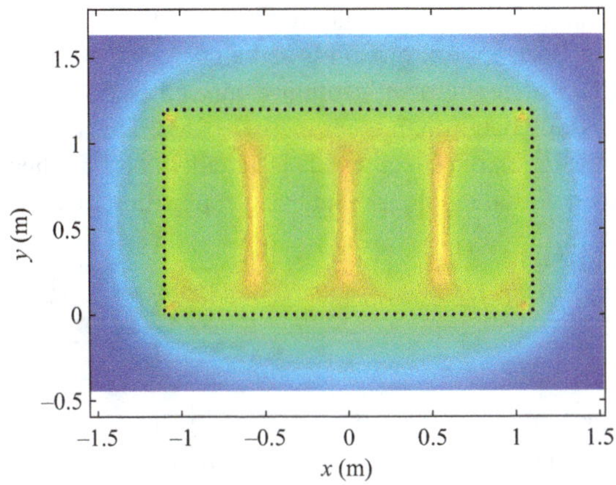

Figure 10.7 Reconstruction of a 2.2 m-by-1.2 m PEC rectangle at 300 MHz using the regularized sampling method

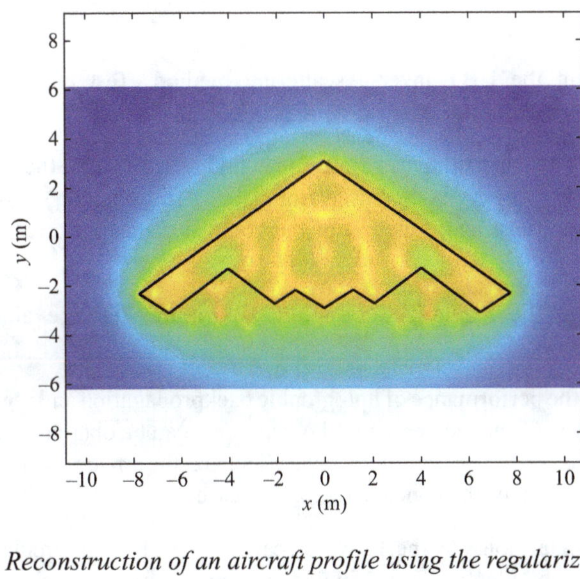

Figure 10.8 Reconstruction of an aircraft profile using the regularized sampling method

A reconstruction of an aircraft profile at 100 MHz is shown in Figure 10.8. The rough shape of the scatterer is represented in the image, but the reconstruction fails to capture fine details in the wing and tail structures. The regularized sampling method does not invoke a linearizing or single scattering approximation, so it might be expected that regularized sampling could provide a better reconstruction of concave object features than Born approximation-based algorithms, but it has been observed that the performance of regularized sampling for some objects with concave features is not substantially better than other methods [13]. This behavior is apparent in Figure 10.8. The resolution of object details is actually better in the reconstruction in Figure 10.5 obtained with the holographic backpropagation method.

Problems

10.1 For the integral operator of Problem 6.1, comment on the conditioning of the resulting numerical method. How does the condition number of the matrix vary with respect to the parameter a? Is regularization needed?

10.2 Implement the method of diffraction tomography for inverse scattering. (a) Verify the algorithms for a circular dielectric and PEC cylinder. (b) Generate forward scattering data for a more complex object using one of the numerical methods of earlier chapters. Comment on the quality of the reconstructed image for various scatterer geometries.

10.3 Implement the HBT inverse scattering method. Test the algorithm as in Problem 10.2.

10.4 Apply the HBT algorithm to a dielectric cylinder and observe the image behavior as the relative permittivity is increased from $\varepsilon_r = 1$ (no scatterer) to $\varepsilon_r = 10$. How does the image change as a function of the contrast of the scatterer relative to the background medium?

10.5 Implement the regularized sampling algorithm. Test the algorithm as in Problem 10.2.

10.6 Compare the performance of holographic backpropagation and regularized sampling for a convex C-shaped geometry. Forward data can be obtained with the FDTD method, surface MoM, or volume MoM. Compare the image quality for the two methods as the depth of the concave region increases.

10.7 Add artificial noise to the forward scattering data for a circular cylinder by adding independent, identically distributed circular complex Gaussian random noise with a given standard deviation to each scattering amplitude. Form an image using an inverse scattering algorithm for the noisy data and compute the contrast ratio of the reconstructed image from the ratio of the root mean square (RMS) value of pixels inside the known scatterer region of support to the RMS pixel value outside. Plot the contrast ratio on a log scale as a function of the signal-to-noise ratio (SNR) at one receiver in dB for (a) holographic backpropagation, and (b) the regularized sampling method. How do the noise sensitivities of the two algorithms compare?

References

[1] D. Colton and R. Kress, *Inverse Acoustic and Electromagnetic Scattering Theory*. Berlin: Springer, 1992.

[2] R. Potthast, *Point Sources and Multipoles in Inverse Scattering*. Boca Raton, FL: Chapman & Hall/CRC, 2001.

[3] A. Kirsch and R. Kress, "Uniqueness in inverse obstacle scattering," Inverse Prob., vol. 9, pp. 285–299, 1993.

[4] R. Potthast, "On a concept of uniqueness in inverse scattering for a finite number of incident waves," SIAM J. Appl. Math., vol. 58, no. 2, pp. 666–682, 1998.

[5] A. C. Kak, "Computerized tomography with X-ray, emission, and ultrasound sources," Proc. IEEE, vol. 67, pp. 1245–1272, 1979.

[6] F. Cakoni, D. Colton, and P. Monk, *The Linear Sampling Method in Inverse Electromagnetic Scattering*. Philadelphia, PA: SIAM, 2011, vol. 80.

[7] F.-C. Chen and W. C. Chew, "Experimental verification of super resolution in nonlinear inverse scattering," Appl. Phys. Lett., vol. 72, no. 23, pp. 3080–3082, 1998.

[8] T. J. Cui, W. C. Chew, X. X. Yin, and W. Hong, "Study of resolution and superresolution in electromagnetic imaging for half-space problems," IEEE Trans. Antennas Propag., vol. 52, no. 6, pp. 1398–1411, 2004.

[9] A. A. Aydiner and W. C. Chew, "On the nature of super-resolution in inverse scattering," in *Proc. Antennas and Propagation Society International Symposium*, Columbus, Ohio, June 22–27, 2003, Piscataway, NJ: IEEE, vol. 4, pp. 776–779, 2003.

[10] P. Blomgren, G. Papanicolaou, and H. Zhao, "Super-resolution in time-reversal acoustics," J. Acoust. Soc. Am., vol. 111, no. 1, pp. 230–248, 2002.

[11] A. J. Devaney, "A filtered backpropagation algorithm for diffraction tomography," Ultrason. Imaging, vol. 4, pp. 336–360, 1982.

[12] T. J. Cui, W. C. Chew, A. A. Aydiner, and S. Chen, "Inverse scattering of two-dimensional dielectric objects buried in a lossy earth using the distorted Born iterative method," IEEE Trans. Geosci. Remote Sens., vol. 39, no. 2, pp. 339–346, 2001.

[13] M. Brandfass, A. Lanterman, and K. F. Warnick, "A comparison of the Colton-Kirsch inverse scattering method with linearized tomographic inverse scattering," Inverse Prob., vol. 17, no. 6, pp. 1797–1816, 2001.

[14] D. Colton and A. Kirsch, "A simple method for solving inverse scattering problems in the resonance region," Inverse Prob., vol. 12, pp. 383–393, 1996.

[15] D. Colton, M. Piana, and R. Potthast, "A simple method using Morozov's discrepancy principle for solving inverse scattering problems," Inverse Prob., vol. 13, pp. 1477–1493, 1997.

[16] D. Colton and P. Monk, "A linear sampling method for the detection of leukemia using microwaves," SIAM J. Appl. Math., vol. 58, pp. 926–941, 1998.

[17] R. Kress and L. Kühn, "Linear sampling methods for inverse boundary value problems in potential theory," Appl. Numer. Math., vol. 43, pp. 143–155, 2002.

[18] F. Cakoni and D. Colton, "On the mathematical basis of the linear sampling method," Georgian Math. J., vol. 10, no. 3, pp. 411–425, 2003.
[19] D. Colton, K. Giebermann, and P. Monk, "A regularized sampling method for solving three dimensional inverse scattering problems," SIAM J. Sci. Comput., vol. 21, no. 6, pp. 2316–2330, 2000.
[20] N. Shelton and K. F. Warnick, "Behavior of the regularized sampling inverse scattering method at internal resonance frequencies," Prog. Electromag. Res., vol. 38, pp. 29–45, 2002.

Index

www.ingramcontent.com/pod-product-compliance
Lightning Source LLC
LaVergne TN
LVHW011839200926
848484LV00002B/376
* 9 7 8 1 8 3 9 5 3 0 7 3 9 *